Microfluidic Chip-Capillary Electrophoresis Devices

Microfluidic Chip-Capillary Electrophoresis Devices

Edited by **Ying Sing Fung**

The University of Hong Kong
Hong Kong, China

Co-Editors: Qidan Chen, Fuying Du,
Wenpeng Guo, Tongmei Ma, Zhou Nie,
Hui Sun, Ruige Wu, and Wenfeng Zhao

CRC Press
Taylor & Francis Group
Boca Raton London New York

CRC Press is an imprint of the
Taylor & Francis Group, an **informa** business

CRC Press
Taylor & Francis Group
6000 Broken Sound Parkway NW, Suite 300
Boca Raton, FL 33487-2742

First issued in paperback 2019

© 2016 by Taylor & Francis Group, LLC
CRC Press is an imprint of Taylor & Francis Group, an Informa business

No claim to original U.S. Government works

ISBN-13: 978-1-4822-5164-7 (hbk)
ISBN-13: 978-0-367-37748-9 (pbk)

Visit the Taylor & Francis Web site at
http://www.taylorandfrancis.com

and the CRC Press Web site at
http://www.crcpress.com

Contents

Section I   Background

Section II   Integration to Improve Sample
Preparation and Cleanup

Section V   Summary and Outlook

About the Editors

Dr. Ying Sing Fung received his BSc and MPhil from The University of Hong Kong, Hong Kong SAR, China, in 1975 and 1977, respectively, and PhD from the Imperial College of Science, Technology and Medicine, University of London, United Kingdom, in 1980. He is currently associate professor in the Department of Chemistry, the University of Hong Kong. He has been a member of the editorial board of *Journal of Chemical Education* (Chinese Chemical Society), *Journal of Biochemical and Biophysical Methods* (Elsevier Science, the Netherlands), and also Scientific Advisory Council member for Separation Science (Eclipse, UK).

He was an academic advisory committee member for Asian Conference on Analytical Chemistry, Asia-Pacific International Symposium on Microscale Separation and Analysis, and China-Japan Molten Salt Chemistry and Technology Symposium, and was appointed as guest professor by the following institutions in China: Open Laboratory of Electroanalytical Chemistry (Changchun Institute of Applied Chemistry), Institute of Material Science and Engineering (Jilin University), and Dongguan Institute of Technology.

He was appointed as advisor for various Hong Kong SAR government departments in different capacities such as assessor for the Hong Kong Laboratory Accreditation Scheme for the Innovation and Technology Commission, member of the Working Group on Regulatory Control of Volatile Organic Compounds for the Environmental Protection Department, and consultant on technical matters for the Drainage Services Department, Architectural Services Department, and Corrective Services Department.

He is active in professional societies such as the Council/Executive Board of the Hong Kong Air & Waste Management Association (USA), the Hong Kong Association for the Advancement of Science and Technology, and the Hong Kong Chemical Society, and is technical advisor for industrial and public organizations such as Hong Kong Green Council, Hong Kong Architectural Coating Association, Hong Kong Critical Components Manufacturers Association, and the Hong Kong Metal Finishing Society.

His research interests include microfluidic chip-capillary electrophoresis devices for environmental, biomedical, and food safety application, and chemical sensors and biosensors, based on the piezoelectric quartz crystal technology. He holds 4 patents, and has authored or co-authored 12 books and reviews, more than 130 journal articles, and 290 conference papers. He is the chief editor of this book and also the principal author of Chapters 1, 3, and 17.

Dr. Qidan Chen received her BSc and MSc degree from Jilin University, Changchun, People's Republic of China, in 2005, and PhD from The University of Hong Kong, Hong Kong SAR, China, in 2010. She is currently associate professor at Zhuhai College, Jilin University. Dr. Chen was appointed as the Honorary Research Associate of the Science Faculty at The Hong Kong University, in 2011. In 2015, she was appointed as a teaching advisory committee member of higher education in Guangdong, People's Republic of China.

Her research interests include the application of nanomaterial in microfluidic-chip capillary electrophoresis (MC-CE) devices and in food testing research, using various international advanced technologies based on semiconductor nanoparticles, microfluidic electrophoresis devices, and laser fabrication techniques, and the development of novel analytical methodology for the determination of toxic environmental pollutants and new analytical methods for the determination of contaminants in food. She is currently a principal investigator of three research grants, the National Natural Science Foundation of China (NSFC) (Grant No. 21307039), foundation for University Excellent Young Teachers Program of Guangdong, and foundation for Distinguished Young Talents in Higher Education of Guangdong (Grant No. 2013LYM_0120). Her research work has led to more than 20 publications in journals such as *Electrophoresis* and presentations at international conference proceedings. She is a co-editor of this book and also principal author of Chapters 10 and 11 of this book.

Dr. Fuying Du received her BSc and MPhil degrees from the Wuhan University, Wuhan, People's Republic of China, in 2003 and 2006, respectively, following by awarding a PhD degree from The University of Hong Kong, Hong Kong SAR, China, in 2012. After graduation, she has taken a postdoctoral position in the Department of Water Quality Engineering, Wuhan University for two years prior to appointment for the current position as a lecturer in the Department of Water Quality Engineering, Wuhan University.

Her research interests include electrochemical sensor and electrochemical biosensor for water analysis, and semiconductor photocatalysts for hydrogen evolution from water splitting. She is a co-editor of this book and also the principal author of Chapters 8 and 9.

Dr. Wenpeng Guo received his BSc from the Sichuan University, Chengdu, People's Republic of China, in 2003, MPhil from the Beijing University of Technology, in 2006, and PhD from The University of Hong Kong, Hong Kong SAR, China, in 2011. He is currently associate professor in The First Affiliated Hospital of Shenzhen University (Shenzhen Second People's Hospital), Shenzhen, People's Republic of China, and also academic advisor to Hubei University of Chinese Medicine, Wuhan. He was appointed as committee member of Guangdong Association of the Integrative Traditional and Western Medicine and as the English editor of *Shenzhen Journal of Integrated Traditional Chinese and Western Medicine*.

His research interests include the development of microfluidic chip-capillary electrophoresis devices for biomedical and clinical application and on clinical integrative research of traditional and Western medicine. He holds 18 patents, and has authored or co-authored 2 books and reviews, and more than 15 journal articles. He is a co-editor of this book and also the principal author of Chapters 2, 4, and 5.

Dr. Gloria Kwan-lok SZE received her PhD from The University of Hong Kong (Hong Kong, SAR) in 2009. She took up a postdoctoral research work for the lithium ion battery project funded by the Hong Kong Innovation and Technology Fund for a year. She then worked in the R&D Department at Bureau Veritas HK Ltd. (Hong Kong SAR) in 2010 prior to joining the Government Laboratory in 2012 as a Chemist.

Her research interest is focused on protein characterization using MC-CE devices integrated with Gas-phase Electrophoretic Mobility Molecular Analyzer (GEMMA). She is the author for Chapter 16 of the book entitled "Protein Characterization and Quantitation – Integrating Microfluidic-chip Capillary Electrophoresis (MC-CE) Device with Gas-phase Electrophoretic Mobility Molecular Analyzer (GEMMA)".

Dr. Tongmei Ma received her PhD from The University of Hong Kong, Hong Kong SAR, China, in 2005, and took up a postdoc position in Arizona State University, Tempe, Arizona, after PhD graduation. Since 2009, she has joined the South China University of Technology, Guangzhou, China.

Her research interests include spectral studies on transition metal containing molecules, transition metal clusters, protein molecules, atmospheric pollutants, and others by employing various modern experimental methods combined with theoretical calculations. She is the author of more than 50 journal and conference papers. She is a co-editor of this book and also the principal author of Chapter 16.

Dr. Zhou Nie received his BSc degree in 1998 and Master of Analytical Chemistry in 2003 from Wuhan University, China. He was then worked as a Chemist in SGS OGC Lab (Shenzhen, Guang Dong Province) from 2003 to 2004 for testing petrochemicals and related petroleum products. After working in industry for a year, he took up a PhD study at the Chemistry Department, University of Hong Kong (Hong Kong SAR) and received his PhD degree in 2008.

After graduation he joined SGS R&D Department (Shatin, Hong Kong) from 2009 to 2010. During this period, he had developed and validated new testing methods to meet requirements for environment and safety regulation. From 2010 till present, he works in Amway China (Guangzhou, Guangdong Province, China) focusing on quality control of botanical extracts in health foods for the Chinese market. His current interest is developing microfluidic chip-capillary electrophoresis devices for on-site quality assessment of Traditional Chinese Medicine and related herbal extract and for determining the binding capacity of bilirubin with human serum proteins. He is a co-editor and principal author of Chapter 12 and 13 of this book.

Dr. Hui Sun received her BSc from the Wuhan University, Wuhan, People's Republic of China, in 1994, and MPhil from the Nankai University, Tianjin, People's Republic of China, in 2002, and PhD from The University of Hong Kong, Hong Kong SAR, China, in 2006. From 2006 to 2010, she worked as research associate for post-doctoral research work in the Department of Chemistry, The University of Hong Kong. She joined the Guangzhou University in 2011. She is currently associate professor in the College of Environmental Science & Engineering, Guangzhou University, Guangzhou, People's Republic of China.

Her research interests include microfluidic chip-capillary electrophoresis devices for environmental, biomedical, and food safety application, and chemical sensors and biosensors based on the piezoelectric quartz crystal technology. She holds 3 patents, and has authored 1 book and more than 50 journal and conference papers. She is a co-editor and principal author of Chapter 6 of this book.

Dr. Ruige Wu received her BSc and MSc, respectively, from Nankai University, Tianjin, People's Republic of China, in 1994, and Peking University, Beijing, People's Republic of China, in 1998. She obtained her PhD on microfluidics applications in analytical chemistry from The University of Hong Kong Hong Kong SAR, China, in 2011. She worked in CapitalBio, Beijing, as a research scientist on microarray technology from 2001 to 2006. She is currently a scientist of Singapore Institute of Manufacturing Technology (SIMTech, A*Star), Singapore.

Her research interests include the applications of microfluidic technology on biomedical and biological areas. She is a co-editor and co-principal author of Chapters 3 and 7 of this book.

Dr. Wenfeng Zhao received her MPhil from Nankai University, Tianjin, People's Republic of China, in 2006, and PhD from The University of Hong Kong, Hong Kong SAR, China, in 2010. She is currently associate professor in the School of Chemical and Engineering, Jiangsu Normal University, Xuzhou, People's Republic of China.

Her research interests include microfluidic chip-capillary electrophoresis devices for biomedical nanoparticles analysis and single molecule detection. She holds a patent, and has authored more than 20 journal and conference papers. She is a co-editor and principal author of Chapters 14 and 15 of this book.

Contributors

Qidan Chen
Department of Chemistry and
  Pharmacy
Jilin University, Zhuhai College
Zhuhai, People's Republic of China

Fuying Du
Department of Water Quality
  Engineering
Wuhan University
Wuhan, People's Republic of China

Ying Sing Fung
Department of Chemistry
The University of Hong Kong
Hong Kong SAR, People's Republic
  of China

Wenpeng Guo
Shenzhen Second People's Hospital
First Affiliated Hospital of
  Shenzhen University
Shenzhen, People's Republic of China

Gloria Kwan-lok SZE
Government Laboratory
Ho Man Tin Government Offices,
  Kowloon
Hong Kong SAR, People's Republic
  of China

Tongmei Ma
School of Chemistry and
  Chemical Engineering
South China University of Technology
Guangzhou, People's Republic
  of China

Zhou Nie
Guangzhou Amway (China) Co., Ltd.
Guangzhou, People's Republic
  of China

Hui Sun
College of Environmental Science &
  Engineering
Guangzhou University
Guangzhou, People's Republic
  of China

Ruige Wu
Singapore Institute of
  Manufacturing Technology
Singapore

Wenfeng Zhao
School of Chemistry and Chemical
  Engineering
Jiangsu Normal University
Xuzhou, People's Republic
  of China

Section I

Background

1

Microfluidic Chip-Capillary Electrophoresis: Expanding the Scope of Application for On-Site Analysis of Difficult Samples

Ying Sing Fung

The University of Hong Kong
Hong Kong SAR, People's Republic of China

CONTENTS

1.1 Needs and Approaches for On-Site Analysis of Difficult Samples Containing Interfering Substances

Due to the rising demand for on-site analysis to deliver timely results in the diagnosis of infective agents and treatment of patients under critical conditions, assay of drugs and metabolites at point-of-care, on-site toxic agent detection for home security, and monitoring of environmental pollutants and food contaminants, various methods capable of on-site operation have been developed in the past 20 years. There are two major methods adopted in various analytical methodologies: direct methods, via specific biochemical reactions with target analytes normally present at high levels in a simple sample matrix; and separation methods, via a separation scheme to separate analytes normally present at low levels in complex samples containing interfering substances [1–3].

Direct methods can be divided further into two groups: low sample throughput laboratory-based methods and high sample throughput clinical-based methods. High sample throughput platforms have been developed for routine monitoring of important clinical parameters. However, the automated system is often targeted to a single parameter, and only a limited number of automated analyzers are available for frequently analyzed parameters because of the high development cost for the automated analyzer.

Laboratory-based low sample throughput platforms are developed to meet difficult analyses requiring sample cleanup and analyte enrichment prior to direct determination by specific biochemical reagents. Due to the requirement of expensive instrumentation and supporting facilities, high capital and running costs are often needed. Thus, this approach is not likely to be useful for on-site monitoring applications.

Procedures utilizing various separation schemes are essential in methods for the determination of several analytes in complex samples containing interfering substances, such as the assay of urinary proteins and metabolites [4]. However, for samples with analytes present in widely different concentrations, dilution or preconcentration has to be carried out for the determination of an individual analyte to match its dynamic working-range requirement. For profile analysis, a large number of analytes have to be determined in each assay, thus requiring a large number of dilutions to be carried out for each sample to meet the quantitation requirement for an individual analyte.

To deal with the problem of determining trace analytes in complex samples, appropriate sample pretreatment procedures have to be carried out to remove interfering compounds. For example, a high-salt environment in a urine sample can interfere with the determination of urinary metabolites when using mass spectrometry (MS), and it also affects the separation of urinary proteins by capillary electrophoresis (CE) or other separation techniques [5,6].

Pretreatment procedures to clean up complex samples are often tedious and time-consuming. Thus, alternative procedures with online sample pretreatment are needed.

1.2 MC Devices for On-Site Analysis

The emergence and value of microfluidic chip (MC) devices for on-site analysis are indicated by the increasing number of publications under the term "microfluidic chip" since its first appearance in the literature in 1994 (Figure 1.1). The pace of increase was slow at the start (1994–2005) but then showed a rapid increase to present.

The various MC devices developed in the past 20 years are listed in Table 1.1, with a brief description on their modes of operation. Although there are various terms used to describe types of MCs; they generally can be classified into four major types: 1) MC as a general term for any device using microfluidic operation; 2) MC or biochip, which utilizes direct detection by specific biochemical reagents with the analyte for qualitative and semiquantitative analysis; 3) microfluidic chip electrophoresis (MCE) or microchip-CE, which incorporates a separation scheme to handle complex samples for quantitative determination of analytes; and 4) microfluidic chip-capillary

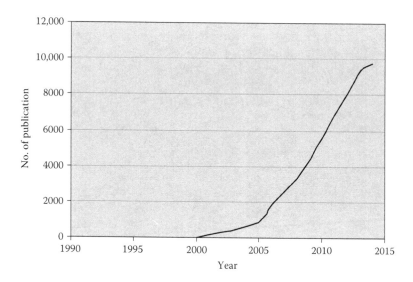

FIGURE 1.1
The cumulative number of publications under microfluidic chip since 1994 to May 2014. (Data from ISI Web of Science.)

TABLE 1.1

Various Microfluidic Chip Devices Developed in the Past 20 Years Since 1994

Microfluidic Chip Device	Year Found	No. of Publications up to 2014[a]	Mode of Operation
Microfluidic chip	1994	9778	Microchip with samples or reagents moved by pumps or electrophoresis with direct analyte detection or analyte separation before detection
Microarray chip/biochip	1994	5364	Samples delivered to specific reagents placed in a given static vial for specific analyte detection
Microfluidic chip electrophoresis or microchip-capillary electrophoresis	1996	2410	Samples or reagents moved by pump or electrophoresis; analyte separated on chip before detection
Microfluidic chip-capillary electrophoresis[b]	1998	90	Samples or reagents moved by electrophoresis in microchip with analyte separated in capillary column; mixing up in terminology with microfluidic chip electrophoresis in literature

[a] Data from ISI Web of Science (1994–2014).
[b] Data from ISI Web of Science (1994–2014) and author's database (2003–2014).

electrophoresis (MC-CE), which integrates a much longer quartz capillary to perform different modes of CE separation and/or incorporates new detection modes to extend the scope of application to difficult samples using on-chip operations, such as sample cleanup, analyte enrichment, high CE separation efficiency, and high detection sensitivity and selectivity for new detection modes. Full definitions of the four forms of microfluidic devices are given in Appendix I.

1.2.1 Microarray Chip/Biochip with Direct Detection

The early form of MC used a pump to transfer samples and reagents to vials fabricated on microchips loaded with chemical or biochemical reagents with specific interaction with the target analytes in the sample for their identification and semiquantitative determination. The number of vials on each chip can reach a few million in the form of a microarray of vials fabricated on the microchip surface, commonly known as a microarray chip or biochip for target sample application. The expensive biochemical reagents needed and the high production costs involved in making microarray chips (a chip can have a few million vials, with each vial containing a different reagent makeup) lead to a high operation cost for using the microarray chip for each assay. In addition, it can only be used once before disposal. As a result of the difficult-to-avoid production error for filling up a few of the vials with the incorrect reagent makeup, false positives and false negatives have been reported in the use of microarray chips, thus affecting interpretation of the results. The high running cost of the microarray chip eliminates its application for general clinical assays and restricts its application to special areas requiring a lot of information from very precious and limited samples, such as DNA assays for exploratory studies of research subjects.

1.2.2 MCE/Microchip-CE with a Separation Scheme

Although the use of a pump to move microfluids provides a nonbiased movement for all microfluids, it can only deliver microfluids from one position to another at a time by applying differential pressure between the two positions. For an interconnecting microchannel pattern, the control of pressure at different parts of a complex microchannel pattern is complicated and expensive, because many pumps have to be employed. Thus, the pressure-driven channel pattern is limited to a very simple design for each operation. In contrast, the use of electrophoresis as the driving force to move samples and reagents is easy, because only one high-voltage power source is needed. Different potentials can be imposed to different vials fabricated on a microfludic chip by imposing differential voltages at each vial and by switching voltages at a desired program. For complicated operations at an open complex microchannel pattern, high-voltage-driven control of microfluids is highly preferred and frequently used for analytical application.

In addition, separation is required for difficult samples for the determination of analytes at trace levels in an interfering sample matrix, and the procedure can be used for the determination of several related analytes in each assay. As electrophoretic separation can provide both high-separation efficiency for analytes and for moving the samples within the MC to a desirable position, it has been adopted as the driving force for moving microfluids and for analyte separation. MCs using electrophoresis for separation of analytes are known as MCE or microchip-CE. Both MCE and microchip-CE refer to the same MCs using electrophoresis in the separation scheme.

The term microchip-CE can sometimes cause confusion. A considerable number of papers reported in the literature [7–9] do not use a quartz capillary to operate in a given mode of CE for analyte separation. Instead, separation is executed for analyte separation in a segment of microchannel a few micros in length is commercially available MCs. Due to the short separation distance in the MC, only a limited separation can be achieved. Thus, a sample with a simple matrix is often used to show the increase in detection sensitivity for the determination of drugs by using a newly developed electrochemical detector for drugs and associated metabolites [9–11].

Despite the increasing number of publications since 1994 covering many application areas, relatively few reports were found using real samples for testing. This discrepancy may due to 1) interference from the complex sample matrixes; 2) use of standard, commercially available MCs with fixed channel configuration to test new ideas, such as new detection principle, and use of a standard mixture to illustrate the enhancement in detection sensitivity; and 3) use of MCs that require expensive research instrumentation, making it difficult for other researchers to follow up on previous work for real sample assay.

In summary, for MCs using direct biochemical detection for point-of-care application, sample matrix interference is a major problem that must be tackled to enable the success of MC for real sample application. For MCE and microchip-CE with separation scheme, the relatively short microchannel (a few centimeters in length) fabricated on the microchip for separation is not sufficient for handling complex samples containing interfering substances. The integration of a quartz separation capillary for high-separation efficiency for analyte coupled with MC for sample preparation provides a promising approach for real sample application, with details given in Section 1.2.3.

1.2.3 Integrated MC-CE Device with a Silica Separation Capillary

The first paper incorporating a capillary in the MC was found in 1998 [12] using a silica capillary to collect cells and deliver the contents for reaction with fluorescein on a glass chip with an etched channel for microscopic observation. There are very few papers published on MCs integrated with a separation capillary after 1996, compared with microarray chip/biochip and MCE/microchip-CE development in the subsequent 10 years, possibly because of the additional work involved in incorporating a capillary with a MC. The interest in online integration of an MC with on-chip sample pretreatment capability to a capillary column with high-separation efficiency under electrophoretic run had initially started in Fung's group in 2000, with an aim to solve problems in handling environmental samples containing interfering substances for trace analysis [13]. In view of the slow progress to develop portable devices to deliver urgently needed results on demand and the lack of affordable commercial instrumentation in support of MC devices for on-site analysis, the approach has later been incorporated into devices showing high potential for biomedical application. The advantages gained in integrating CE with MC devices are found to be far beyond the additional work for bonding a quartz capillary to the MC to fabricate the MC-CE device.

CE is a well-established separation technique that provides an efficient separation with a short assay time, reduced reagent consumption, and low cost. Different separation modes, such as isoelectric focusing (IEF), micellar electrokinetic chromatography (MEKC), and capillary zone electrophoresis (CZE), have been established for the separation of molecules with biomedical significance. For example, CZE separates charged analytes with high resolving power by mass-to-charge ratio (m/z). MEKC provides an effective separating method for both neutral and ionic compounds. IEF produces a pH gradient from the focusing of ampholytes according to their isoelectric points, which can be used to isolate protein fractions for subsequent separation [14,15]. The different separation modes provide flexibility for selecting a desired mode for a given analytical separation. In addition, other sample introduction techniques, such as analyte stacking and isotachophoresis (ITP), have been developed for analyte enrichment and sample cleanup, such as desalting, which can enhance the detection sensitivity.

The integration of a separation capillary with the MC in the MC-CE device has quickened the pace for developing devices for target application in the following ways. First, commercial instrumentation with affordable cost is available from CE manufacturers who have developed CE instrumentation, such as detectors and high-voltage sources, for portable application. Second, various fully developed CE separation modes (MEKC, CZE, and IEF) can be used to assist in the separation of targeted analytes from given samples within the separation capillary. Third, the various analytical enrichment and sample cleanup procedures developed in the microchip can be coupled with CE separation to meet demand for a given task. Fourth, the databases and knowledge accumulated from existing CE separation procedures and buffer systems developed over the years for biomedical, environmental, food safety, security protection, and other areas can be utilized to tackle difficult separation. Fifth, microfluidic techniques developed in MCs, such as mixing, valving, and microflow patterning, can be adopted for on-chip operations prior to CE separation. The integration of two microscale techniques, performing different tasks leads to enhanced capabilities that exceed the use of either technique alone, making it possible to analyze complex real samples for target analytes as shown in Sections II–IV of this book.

The first area of integration of MC-CE devices started in 2000. It incorporated various sample pretreatment procedures, such as sample cleanup and analyte enrichment [13,16,17], with CE technology by using established separation modes for specific biomedical molecules [18–23] and a high-separation efficient CE column to reduce interference from the sample matrix. The second area involved integration with a sensitive and selective detection mode to expand the scope of application to new areas, such as the determination of a non-ultraviolet (UV)-absorptive analyte and nonelectroactive analyte by using optical and electrochemical detectors, respectively.

The success reported in the literature since the mid-2000s using MC-CE devices for the determination of trace analytes in difficult samples covers widely different application areas, including biomedical, environmental, and food safety areas. Wider application was made possible by the availability of affordable facilities and instrumentation at the time to enable the operator to design and fabricate desirable microchannel patterns on MCs for intended application, as ascribed in Section 1.3.

1.3 Fabrication and Control Instrumentation for MC-CE Devices for Intended Application

1.3.1 Instrumentation for Fabrication and Control of MC-CE Devices

There are two major forces driving the development of microchip technology. The first is the need for analyzing small samples for the determination of multi-analytes at low levels, for example, the assay of metabolites in urine. The second is the micromachining capability for making MCs with desired microchannel patterns and electrode vials fabricated at suitable locations to control the movement of samples and analytes for following the procedure of the required analytical task. Personnel with analytical background are mostly chemists by training, and those who are experts in microchip fabrication are engineers, with little chemistry background. This distinction provides a bottleneck for the development of a specific microchip for a target application, as most chemists are often forced to work on a prefabricated chip with standard channel patterns made by engineers, with no consideration for a given analytical task.

The recent advancement in laser fabrication technology driven by industrial prototyping and micromachining allows MC fabrication by computer-aided manufacturing under software control. This technological approach enables the chemist to fabricate a self-designed channel pattern targeting the demand of a given analytical task, leading to a rapid pace of development in recent years of microchip-based methods. The lack of commercially available instrumentation for high-voltage control and sensitive detection for target analytes that are often present in trace levels in complex sample matrixes hinders the pace of development toward real sample application. Recently, there is commercially available instrumentation from Taiwan and Singapore for an economic CO_2 laser ablation unit and programmable high-voltage switching instrumentation. Details are given in Chapters 2 and 3.

The availability of an economic, computer-aided laser ablation unit makes the development of an MC-CE device possible by a chemist to test the desired MC with a self-designed microchannel pattern for analyzing samples with complicated matrixes. The capabilities and advantages for integration of MC with CE for sample pretreatment as well as for the incorporation of sensitive detection modes are discussed in Section 1.3.2, followed by a case study for the application of MC-CE devices to tackle problems in real samples (see Section 1.4).

1.3.2 Fabrication of MC-CE Devices by CO_2 Laser Ablation on PMMA

A commercially available CO_2 laser engraver (V-series, Pinnacle, Great Computer Corporation, Taipei) at an affordable cost was used to fabricate a desired channel pattern onto a poly(methyl methacrylate) (PMMA) polymer surface under the control of CorelDRAW 10 (Corel Corporation, Ottawa,

Ontario) computer software. To assess its capability for repeatable profiling of a microchannel, the fabricated channel was investigated by scanning electron microscopy (SEM); the SEM micrograph in Figure 1.2 shows a cross section of the microchannel produced by CO_2 laser ablation [24]. The profile shows a good match to an overlaid Gaussian curve, with normal distribution of the laser energy for vaporization of PMMA [25]. For laser ablation to a desired depth and width of the microchannels, careful control of the laser power and its velocity drawn across the PMMA surface is required. Details on the effect of laser power and laser speed for fabricating MC-CE devices are given in Chapter 2.

One distinct advantage for laser ablation is the repeatability for fabricating microchannels with a given width and depth, as well as fabricating a specific length of a given segment of the desired microchannel pattern. Thus, one can use different numbers of microchannel segments for mixing of samples and a standard and for calculating the mixing ratio for the on-chip standard addition operation. An example is given below to illustrate the fabrication of an MC-CE device for emergency assay for free bilirubin in the blood of newborn babies. The assay can be carried out at any time of the day by on-chip titration to determine the binding capacity of a given blood sample. Four bilirubin/albumin titrations are needed for each sample with varying mixing ratios between standard bilirubin and the human sera sample to assess its binding capacity. The assay is needed

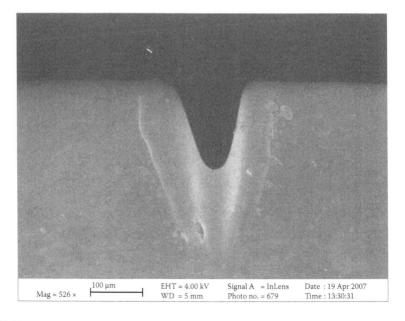

Mag = 526 ×	100 μm	EHT = 4.00 kV	Signal A = InLens	Date : 19 Apr 2007
		WD = 5 mm	Photo no. = 679	Time : 13:30:31

FIGURE 1.2
Cross section of a PMMA microchannel produced by CO_2 laser ablation as shown by the SEM micrograph. (From Nie, Z. and Fung, Y.S. *Electrophoresis* 2008, 29, 1924–1931.)

to be operable by clinical staff because the assay may be urgently needed in the middle of the night. Thus, the MC-CE devices are preloaded with bilirubin at different concentrations for mixing with a specific sera sample injected into the four sample vials for operation to be carried out automatically when the assay is needed.

For fabrication of the MC-CE device (Figure 1.3) for the determination of free bilirubin by using the frontal analysis/CE mode, four pairs of double-T injector with a common connection to the inlet of the fused silica separation capillary are fabricated by CO_2 laser with a wavelength at 10.6 μm [24]. All channels are ablated to a standard configuration with 100 μm depth and 150 μm width at the top of all channels. For the four pairs of double-T injectors, identical dimensions are ablated onto a 30 mm × 40 mm × 0.15 mm Slide (PMMA, Ensinger Ltd, Mid Glamorgan). The channel segment with a fixed distance between each T section thus provides an equal volume of test samples. Desirable sera/bilirubin volume ratios can be produced by fabricating channel segments with calculated distance between each T section.

The 3-mm double-T injector is used to introduce a large sample plug, and the 8-mm length channel is used to connect each of the four double-T injectors to the embedded fused silica separation capillary at equal distance. The 50 μm inside diameter and 13-cm capillary (detection window 4 cm from capillary end) with an effective length of 9.8 cm is used for separation. The capillary is sandwiched between two PMMA plates at the intersection of the four double-T injectors.

The PMMA MC with prefabricated channel configuration is bonded by a press with hot plate to the quartz separation capillary under constant pressure and temperature for 15 min at 0.6 MPa and 92°C, respectively.

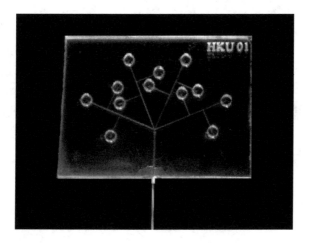

FIGURE 1.3

Intersection between four double-T injectors and the inlet of the separation capillary (60× magnification). (From Nie, Z. and Fung, Y.S. *Electrophoresis* 2008, 29, 1924–1931.)

Satisfactory bonding is shown between the silica capillary and the PMMA plate. After bonding, the MC-CE device is cooled in air to room temperature, washed in distilled water inside an ultrasonic bath, and dried before use.

1.4 Integration of MC-CE Devices for Sample Preparation and Analyte Detection

Integration of the MC-CE device for sample preparation and analyte detection expands its scope to handle complicated sample matrixes, such as determining proteins and metabolites in urine, pesticides in food and environmental samples, and monitoring of drugs and associated metabolites in blood and sera. The simplest integration follows an online format with sequential arrangement for consecutive operations, such as first passing of the sample through the MC for sample cleanup, followed by analyte enrichment by stacking, separation of analytes in the CE column, and finally their quantitation by a detector.

The advantages for sequential operation are as follows: (1) a similarly developed CE procedure can be adopted for operation in MC-CE devices; (2) there is no contamination during operation because the entire procedure is performed in an enclosed MC-CE device; (3) automation of the operation procedure can be done by high-voltage control at electrodes placed in vials fabricated on the MC-CE device; and (4) the MC-CE device can be operated on-site to deliver results on demand.

The disadvantage of the sequential arrangement is that the total analysis time is the sum of the time for all operations to be performed in the procedure. An off-line integration offers an alternative time management. It is more efficient in cost for MC-CE devices for procedures using multiple relative slow sample preparation units to integrate with a fast but expensive detection unit in subsequent operation. An off-line development is necessary to obtain results needed for design of MC-CE devices using online procedures. In summary, the online integration is preferred for on-site analysis. Examples for various forms of integration are given in Sections II and III of this book.

There are two major lines of integration. The first line is based on the lab-on-a-chip approach, such as the incorporation of on-chip sample preparation procedures before analyte separation in the CE separation column. The sample preparation procedures include on-chip dilution, sample cleanup, dual-channel mixing, and online standard addition. The various on-chip operations can be achieved by a controlled manipulation of the microfluids via a preprogrammed high-voltage switching at designated vials, making the MC-CE device a lab-on-a-chip device for field operation. The integration with sample preparation provides the first batch of successful cases to demonstrate the capability of MC-CE devices using commercially available portable and economic instrumentation for tackling difficult samples for determination of trace levels of analytes.

The second area for integration is the incorporation of new sensitive detection modes on-chip to determine target analytes eluted out from the separation capillary. MCs with desired microchannels can be fabricated to guide the positioning of the microelectrode detector to the desired position at the middle of the exit of the separation capillary, which is known to give the highest detection signal. To enhance the use of quantum dot (QDs)-assisted detection for analytes with no detectable functional groups, the detection zone is guided by the MC to a suitable location within the detection zone of the capillary at the focus of the optical system to reduce a stray-light effect and to lower background noise. Another extremely useful mode to enable the detection of analytes with no detectable functional groups is fabricating a specially designed MC to supply the chemical reagent through the crack of the capillary to react with the analyte after its exit from the separation capillary, making possible the detection of analyte with no detectable function to be detected by sensitive detection mode based on dual-electrode detection as well as the extremely sensitive detection mode based on QDs-assisted laser-induced fluorescence (LIF) detection.

Using a carefully designed MC-CE device with a suitable combination of procedures, such as sample cleanup, analyte enrichment, CE separation mode, and selective and sensitive detection of the analyte eluted out from the capillary column, the procedure developed for the fabricated MC-CE device can be optimized with results to demonstrate its capability to remove interfering substances from difficult sample matrixes for the quantitation of analytes up to the required level for the intended application. Successful cases for integrated MC-CE devices for sample preparation are given in Section 1.4.1 and for integration with new detection modes in Section 1.4.2.

1.4.1 Integrating MC-CE Devices for Sample Preparation

The various MC-CE devices integrated with different sample preparation and analyte enrichment procedures are listed in Table 1.2. The examples listed are cases to handle difficult analytical tasks, such as the determination of minor protein in a high-protein sample, determination of trace levels of protein in urine, and on-chip binding assay incorporating several laboratory operations to be carried out on MC-CE devices to deliver timely results. Commercially available instrumentation is used in the examples listed, such as a UV detector from CE equipment, for portable CE equipment, and high-voltage supply, which can be purchased from CE equipment manufacturers. The only research and developed work required are the design and fabrication of MC-CE device with a desirable microchannel pattern and the optimization of related operation procedures.

Although CE has been well-established in laboratories, it has not been extensively used for online clinical analysis of urinary metabolites

TABLE 1.2

Integrating of Various Sample Preparation Techniques with Microfluidic Chip-Capillary Electrophoresis Devices

Sample	Analyte	Sample Preparation before Capillary Electrophoresis Separation	References
Milk, baby formula, dairy products	Minor proteins	On-chip isoelectric focusing fraction isolation and cleanup On-chip multidimensional separation and analyte enrichment by precolumn stacking	[26–30]
Human sera	Free bilirubin	On-chip titration for binding assay On-chip multisegment mixing	[25,31–33]
Urine	Metabolites	On-chip dilution and standard addition	[16,34–36]
Urine	Trace proteins	On-chip two-dimensional separation and transient isotachophoresis analyte enrichment	[16,37,38]
Ambient air	Carbonyls	Solid phase extraction	[39]
Vegetable	Pesticides	Microextraction and microevaporation	[40,41]

and proteins. One reason is its relatively poor concentration detection limit (cLOD) due to the small volume of the injected sample plug (nanoliter level) and the very short optical path length (micrometers) for absorbance detection. However, the cLOD can be lowered using highly sensitive detection modes such as MS or LIF. Ramautar et al. [42] reviewed CE-MS for its application in metabolomics, covering different capillary coatings, separation modes, data processing methods, and other techniques on intended applications. The use of appropriate sample pretreatment methods is important for improving the detection sensitivity for real samples. For example, solid phase extraction (SPE)-related technique has been employed for selective isolation and enrichment of targeted metabolites, such as nucleosides from urine samples [43]. However, the use of off-line sample pretreatment procedures leads to undesirable lengthening of the analysis time, the requirement for a larger sample volume, and the need of laboratory facilities and environments for the operation of the pretreatment procedures and CE instruments. Thus, the application of CE for routine clinical analysis is currently limited to DNA assays [44,45].

Various MC-CE devices listed in Table 1.2 have integrated analyte enrichment procedure on-chip for sample preparation and are used to determine analytes presented at low levels in difficult samples, such as the determination of urinary proteins and assay of minor proteins in milk, baby formula, and dairy products. The MC-CE devices for the aforementioned applications are all using UV-visible detectors available commercially from CE manufacturers. The results from using MC-CE devices integrating with on-chip analyte enrichment procedures with UV-visible detection are found to be sufficient to meet the required detection sensitivity. The outcome demonstrates that the on-chip integration of sample preparation techniques in the MC-CE devices to match the need of the user is essential for the success the MC-CE device for the intended application. The use of commercially available standard MC with double-T configuration has a very limited scope of application for real sample analysis. The approach for fabricating specific MC-CE devices with target for intended application provides the most promising path for the proof-of-concept stage to demonstrate the capability of the device fabricated to meet the required the specifications for real sample application prior to its commercialization.

1.4.2 Integrating MC-CE Devices to Enhance Detection Sensitivity and Selectivity for Analyte Detection

Most of the work reported by MC-CE devices are using UV detectors commercially available from CE instrument manufacturers as received. The areas covered include urine analysis of metabolites by UV absorbance detection [16,17,46–48]. The capillary column can be easily integrated with PMMA chips by thermal bonding to enable sample pretreatment and analyte preconcentration to be carried out on a PMMA chip with a desired microchannel pattern [16,17] prior to CE separation and UV detection directly using commercially available CE equipment. With the recent availability of other sensitive detectors at an affordable cost from CE manufacturers, such as LIF detectors based on semiconductor laser and other high-sensitivity optical detectors, including fluorescence detection [49–51], bioluminescence [52], and electrochemiluminescence (ECL) detection [53], the scope of application for MC-CE devices is expected to expand rapidly in near future.

The various sensitive detection modes integrated with MC-CE devices are shown in Table 1.3, including sensitive detection mode based on LIF, electrochemical detection (ECD) (dual-electrode detection), and the mass-based gas-phase electrophoretic mobility molecular analyzer (GEMMA). Details on the two commonly available highly sensitive detection modes, QDs-assisted LIF detection and the dual-electrode detection are given in Section 1.4.2.1.

1.4.2.1 ECD/Dual-Electrode Detection

ECD is commonly used for MC-CE detection, particularly in early publications [58–63]. It is mainly due to the easy integration of miniaturized ECD mode online with MC-CE for detection after separation. ECD is sensitive, portable, and low in capital and running costs. The problem involving the use of ECD as a detector for MC is that the detection sensitivity is critically dependent on the positioning of the microelectrode to the separation capillary for detecting exiting analytes. In addition, easy replacement of a fouled microelectrode and plugged separation capillary are required during the operation for assay of metabolites in urine.

Du and Fung [46] have developed a dual opposite carbon microdisk electrode (DOCME) detection cell using microchannels fabricated on a PMMA

TABLE 1.3

Integrated Microfluidic Chip-Capillary Electrophoresis Device with Sensitive and Selective Detection Modes

Detection Method	Sample	Analyte	Special Feature	References
Electrochemical Detector				
Opposite dual microelectrodes	Wine	Polyphenols	Improve detection selectivity	[46–48]
Serial dual microelectrodes	Urine	Metabolites, proteins	Direct detection of electroactive analytes	[49]
	Tear fluid	Amino acids	Indirect detection of nonelectroactive analytes	[50]
Electrochemiluminescence	Urine	Racemic drugs	High detection sensitivity	[51]
LIF Detector				
QD-mediated LIF detection	Potato chips	Acrylamide	Indirect detection of analyte without detectable groups	[52,53]
Immobilized QD/LIF detection	Vegetables	Organophosphorus pesticides	High detection sensitivity / High detection selectivity	[54,55]
Mass-Based Detector				
GEMMA	Milk	Proteins	Analyte identification	[56,57]

Note: LIF, laser-induced fluorescence; QD, quantum dot; GEMMA, gas-phase electrophoretic mobility molecular analyzer.

chip to guide the positioning of two carbon microdisk electrodes placed at opposite sides of the exit of the separation capillary for the determination of polyphenols in wine. Different constant potentials were imposed on the two electrodes placed in an identical environment facing the exit of eluted analytes with a lower potential to detect the background current and a working potential to detect analytes separated by CE. The use of the current ratio of two identical and opposite microelectrodes held at different potentials has demonstrated its capability to differentiate analytes from electroactive impurities coeluted from the CE column at the same migration time. The setup and fabrication of DOCME are given in Chapter 8. As both the separation capillary and DOCME are placed in positions guided by microchannels fabricated onto the PMMA chip, repeatable replacement for fouled DOCME and plugged separation capillary can be easily performed. Details on the fabrication and performance of the DOCME detector are given in Chapter 8.

1.4.2.2 QD-Assisted Detection

QDs were utilized to enhance the detection sensitivity for fluorescence detection [52,54]. Chen and Fung [54] have applied QDs to assist the detection of acrylamide in potato crisp and organophosphorus pesticides (OPs) in vegetables, because both of them do not possess desirable functional groups for UV and ECD. The former provides an indirect detection and the latter a more sensitive direct detection. Using the MC-CE device with QD-assisted LIF detection, a selective method using a simple procedure is developed for the rapid determination of four OPs in vegetable samples: mevinphos, phosalone, methidathion, and diazinon. CdTe/Cds core-shell QDs were immobilized at the inside surface of the separation capillary for selective binding of OPs with enhanced fluorescence signal. For the four OPs investigated by the fabricated MC-CE device, working ranges from 0.1 to 30 mg/kg and cLODs from 50 to 180 μg/kg had been achieved. The total assay time of 12 min is highly desirable compared to the lengthy solvent extraction and time-consuming SPE step for sample preparation, which takes hours to perform. MC-CE devices provide a promising highly sensitive detection mode based on QDs-assisted LIF indirect detection to determine metabolites present at low levels in complex urine samples with no strong detectable functional groups. Details are given in Chapters 10 and 11.

1.5 Application of MC-CE Devices

Noticeable advancement of MC-CE devices has been made in the past 10 years, in particular, for their success demonstrated for assay of proteins and metabolites in complex urine samples. Based on the integration of MC, with enrichment and cleanup capability, to CE, with matured separation and detection modes, the scope of application of MC-CE devices is found exceeding the use of either technique alone. For example, despite using the less sensitive UV detector in MC-CE compared to the use of LIF detectors in commercially available MCE devices, the detection limits of four urinary proteins investigated by the MC-CE device developed have shown to deliver a better performance, reflecting the beneficial effect of desalting and analyte enrichment carried out on-chip prior to CE separation.

The application of MC-CE devices developed for biomedical, food, environmental, and pharmaceutical or herbal medicine analysis is listed in Table 1.4. MC-CE has been shown to cover a wide range of application areas with results from the integration of these complementary techniques by the MC-CE device with target on the intended application. Compared to existing methods, MC-CE devices offer similar analytical performance. In addition, the flexibility of the MC-CE device for handling difficult urine samples using on-chip sample dilution up to 500-fold and online standard addition to handle difficult sample matrixes is much preferred to commercially available MCE chips. For the determination of drugs and associated metabolites, the separation of chiral drugs has been shown to be achievable in urine with the incorporation of a highly efficient separation capillary with sufficient column length in the MC-CE device [51]. Coupling with on-column sensitive ECL detection, the chiral compounds separated can be detected at required concentrations. The on-chip integration of sensitive detection modes such as

TABLE 1.4

Application of Microfluidic Chip-Capillary Electrophoresis Devices in Biomedical, Food, Environmental, and Pharmaceutical Areas

Application Area	Analyte and Sample	References
Medical assay	General	[49,64–72]
	Diagnostic application	[31,40,44,73–76]
	Bilirubin binding assay	[24,32,33,53]
	General metabolite assay	[17,36,77,78]
	Urine metabolites and biomarkers	[34,35,79,80]
	Urinary protein	[16,37]
Food analysis	General protein analysis	[28,38,56,57,81–84]
	Proteins in milk and dairy products	[26,27,85]
	Food quality assessment	[30,86,87]
	Polyphenols in wine	[46,47]
	Acrylamide in potato crisps	[52]
Environment analysis	Pesticides in vegetables	[40,44,54,55,85]
	Carbonyl in air	[39,68]
Biological analysis	Organelles in cell extract	[88–90]
	Amino acids, proteins in biofluids	[50,91]
Pharmaceutical analysis	Active Ingredients in herbal medicine	[51,92,93]

ECL, dual-microelectrode detection, and QDs-assisted LIF detection provides additional drive for further development of MC-CE devices to explore uncharted areas for determining ultratrace levels of metabolites in complex urine samples.

1.6 Aim and Outline of This Book

This book aims to introduce readers to the methodology and benefits gained by integration of two complementary technologies, MC and CE, with results to demonstrate their capability to expand the scope of application to tackle difficult analytical tasks for the determination trace levels of analyte in samples containing potential interfering substances.

This integration is made possible by the availability of affordable equipment for computer-aided fabrication of MC using a CO_2 laser, as it enables the user to design microchannel patterns to solve the problem for difficult samples in an intended application area. With the user fabricating his or her own MC, with application built into the design, and coupled with the use of a highly efficient CE separation and sensitive detection mode, the pace for commercialization of MC-based devices will be quicker, as the device is designed and tested with the need of the market in mind right at the beginning. With this in consideration, this book is divided into the following sections, with key points shown.

Section I Background

- To present a brief account of the historical development of MC, MCE, Microchip-CE, and MC-CE devices
- To highlight the recent development of instrumentation and facilities which makes possible the fabrication of MC-CE devices with designed microchannel patterns by the user for the intended application
- To illustrate the integration of MC with CE for expanding the scope of application in MC-CE devices and give a brief survey on the success made in various application areas

Section II Integration to Improve Sample Preparation and Cleanup

- To explain the need for integrating sample preparation in MC-to-CE separation for real sample analysis
- To give a brief description on the operations for sample cleanup and analyte enrichment by integrated MC-CE devices for the intended application

- To illustrate the salient features and successes of MC-CE devices fabricated to handle difficult samples in demanding application areas as shown in each chapter

Section III Integration to Enhance Analyte Detection

- To introduce the principle and methodology for integrating new detection modes in MC-CE devices
- To explain the operation and benefits gained by integration of sensitive and selective analyte detection modes in MC-CE devices
- To highlight the special features and advantages gained by special MC-CE devices fabricated to determine analytes not possible to be detected by conventional detection modes as shown in each chapter

Section IV Integration to Achieve Intended Application

- To present an overview on integrating MC-to-CE for intended application in biomedical, environmental, and food safety areas
- To explain the design of MC-CE devices and how they can solve the problems for tackling difficult samples for determination of trace levels of analyte in complex sample matrixes containing interfering substances
- To discuss the optimization of the operation procedure for the MC-CE devices fabricated and the successes achieved by MC-CE devices fabricated for intended application

Section V Summary and Outlook

- To summarize on the achievement made by MC-CE devices to date
- To discuss the obstacles to be overcome for commercialization of MC-CE devices for intended application
- To highlight the path to be taken and tasks to be performed for future development of MC-CE devices

References

1. Assadi, F.K. Quantitation of microalbuminuria using random urine samples. *Pediatr Nephrol* 2002, 17, 107–110.
2. Ebbels, T.M., Holmes, E., Lindon, J.C. and Nicholson, J.K. Evaluation of metabolic variation in normal rat strains from a statistical analysis of ^{1}H NMR spectra of urine. *J Pharmaceut Biomed Anal* 2004, 36, 823–833.

3. Iadarola, P., Cetta, G., Luisetti, M., Annovazzi, L., Casado, B., Baraniuk, J., Zanone, C. and Viglio, S. Micellar electrokinetic chromatographic and capillary zone electrophoretic methods for screening urinary biomarkers of human disorders: A critical review of the state-of-the-art. *Electrophoresis* 2005, 26, 752–766.

4. German, J.B., Hammock, B.D. and Watkins, S.M. Metabolomics: Building on a century of biochemistry to guide human health. *Metabolomics* 2005, 1, 3–9.

5. Issaq, H.J., Abbott, E. and Veenstra, T.D. Utility of separation science in metabolomic studies. *J Sep Sci* 2008, 31, 1936–1947.

6. Wood, S.L., Knowles, M.A., Thompson, D., Selby, P.J. and Banks, R.E. Proteomic studies of urinary biomarkers for prostate, bladder and kidney cancers. *Nat Rev Urol* 2013, 10, 206–218.

7. Chandra, P., Zaidi, S.A., Noh, H.B., and Shim, Y.B. Separation and simultaneous detection of anticancer drugs in a microfluidic device with an amperometric biosensor. *Biosens Bioelectron* 2011, 28, 326–332.

8. Zhao, X., You, T., Qiu, H., Yan, J.L., Yang, X.R. and Wang, E.K. Electrochemiluminescence detection with integrated indium tin oxide electrode on electrophoretic microchip for direct bioanalysis of lincomycin in the urine. *J Chromatogr* 2004, 810, 137–142.

9. Zhang, Q.L., Lian, H.Z., Wang, W.H., and Chen, H.Y. Separation of caffeine and theophylline in poly(dimethylsiloxane) microchannel electrophoresis with electrochemical detection. *J Chromatogr A* 2005, 1098, 172–176.

10. Zhang, Q.L., Xu, J.J., Li, X.Y., Lian, H.Z. and Chen, H.Y. Determination of morphine and codeine in urine using poly(dimethylsiloxane) microchip electrophoresis with electrochemical detection. *J Pharmaceut Biomed Anal* 2007, 43, 237–242.

11. Ding, Y., Qi, Y. and Suo, X. Rapid determination of beta(2)-agonists in urine samples by microchip micellar electrokinetic chromatography with pulsed electrochemical detection. *Anal Methods* 2013, 5, 2623–2629.

12. Ocvirk, G., Tang, T. and Harrison, D.J. Optimization of confocal epifluorescence microscopy for microchip-based miniaturized total analysis systems. *Analyst* 1998, 123, 1429–1434.

13. Fung, Y.S., Capillary electrophoresis for environmental analysis. *Abstract of 4th Chinese Capillary Electrophoresis and Related Microscale Techniques*, Guangzhou, China, June 11–13, 2000, p. 18.

14. Kuhn, R. and Hoffstetter-Kuhn, S. *Capillary Electrophoresis: Principles and Practice.* Springer-Verlag, Berlin, 1993, pp. 5–36.

15. Wehr, T., Rodríguez-Díaz, R. and Zhu, M. *Capillary Electrophoresis of Proteins.* Marcel Dekker, New York, 1999, pp. 131–233.

16. Wu, R., Yeung, W.S.B. and Fung, Y.S. 2D t-ITP/CZE determination of clinical urinary proteins using microfluidic-chip capillary electrophoresis device. *Electrophoresis* 2011, 32, 3406–3414.

17. Guo, W.P., Lau, K.M. and Fung, Y.S. Microfluidic chip-capillary electrophoresis for two orders extension of adjustable upper working range for profiling of inorganic and organic anions in urine. *Electrophoresis* 2010, 31, 3044–3052.

18. Liu, Y.J., Foote, R.S., Jacobson, S.C., Ramsey, R.S. and Ramsey, J.M. Electrophoretic separation of proteins on a microchip with noncovalent, postcolumn labeling. *Anal Chem* 2000, 72, 4608–4613.

19. Lacher, N.A., de Rooij, N.F., Verpoorte, E. and Lunte, S.M. Comparison of the performance characteristics of poly(dimethylsiloxane) and Pyrex microchip electrophoresis devices for peptide separations. *J Chromatogr A* 2003, 1004, 225–235.

20. Bousse, L., Mouradian, S., Minalla, A., Yee, H., Williams, K. and Dubrow, R. Protein sizing on a microchip. *Anal Chem* 2001, 73, 1207–1212.
21. Jin, L.J., Giordano, B.C. and Landers, J.P. Dynamic labeling during capillary or microchip electrophoresis for laser-induced fluorescence detection of protein-SDS complexes without pre- or postcolumn. labeling *Anal Chem* 2001, 73, 4994–4999.
22. Giordano, B., Jin, L., Couch, A.J., Ferrance, J.P. and Landers, J.P. Microchip laser-induced fluorescence detection of proteins at submicrogram per milliliter levels mediated by dynamic Labeling under pseudonative conditions. *Anal Chem* 2004, 76, 4705–4714.
23. Hatch, A.V., Herr, A.E., Throckmorton, D.J., Brennan, J.S. and Singh, A.K. Integrated preconcentration SDS-PAGE of proteins in microchips using photo-patterned cross-linked polyacrylamide gels. *Anal Chem* 2006, 78, 4976–4984.
24. Nie, Z. and Fung, Y.S. Microchip Capillary Electrophoresis for frontal analysis of free bilirubin and study of its interaction with-human serum albumin. *Electrophoresis* 2008, 29, 1924–1931.
25. Klank, H., Kutter, J.P. and Geschke, O. CO_2-laser micromachining and back-end processing for rapid production of PMMA-based microfluidic systems. *Lab Chip* 2002, 2, 242–246.
26. Wu, R.G., Fung, Y.S. and Yeung, W.S.B. Determination of lactoferrin and β-lactoglobulin in dairy products by microfluidic-chip capillary electrophoresis. *Abstract P4-10, 9th Asia-Pacific International Symposium on Microscale Separation and Analysis (APCE 2009) and 1st Asian-Pacific International Symposium on Lab-on-Chip (APLOC2009)*, Shanghai, China, October 28–31, 2009, p. 238.
27. Wu, R.G., Fung, Y.S. and Yeung, W.S.B. Determination of lactoferrin and immunoglobulin G in baby formula by microfluidic-chip capillary electrophoresis. *Abstract PB8, 10th Asian-Pacific International Symposium on Microscale Separations and Analysis (APCE 2010)*, Hong Kong, China, December 10–13, 2010, p. 99.
28. Wu, R.G., Fung, Y.S. and Yeung, W.S.B. Microfluidic chip-capillary electrophoresis for separation and determination of low abundance proteins. *Abstract, Microfluidic and Nanofluidic Devices for Chemical and Biochemical Experimentation Symposium, 2010 International Chemical Congress of Pacific Basin Societies (Pacifichem 2010)*, Honolulu, HI, December 15–20, 2010, TECH Paper 547.
29. Wu, R.G., Wang, Z.P., Zhao, W.F., Yeung, W.S.B. and Fung, Y.S. Multi-dimension microchip-capillary electrophoresis device for determination of functional proteins in infant milk formula. *J Chromatogr A* 2013, 1304, 220–226.
30. Wu, R.G., Wang, Z., Fung, Y.S., Seah, D.Y. P. and Yeung, W.S.B. Assessment of adulteration of soybean proteins in dairy products by 2D microchip-CE device. *Electrophoresis* 2014, 35, 1728–1734.
31. Nie, Z. and Fung, Y.S. Determination of binding capacity of albumin for bilirubin by insitu titration at a circular ferrofluid driven micromixer in microfluidic chip-capillary electrophoresis. *Abstract P01-05, 8th Asia-Pacific International Symposium on Microscale Separation and Analysis (APCE 2008)*, Kaohsiung, Taiwan, November 2–5, 2008, p. 99.
32. Nie, Z., Sun, H. and Fung, Y.S. Determination of binding capacity of albumin for bilirubin by microfluidic chip-capillary electrophoresis based on insitu titration at a circular ferrofluid driven micromixer. *Abstract P2-02, 9th Asia-Pacific International Symposium on Microscale Separation and Analysis (APCE 2009) and 1st Asian-Pacific International Symposium on Lab-on-Chip (APLOC2009)*, Shanghai, China, October 28–31, 2009, p. 155.

33. Sun, H. and Nie, Z., Fung, Y.S. Determination of free bilirubin and its binding capacity by albumin using a microfluidic chip-capillary electrophoresis device with a multi-segment circular-ferrofluid driven micromixing injection. *Electrophoresis* 2010, 31, 3061–3069.

34. Guo, W.P. and Fung, Y.S. Microfluidic chip-capillary electrophoresis with adjustable on-chip sample dilution for profiling of urinary markers. *Proceedings, 14th International Conference on Miniaturized Systems for Chemistry and Life Sciences (μTAS 2010)*, Groningen, The Netherlands, October 3–7, 2010, Paper W11A, pp. 1481–1483.

35. Guo, W.P. and Fung, Y.S. Microfluidic chip-capillary electrophoresis with adjustable on-chip sample dilution for profiling of urinary markers. *Abstract PB11, 10th Asian-Pacific International Symposium on Microscale Separations and Analysis (APCE 2010)*, Hong Kong, China, December 10–13, 2010, p. 102.

36. Guo, W.P. and Fung, Y.S. Microfluidic chip-capillary electrophoresis with dynamic multi-segment standards addition for rapidly identifying nephrolithiasis markers in urine. *Electrophoresis* 2011, 32, 3437–3445.

37. Wu, R.G., Fung, Y.S. and Yeung, W.S.B. Microfluidic-chip capillary electrophoresis for determination of clinical urinary proteins. *Abstract PB9, 10th Asian-Pacific International Symposium on Microscale Separations and Analysis (APCE 2010)*, Hong Kong, China, December 10–13, 2010, p. 100.

38. Wu, R.G., Fung, Y.S. and Yeung, W.S.B. Microfluidic-chip capillary electrophoresis for analysis of clinical urinary proteins. *Abstract P136, 25th International Symposium on Microscale Bioseparations (MSB 2010)*, Prague, Czech Republic, March 21–25, 2010, p. 129.

39. Sun, H. and Fung, Y.S. Integrating micellar electrokinetic capillary chromatography with molecular imprinting polymer-solid phase extraction for ambient carbonyl determination. *Abstract, the 12th Asian-Pacific International Symposium on Microscale Separations and Analysis (APCE 2012)*, Singapore, December 16–19, 2012, SICC7-A-0240, p. 1.

40. Chen, Q.D. and Fung, Y.S. Determination of organophosphate pesticides in vegetables by integrating microextraction with microfluidic-chip capillary electrophoresis. *Abstract P4-01, 9th Asia-Pacific International Symposium on Microscale Separation and Analysis (APCE 2009) and 1st Asian-Pacific International Symposium on Lab-on-Chip (APLOC2009)*, Shanghai, China, October 28–31, 2009, p. 226.

41. Chen, Q.D., Ma, T.M., Du, F.Y. and Fung, Y.S. Microchip-CE device with on-chip micro-evaporator for determination of organophosphorus pesticides in vegetable. *Abstract, the 12th Asian-Pacific International Symposium on Microscale Separations and Analysis (APCE 2012)*, Singapore, December 16–19, 2012, SICC7-A-0232, p. 1.

42. Ramautar, R., Somsen, G.W. and de Jong, G.J. CE-MS in metabolomics. *Electrophoresis* 2009, 30, 276–291.

43. La, S., Cho, J., Kim, J. and Kim, K.R. Capillary electrophoretic profiling and pattern recognition analysis of urinary nucleosides from thyroid cancer patients. *Anal Chim Acta* 2003, 486, 171–182.

44. Qin, J., Fung, Y.S. and Lin, B.C. DNA diagnosis by capillary electrophoresis and microfabricated electrophoretic devices. *Expert Rev Mol Diagn* 2003, 3, 387–394.

45. Chang, P.L., Kuo, I.T., Chiu, T.C. and Chang, H.T. Fast and sensitive diagnosis of thalassemia by capillary electrophoresis. *Anal Bioanal Chem* 2004, 379, 404–410.

46. Du, F.Y. and Fung, Y.S. Development of CE-dual opposite carbon-fibre micro-disk electrode detection for peak purity assessment of polyphenols in red wine. *Electrophoresis* 2010, 31, 2192–2199.

47. Du, F.Y. and Fung, Y.S. Differential amperometric dual electrode detection for peak purity assessment of polyphenols in red wine after separation by microfluidic chip-capillary electrophoresis. *Abstract P3-01, 9th Asia-Pacific International Symposium on Microscale Separation and Analysis (APCE 2009) and 1st Asian-Pacific International Symposium on Lab-on-Chip (APLOC2009)*, Shanghai, China, October 28–31, 2009, p. 212.

48. Fung, Y.S. and Du, F.Y. Microfluidic chip-capillary electrophoresis devices—Dual electrode detectors for direct and indirect determination of analytes in complicated sample matrixes. *Abstract IL2, 11th Asian-Pacific International Symposium on Microscale Separations and Analysis (APCE 2011)*, Hobart, Australia, November 27–30, 2011, p. 21.

49. Du, F.Y., Chen, Q.D., Zhao, W.F. and Fung, Y.S. A new microfluidic-chip capillary electrophoresis device with a serial dual-electrode detector for biomedical analysis. *Abstract T21, 4th International Symposium on Microchemistry and Microsystems (ISMM 2012)*, Zhubei City, Taiwan, June 10–13, 2012, pp. 326–327.

50. Du, F.Y., Wu, R.G. and Fung, Y.S. Microfluidic chip-capillary electrophoresis devices for determination of proteins and amino acids in biofluids. *Abstract, the 13th Asian-Pacific International Symposium on Microscale Separations and Analysis (APCE 2013)*, Jeju, November 3–6, 2013, Korea Paper LOC-WK02, p. 1.

51. Guo, W.P., Rong, Z.B., Li, Y.H., Fung, Y.S., Gao, G.Q. and Cai, Z.M. Microfluidic chip capillary electrophoresis coupled with electrochemiluminescence for enantioseparation of racemic drugs using central composite design optimization. *Electrophoresis* 2013, 34, 2962–2969.

52. Chen, Q.D., Zhao, W.F. and Fung, Y.S. Determination of acrylamide in potato crisps by capillary electrophoresis with quantum dot-mediated LIF detection. *Electrophoresis* 2011, 32, 1252–1257.

53. Mo, S.L. and Fung, Y.S. Quantum-dots mediated LIF detection of free bilirubin by frontal analysis using microfluidic chip-capillary electrophoresis device. *Abstract P32, 11th Asian-Pacific International Symposium on Microscale Separations and Analysis (APCE 2011)*, Hobart, Australia, November 27–30, 2011, p. 95.

54. Chen, Q.D. and Fung, Y.S. Capillary electrophoresis with immobilized quantum dots fluorescence detection for rapid determination of organophosphorus pesticides in vegetables. *Electrophoresis* 2010, 31, 3107–3114.

55. Chen, Q.D. and Fung, Y.S. Microfluidic-chip capillary electrophoresis with immobilized quantum dots detection for the separation and determination of organophosphorus pesticides in contaminated vegetables. *Abstract P01-07, 8th Asia-Pacific International Symposium on Microscale Separation and Analysis (APCE 2008)*, Kaohsiung, Taiwan, November 2–5, 2008, p. 101.

56. Ma, T.M., Wu, R.G. and Fung, Y.S. Hyphenation of microfluidic-chip capillary electrophoresis with gas-phase electrophoretic mobility molecular analyzer (GEMMA) for protein characterization. *Abstract PB10, 10th Asian-Pacific International Symposium on Microscale Separations and Analysis (APCE 2010)*, Hong Kong, China, December 10–13, 2010, p. 101.

57. Ma, T.M. and Fung, Y.S. Microchip CE device for coupling gas-phase electro-phoretic mobility molecular analyzer with capillary electrophoresis for study of metal-protein binding in milk. *Abstract M20, 4th International Symposium on Microchemistry and Microsystems (ISMM 2012)*, Zhubei City, Taiwan, June 10–13, 2012, pp. 134–135.

58. Horng, R.H., Han, P., Chen, H.Y., Lin, K.W., Tsai, T.M. and Zen, J.M. PMMA-based capillary electrophoresis electrochemical detection microchip fabrication. *J Micromech Microeng* 2005, 15, 6–10.

59. Zúborová, M., Masár, M., Kaniansky, D., Johnck, M. and Stanislawski, B. Determination of oxalate in urine by zone electrophoresis on a chip with con-ductivity detection. *Electrophoresis* 2002, 23, 774–781.

60. Zeng, Y., Chen, H., Pang, D.W., Wang, Z.L. and Cheng, J.K. Microchip cap-illary electrophoresis with electrochemical detection. *Anal Chem* 2002, 74, 2441–2445.

61. Martin, R.S., Ratzlaff, K.L., Huynh, B.H. and Lunte, S.M. In-channel electro-chemical detection for microchip capillary electrophoresis using an electrically isolated potentiostat. *Anal Chem* 2002, 74, 1136–1143.

62. Wu, C.C., Wu, R.G., Huang, J.G., Lin, Y.C. and Chang, H.C. Three-electrode electrochemical detector and platinum film decoupler integrated with a capil-lary electrophoresis microchip for amperometric detection. *Anal Chem* 2003, 75, 947–952.

63. Jiang, L., Lu, Y., Dai, Z., Xia M.H. and Lin, B.C. Mini-electrochemical detector for microchip electrophoresis. *Lab Chip* 2005, 5, 930–934.

64. Fung, Y.S. and Nie, Z. Microfluidic chip-capillary electrophoresis for biomedical analysis. *Sep Sci* 2009, 1, 3–8.

65. Fung, Y.S. and Nie, Z. Microfluidic chip-capillary electrophoresis for biomedical analysis. *Sep Sci China* 2009, 1, 18–23.

66. Guo, W.P., Lau, K.M., Sun, H. and Fung, Y.S. Microfluidic chip-capillary electro-phoresis for emergency onsite biomedical analysis. *Sep Sci* 2010, 2, 3–10.

67. Fung, Y.S. Microchip-capillary electrophoresis for biomedical application in Hong Kong. *Abstract, International Symposium on Smart Materials and Devices/ Workshop on "Microfluidics and Its Application,"* Hong Kong, December 10–13, 2007, Paper WP.04, p. 114.

68. Fung, Y.S. Microfluidic chip—Capillary electrophoresis for biomedical and environmental analysis. *Abstract, 7th Asia-Pacific International Symposium on Microscale Separation and Analysis (APCE 2007)*, Singapore, December 16–19, 2007, Paper 5:14, pp. 5:128.

69. Fung, Y.S. Microfluidic chip-capillary electrophoresis for biomedical applica-tions. *Abstract, Separation Science Singapore*, Biopolis Park, Singapore, August 26–28, 2009, p. 36.

70. Guo, W.P., Lau, K.M. and Fung, Y.S. Microfluidic chip-capillary electrophoresis for emergent biomedical analysis in Hong Kong. *Abstract K27, 9th Asia-Pacific International Symposium on Microscale Separation and Analysis (APCE 2009) and 1st Asian-Pacific International Symposium on Lab-on-Chip (APLOC2009)*, Shanghai, China, October 28–31, 2009, p. 36.

71. Fung, Y.S. Microfluidic chip-capillary electrophoresis devices—From sam-ple pretreatment to analyte detection for clinical diagnosis and monitoring of biomarkers and metabolites. *Abstract KL21, 10th Asian-Pacific International*

Symposium on Microscale Separations and Analysis (APCE 2010), Hong Kong, China, December 10–13, 2010, p. 48.

72. Fung, Y.S. Microfluidic chip—Capillary electrophoresis for biomedical analysis. *Abstract APCAS 2, 2012 Asia Pacific Conference on Analytical Science (APCAS)*, Manila, Philippines, April 11–13, 2012, pp. 26–27.

73. Qin, J.H., Fung, Y.S. and Lin, B.C. *Chinese J Chromatogr* 2003, 21, 464–468.

74. Nie, Z. and Fung, Y.S. Multi-segment incremental sample injection for the determination of bilirubin binding capacity of albumin and study of drug interaction by microchip capillary electrophoresis. *Abstract, 22nd International Symposium on Microscale Bioseparations & Methods for System Biology (MSB 2008)*, Berlin, Germany, March 9–13, 2008, Paper P.284-Tu, p. 417.

75. Fung, Y.S. Microfluidic chip—Capillary electrophoresis for investigation of binding interaction between free bilirubin, trace metals and essential minerals with biological significant proteins. *Abstract I-03, 8th Asia-Pacific International Symposium on Microscale Separation and Analysis (APCE 2008)*, Kaohsiung, Taiwan, November 2–5, 2008, p. 65.

76. Fung, Y.S. Microfluidic chip-capillary electrophoresis for assessing bilirubin-protein interaction and quality control of Chinese medicine. *Abstract, 8th Chinese Symposium on Microscale Bioseparations & Methods for System Biology (CMSB 2008)*, Guilin, China, November 21–24, 2008, p. 4.

77. Mo, S.L. and Fung, Y.S. Device based on microfluidic chip-capillary electrophoresis for the determination of free bilirubin in human serum. *Abstract PB13, 10th Asian-Pacific International Symposium on Microscale Separations and Analysis (APCE 2010)*, Hong Kong, China, December 10–13, 2010, p. 104.

78. Fung, Y.S., Guo, W.P. and Du, F.Y. Microfluidic chip-capillary electrophoresis devices for metabolomics applications. *Abstract L-121, 19th International Symposium on Capillary Electroseparation Techniques (ITP 2012)*, Baltimore, MD, September 30–October 3, 2012, p. 9.

79. Fung, Y.S. Microchip-capillary devices for metabolomics application. *Abstract, 1st International Conference on Urine Omics (URINOMICS)*, Caparica-Almada, Portugal, September 9–11, 2013, Paper O54, pp. 141–142.

80. Fung, Y.S. Microfluidic chip-capillary electrophoresis for emergent clinical diagnosis and monitoring of biomarkers and metabolites. *Abstract L-128, 17th International Symposium on Capillary Electroseparation Techniques (ITP 2010)*, Baltimore, MD, August 29–September 1, 2010, p. 13.

81. Du, F.Y. and Fung, Y.S. Microfluidic-chip capillary electrophoresis device incorporating a new serial dual-microelectrode detector for the determination of amino acids and proteins in food. *Abstract PB12, 10th Asian-Pacific International Symposium on Microscale Separations and Analysis (APCE 2010)*, Hong Kong, China, December 10–13, 2010, p. 103.

82. Wu, R.G., Fung, Y.S. and Yeung, W.S.B. Microfluidic chip-capillary electrophoresis for 2-D separation of proteins. *Abstract, 7th Asia-Pacific International Symposium on Microscale Separation and Analysis (APCE 2007)*, Singapore, December 16–19, 2007, Paper 5:49, pp. 5:141.

83. Wu, R.G., Fung, Y.S. and Yeung, W.S.B. Microfluidic chip-capillary electrophoresis for separation and determination of low abundance proteins. *Abstract P01-02, 8th Asia-Pacific International Symposium on Microscale Separation and Analysis (APCE 2008)*, Kaohsiung, Taiwan, November 2–5, 2008, p. 96.

84. Du, F.Y., Mo, S.Y. and Fung, Y.S. Dual microelectrodes as CE and HPLC post column detector for peak purity assessment and protein determination. *Extended Abstract, 13rd International Symposium on Electroanalytical Chemistry (13th ISEAC)*, Changchun, China, August 19–22, 2011, pp. 63–64.

85. Zhao, W.F., Du, F.Y., Chen, Q.D. and Fung, Y.S. Microchip-capillary electrophoresis devices incorporating magnetically activated microbeads with molecularly imprinted polymer for determination of antibiotics in milk. *Abstract M22, 4th International Symposium on Microchemistry and Microsystems (ISMM 2012)*, Zhubei City, Taiwan, June 10–13, 2012, pp. 138–139.

86. Sze, K.L. and Fung, Y.S. Microfluidic chip-capillary electophoresis for trace metal determination in milk. *Abstract, Analytical Research Forum, Royal Society of Chemistry*, Hull, UK, July 19–22, 2008, p. 33.

87. Sze, K.L. and Fung, Y.S. Microfluidic chip-capillary electrophoresis for the determination of trace metals in milk. *Abstract P01-03, 8th Asia-Pacific International Symposium on Microscale Separation and Analysis (APCE 2008)*, Kaohsiung, Taiwan, November 2–5, 2008, p. 97.

88. Zhao, W.F., Chen, Q.D., Wu, R.G., Wu, H., Fung, Y.S. and O, W.S. Capillary electrophoresis with LIF detection for assessment of mitochondrial number based on the cardiolipin content. *Electrophoresis* 2011, 32, 3025–3033.

89. Zhao, W.F., O, W.S. and Fung, Y.S. Microfluidic chip-capillary electrophoresis with laser induced fluorescence detection for assessing changes in mitochondria from HepG2 cells. *Abstract P01-06, 8th Asia-Pacific International Symposium on Microscale Separation and Analysis (APCE 2008)*, Kaohsiung, Taiwan, November 2–5, 2008, p. 100.

90. Zhao, W.F., Fung, Y.S. and O, W.S. Assessment of the mitochondria number based on cardiolipin by capillary electrophoresis-LIF detection. *Abstract PA17, 10th Asian-Pacific International Symposium on Microscale Separations and Analysis (APCE 2010)*, Hong Kong, China, December 10–13, 2010, p. 89.

91. Guo, W.P. and Fung, Y.S. Microfluidic chip-capillary electrophoresis for the determination of organic acids in biofluids. *Abstract, 8th Chinese Symposium on Microscale Bioseparations & Methods for System Biology (CMSB 2008)*, Guilin, China, November 21–24, 2008, p. 197.

92. Nie, Z. and Fung, Y.S. Microfluidic chip-capillary electrophoresis for quality assessment of a complex herbal preparation. *Abstract, 7th Asia-Pacific International Symposium on Microscale Separation and Analysis (APCE 2007)*, Singapore, December 16–19, 2007, Paper 5:48, pp. 5:141.

93. Mo, S.L. and Fung, Y.S. Development of microchip—Capillary electrophoresis for quality control of herbal medicine. *Abstract, 8th Chinese Symposium on Microscale Bioseparations & Methods for System Biology (CMSB 2008)*, Guilin, China, November 21–24, 2008, p. 187.

2

Instrumentation and Facilities I: Fabrication of Microfluidic Chip-Capillary Electrophoresis Device

Wenpeng Guo

First Affiliated Hospital of Shenzhen University
Shenzhen, People's Republic of China

CONTENTS

2.1 Fabrication of Microfluidic Chips

Initially, most of the processes for fabricating microfluidic chips (MCs) were developed using silicon as the substrate material, with the fabrication process derived from well-established technologies in the semiconductor manufacturing industry [1]. Due to the increasing demand for specific biocompatibility, desirable optical characteristics, faster prototyping, and lower production costs, new materials and new processes have been developed for selective etching of glass, metal, and ceramic substrates in conventional lithography, as well as for deposition and bonding of polydimethyldisiloxane (PDMS) with desired substrates in soft lithography [2]. Other new processes, such as laser ablating for poly(methyl methacrylate) (PMMA) substrates, thick-film, and stereolithography [3], are also developed as well as fast replication methods via electroplating, injection molding, and embossing [4].

2.1.1 Conventional Photolithography Methods

The most common methods for microfabrication of MCs are based on photolithography, using a photomask process developed for selective protection and layer-by-layer removal of the designed parts from a substrate such as silicon or glass (Figure 2.1) [1]. The geometric pattern is transferred by light from a photomask to a photoresist covering a light-sensitive chemical layer previously deposited onto a silicon or glass substrate. The exposure patterns underneath the photoresist are then visualized after applying a series of chemical treatments.

2.1.2 Soft Lithography Methods

Soft lithography was first introduced by Whitesides and coworkers [5] in 1997. The fabrication process used a hard master mold to prepare a PDMS replica. Soft lithography provides a low-cost method for making integrated MCs. Two of the most commonly used soft lithography methods are replica molding and microcontact printing [6,7]. For the replica molding method, an elastomeric mold and ultraviolet (UV) light or thermally sensitive epoxy master is typically fabricated by the photolithographic method to

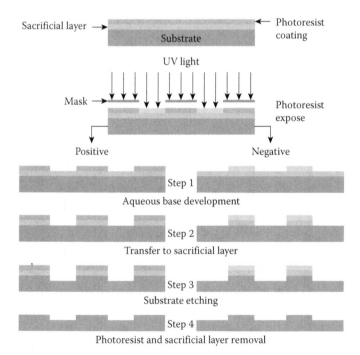

FIGURE 2.1
Schematic diagram showing the photolithographic process for fabrication of silicon or glass chip.

define a stamp pattern as shown in Figure 2.2. Upon placement of the mold onto the polymer coating at the substrate surface, heat energy or UV light irradiation is applied to generate the desired micropattern.

Another method for generating a channel pattern on a PDMS mold is microcontact printing by transferring a patterned self-assembled monolayer (SAM) onto the PDMS mold surface for use as an etching mask [8]. The process is shown in Figure 2.3. The SAM solution is deposited onto the surface

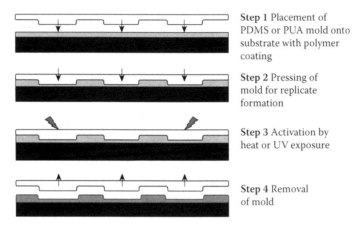

Step 1 Placement of PDMS or PUA mold onto substrate with polymer coating

Step 2 Pressing of mold for replicate formation

Step 3 Activation by heat or UV exposure

Step 4 Removal of mold

FIGURE 2.2
Schematic diagrams showing the operation steps for the soft lithographic process using polyurethane acrylate (PUA) or PDMS molding as replica to fabricate PDMS chip.

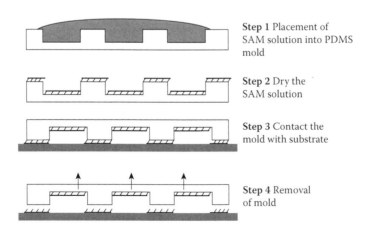

Step 1 Placement of SAM solution into PDMS mold

Step 2 Dry the SAM solution

Step 3 Contact the mold with substrate

Step 4 Removal of mold

FIGURE 2.3
Schematic diagrams showing the soft lithographic process using SAM for microcontact printing to fabricate PDMS chip.

of a PDMS mold. Upon the evaporation of the SAM solution, a pattern is formed at the surface of the mold. The pattern is transferred to the substrate by contacting the mold with the substrate.

2.1.3 Fabrication of MC-CE Devices by Laser Ablation

Microfluidic chip-capillary electrophoresis (MC-CE) devices using laser ablation offer a widely used method for fabricating MCs made of polymers such as PMMA, polystyrene, polycarbonate, polyethylene terephthalate, polyimide, cellulose acetate, and photoresists [9,10]. In the fabricating process, the laser beam that is focused on the solid polymer surface transfers its energy to the solid surface and vaporizes the polymer to create a predetermined channel pattern.

Both UV and CO_2 lasers are frequently used as the source for ablation on the polymeric material. The ablation mechanism of the UV laser is a combination of photothermal and photochemical processes, which break down and remove the target area of the polymer substrate thermally as well as photochemically, simultaneously. In contrast to the UV laser, the ablation of the CO_2 laser is mostly attributed to a photothermal process at the plastic surface under a strong incidence of the infrared radiation at 10.6-μm wavelength [11]. A UV laser with a shorter wavelength is typically operated in pulsed mode, which can produce a finer structure compared to the CO_2 laser. However, the relatively high capital and operation costs as well as the rough surface produced by the pulse mode hinder the use of the UV laser for fabricating polymeric substrates.

In contrast, the CO_2 laser emits high intensity at short pulse radiation, giving rise to a rapid and uniform increase in the temperature of the irradiated spot by the focused laser beam to decompose the polymer material. The decomposition mechanism varies for the different materials, depending on the chemical bonds that make up the polymer and its associated structure. For PMMA, monomers are formed during vaporization at decomposition temperature. A molten polymer pool is created at the spot where the laser beam was focused onto the surface, leading to the breakdown of the PMMA polymer structure and the production of methyl methacrylate (MMA) from the ablated zone. The heated gases exploding from the vaporizing polymer push the liquid polymer away from the hot zone, resolidifying them at the nearby cooler areas. Thus, the moving laser beam is able to remove polymer materials to produce structures such as channels and wells, as illustrated in Figure 2.4, which shows the process for the formation of molten PMMA and the exit of the MMA vapors from the ablated PMMA region [11].

The width of the channels produced depends on the speed of heat dispersion through the polymer material and the distribution of laser intensity within the laser beam. Due to the slow thermal diffusivity in polymers, the intensity

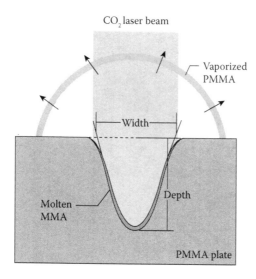

FIGURE 2.4
Channel profile created by laser ablation at the PMMA chip.

distribution of the laser beam determines the channel cross section. Thus, a Gaussian-like profile is normally produced, as shown in Figure 2.4.

2.2 Instrumentation and Facilities for Fabrication of MC-CE Devices by Laser Fabrication

The basic system for laser fabrication includes a computer installed with computer-aided design (CAD) software, a laser machine, and a thermal bonding machine, as shown in Figure 2.5. The designed channel pattern is created by a Pinnacle V-series CO_2 laser with a wavelength of 10.6 μm onto a PMMA polymer plate under the control of CorelDraw 10.0 computer software. The hot press machine is operated up to 500°C under 1 MPa of air pressure and was purchased from Guangju Machinery Company (Dongguan, China).

A laboratory-built chip workstation was used to enable microscopic observation of the microfluidic channels during operation. It has the capability for three-dimensional (X, Y, and Z directions) control of the MC. The high-voltage power supply (CZE 1000R, Spellman, Suzhou, China) was controlled by an electronic control system (see Chapter 3 for details) to deliver high voltages ranging from −10 to 0 kV to the MC-CE system. The analytes separated by CE were detected by an SPD-10A UV-visible spectrophotometric detector (Shimadzu Corporation, Kyoto, Japan).

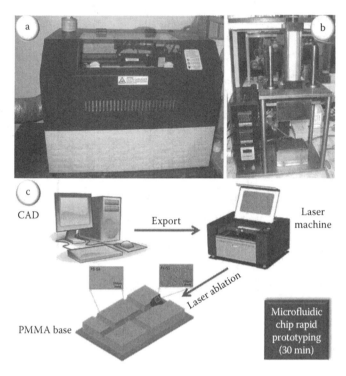

FIGURE 2.5
Basic components required for the fabrication of MC-CE devices by laser ablation. (a) CO_2 laser ablation equipment; (b) thermal bonding machine; (c) computer with CAD software for controlling CO_2 laser fabrication.

2.3 Advantages for Fabrication of MC-CE Devices by Laser Ablation

Although femtosecond laser-based prototyping had been reported for fabrication of MCs by Wu et al. [12] to develop movable devices and functional microvalves, only limited operations can be performed on-chip. With a suitable microchannel pattern fabricated on-chip using a CO_2 laser and programmed high-voltage switching at desired vials, Guo et al. [13–15] developed complex procedures to enable mixing and standard addition to be performed on PMMA chips fabricated using a low-cost laser machine. Examples for single- and dual-channel mixing are shown with special features and discussed in Chapters 4 and 5 to illustrate the flexible design enabled by a fast and one-step laser fabrication, lower fabrication cost, and easy bonding procedure.

A typical MC-CE device developed and applied for biomedical analysis is shown in Figure 2.6 [14]. Different lengths of the microchannel can be

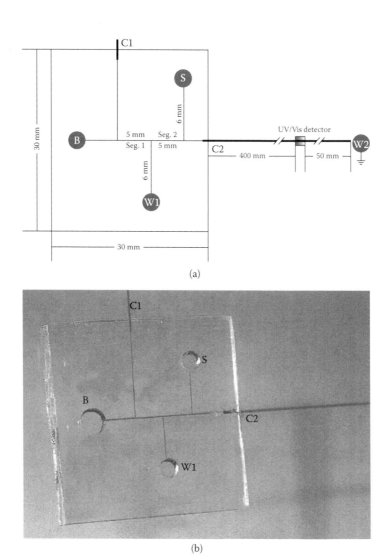

(a)

(b)

FIGURE 2.6
(a) Schematic layout of the MC-CE device. All dimensions shown are given in millimeters.
(b) Photograph showing the MC-CE device fabricated for profiling of organic metabolites
in human serum. B, buffer vial; W1, W2, waste vial 1, waste vial 2; S, standard/sample vial;
C1, sample inlet capillary; C2, separation capillary.

fabricated to produce different volumes for mixing and standard addition,
thereby making it possible to carry out various lab-on-chip operations
on-site. The MC-CE device can also incorporate sample pretreatment and
cleanup prior to highly efficient separation by CE using commercially
available instrumentation.

TABLE 2.1

Comparison of Three Conventional Fabrication Methods for Microfluidic Chips

Fabrication Method	Time Required		Material	Fabrication Cost	Reference
	Design	Fabrication			
Laser fabrication	Short	Short	Polymers (CO_2 laser), glass (UV laser)	Low	[14]
Photolithography	Long	Long	Glass, silicon	High	[1]
Soft lithography	Moderate	Moderate	Polymers	Moderate	[5]

2.4 Summary

Table 2.1 compares the three conventional MC fabrication methods: laser ablation, photolithography, and soft lithography. The fabrication time using laser ablation is much shorter than those of photolithography and soft lithography. Photolithography has the longest time due to complex operation steps. Photolithography can handle both glass and silicon, whereas soft lithography is better for polymer chips. The laser fabrication method can handle both glass and polymer substrates using a UV and CO_2 laser, respectively. The fabrication cost is the lowest for laser fabrication and the highest for photolithography, whereas soft lithography shows a moderate cost compared to the other fabrication methods.

In summary, the laser method provides the optimal procedure for fabrication of MCs, in particular, for polymer chips. The flexibility of CAD enables the user to create the best designed microchannel pattern. This approach enhances the ability to solve demanding applications for complex samples that require extensive and lengthy procedures for sample preparation before the use of high-efficiency CE for separation and quantitation of the targeted analytes.

References

1. Camp, J., Stokol, T. and Shuler, M. Fabrication of a multiple-diameter branched network of microvascular channels with semi-circular cross-sections using xenon difluoride etching. *Biomed Microdevices* 2008, 10, 179–186.
2. Seo, J.H., Leow, P.L., Cho, S.H., Lim, H.W., Kim, J.Y., Patel, B.A., Park, J.G. and Hare, D. Development of inlaid electrodes for whole column electrochemical detection in HPLC. *Lab Chip* 2009, 9, 2238–2244.

3. Truckenmuller, R., Giselbrecht, S., Bilitterswijk, C.V., Dambrowsky, N., Gottwald, E., Mappes, T., Rolletschek, A., et al. Flexible fluidic microchips based on thermoformed and locally modified thin polymer films. *Lab Chip* 2008, 8, 1570–1579.
4. Koesdjojo, M.T., Tennico, Y.H. and Remcho, V.T. Fabrication of a Microfluidic System for Capillary Electrophoresis Using a Two-Stage Embossing Technique and Solvent Welding on Poly(methyl methacrylate) with Water as a Sacrificial Layer. *Anal Chem* 2008, 80, 2311–2318.
5. Zhao, X.M., Xia, Y.N. and Whitesides, G.M. Soft lithographic methods for nano-fabrication. *J Mater Chem* 1997, 7, 1069–1074.
6. Xia, Y. and Whitesides, G.M. Soft Lithography. *Angew Chem Int Ed* 1998, 37, 551–575.
7. Xia, Y., Rogers, J.A., Paul, K.E. and Whitesides, G.M. Unconventional methods for fabricating and patterning nanostructures. *Chem Rev* 1999, 99, 1823–1848.
8. Kim, P., Kwon, K.W., Park, M.C., Lee, S.H., Kim, S.M. and Suh, K.Y. Soft lithography for microfluidics: a review. *Biochip J* 2008, 2, 1–11.
9. Roberts, M.A., Rossier, J.S., Bercier, P. and Girault, H. UV Laser Machined Polymer Substrates for the Development of Microdiagnostic Systems. *Anal Chem* 1997, 69, 2035–2042.
10. Pethig, R., Burt, J.P., Parton, A., Talary, M.S. and Tame, J.A. Development of bio-factory-on-a-chip technology using excimer laser micromachining. *J Micromech Microeng* 1998, 8, 57–63.
11. Klank, H., Kutter, J.P. and Geschke, O. CO_2-laser micromachining and back-end processing for rapid production of PMMA-based microfluidic systems. *Lab Chip* 2002, 2, 242–246.
12. Wu, D., Chen, Q., Niu, L.G., Wang, J.N., Wang, R., Xia, H. and Sun, H. Femtosecond laser rapid prototyping of nanoshells and suspending components towards microfluidic devices. *Lab Chip* 2009, 9, 2391–2394.
13. Guo, W.P., Lau, K.M. and Fung, Y.S. Microfluidic Chip-Capillary Electrophoresis for Emergency Onsite Biomedical Analysis. *Sep Sci* 2009, 2, 3–10.
14. Guo, W.P., Lau, K.M. and Fung, Y.S. Microfluidic chip-Capillary Electrophoresis for two orders extension of adjustable upper working range for profiling of inorganic and organic anions in urine. *Electrophoresis* 2010, 31, 3044–3052.
15. Guo, W.P. and Fung, Y.S. Microfluidic chip-capillary electrophoresis with adjustable on-chip sample dilution for profiling of urinary markers. *Proceedings of the 14th International Conference on Miniaturized Systems for Chemistry and Life Sciences*, Groningen, The Netherlands, 2010, pp. 1481–1483.

3

Instrumentation and Facilities II: Manipulating Nanofluids for Desired Operations to Achieve Intended Application by MC-CE Device

Ruige Wu

Singapore Institute of Manufacturing Technology
Singapore

Zhou Nie

Guangzhou Amway (China) Co., Ltd.
Guangzhou, People's Republic of China

Ying Sing Fung

The University of Hong Kong
Hong Kong SAR, People's Republic of China

CONTENTS

3.1 Introduction

Due to the rising demand for on-site analysis and the recent advancement in research findings such as the discovery of biomarkers and protein
drugs, demand has arisen in recent years for their assay in body fluid. Brief
descriptions and discussion are given on the rising needs and requirements
for protein assay and the current analytical methods for their determination in biofluids. Controlled manipulation of nanofluids by incorporation of
required analytical procedures with MC-CE devices is reviewed and operations discussed to illustrate the advantages for functional integration of MC
with CE to achieve desired applications.

   The recent success of MC-CE device to achieve intended application is
attributed to the availability of required high voltage instrumentation for
electric field–driven nanofluid manipulation and laser fabrication facilities
to create desired microchannel pattern on-chip by users. Thus, the required
high voltage instrumentation for on-demand nanofluid manipulation and
the various microchannel pattern to manipulate nanofluids for desired operations are reviewed in this chapter. The various laser fabrication facilities
and operations are presented and discussed in Chapter 2.

3.1.1 Need, Requirements, and Analytical Approaches
       for On-Site Analysis

The need for analytical devices capable of on-site operation have arisen recently
to meet the increasing worldwide demand on environmental monitoring,

food safety, medical care and clinical diagnosis, public health, and security. Sensitive, economic, and quick analytical techniques are required for on-site monitoring. Currently, there are two major approaches adopted for on-site clinical assays. The first approach is based on the use of a chemical reaction, specific to interact with the target analyte to produce measurable products for detection, such as a change in color for the testing paper strip or solution upon reaction with specific reagents. This approach provides a cheap assay without the need to use expensive equipment, and it is capable of on-site analysis and delivers quick results on-site to meet the demand. It has been adopted in the form of reagent kits for field monitoring in water analysis, and it is frequently used for various clinical assays handled by medical personnel [1]. Its limitations are potential interference from substances present in samples with a complex and highly variable sample matrix, relatively low detection sensitivity, and a limited number of chemical reactions highly specific to target analytes that can be used for analytical application.

The second approach for on-site clinical assays is based on current instrumental methods that often require high capital and operational costs and the use of complex and lengthy procedures. They are used for assay of important clinical parameters to assess for patients under critical conditions in hospitals and clinics as well as for monitoring of toxic environmental pollutants present at trace levels in complex samples, such as toxic metals in air particulates [2,3]. Although it has high capital and running costs, this approach enables the determination of important analytes in complex samples which cannot be handled by the specific chemical approach. However, it is not suitable for routine on-site assay for determining analytes that require frequent monitoring to obtain timely critical results for immediate action.

The recently developed microfluidic chip-capillary electrophoresis (MC-CE) devices provide a promising approach for on-site routine analysis of critical parameters requiring results for immediate action, such as bedside monitoring of parameters for medical management of patients under critical conditions and fast diagnosis of infective diseases as a quick response to implement a timely measure to combat a significant threat for public health [4]. To develop an MC-CE device capable of carrying out required lab-on-a-chip operations for the isolation of analytes prior to their separation and accurate determination using CE procedures, important laboratory operations and procedures integrated with the MC-CE devices are needed, and they are discussed in Section 3.1.2.

3.1.2 Integrating Analytical Procedures with MC-CE Devices for Desired Operations by Nanofluid Manipulation

Various procedures to be integrated with MC-CE devices to enable on-site lab-on-a-chip operations are as follows:

1. Introduction of biofluids at a constant and suitable sample size for the MC-CE device

2. Cleanup of sample introduced into the MC-CE device to remove substances present in the sample matrix which may interfere with the determination of the target analyte

3. Enrichment of analytes from the cleanup samples to a level sufficient to meet the application requirements and be within the detection sensitivity of the detector

4. Separation of analytes from interfering substances with similar chemical properties

5. Quantitation of analytes by the detector, with sufficient detection sensitivity to meet the requirements for the intended application

Adoption of the above-mentioned procedures are illustrated in Sections II–IV of this book for intended applications. Most procedures are based on the arrangement for continuous online integration of suitable operations performed in the device to achieve the target analytical tasks. To complete the required analytical tasks, the following operational steps have to be carried out in suitable sequence in the MC-CE devices to achieve the intended application.

1. Injection of samples with different volumes at the nanofluid scale into MC-CE device for determination of the target analyte present in the samples

2. Dilution of samples on-chip to a suitable ratio to bring samples with highly variable analyte concentrations within the working range of the MC-CE device and reduce the amount of interfering substances to a level capable of being handled by the device

3. Enrichment of analyte in samples with analyte concentration below the detection limit of the detector

4. Addition of analyte standard to sample at different concentration ratios for their determination under the same environment as those originally present in the sample to correct for sample matrix interference

5. Mixing of different nanofluids on-chip at adjustable volume ratios required for standard addition, dilution, and binding assay and to carry out highly sensitive detection schemes by postcolumn chemical reaction

6. Integration of operations for multidimensional separation, sample cleanup, and sensitivity enhancement for determination of analytes at trace levels in highly complex samples

7. Controlled movement of nanofluids to designated places to stay for a specified time duration for sample cleanup, analyte isolation, stacking, and enrichment prior to their detection after CE separation

Successful cases for implementation of the above-mentioned operations are discussed in Sections II–IV of this book for integration of various operations for intended applications. The most basic operation for all procedures to be carried out is a controlled manipulation of nanofluids in the MC-CE devices to move to designated places for desired times. Important operations to control the movement of nanofluids enabled by pressure-driven, electric field–driven, and magnetic-activated modes are discussed in Section 3.2, following by a brief description of required instruments and facilities in Section 3.3 and their application for essential operations to be carried out in MC-CE devices in Section 3.4.

3.2 Modes for Manipulation of Nanofluids for Desired Operations

There are three major modes for manipulation of nanofluids for intended operations: (1) pressure-driven mode, (2) magnetic-activated mode, and (3) electric field–driven mode. The advantages and limitations of each mode for nanofluid manipulation in MC-CE devices are discussed in Sections 3.2.1–3.2.4. To meet the requirements for an intended application, it is often necessary to use two or three modes in a suitable combination to implement the desired operations. The use of desired operations by manipulating nanofluids for sample injection, standard addition, binding assays, multidimensional separation, and other more complex procedures as implemented in MC-CE devices for specific applications is discussed in Section 3.4.

3.2.1 Pressure-Driven Mode

Pressure-driven modes to manipulate nanofluids, such as the use of syringe pump, vacuum suction, and gravity feed operations, are commonly used to control the movement of nanofluids, and they show satisfactory repeatability for sample introduction prior to running the CE procedures for separation and determination of analytes. The advantages of the pressure-driven mode are the equal rate for introduction of analytes in the sample to the microscale devices with no biases among different analytes; capability to introduce different sample volumes to the device as required in the analytical procedure; and supporting proven interface with CE, MC-CE, and other microscale devices.

Gravity feed mode is commonly used to introduce samples for MC-CE devices, as well as for feeding reagents for postcolumn reaction to enhance the detection sensitivity for MC-CE devices [5–7]. It has a limited range for control of nanofluid flowrate, depending on the limited height difference for the liquid reservoir. Moreover, the action to raise the liquid reservoir to different heights is slow, hindering its implementation for procedure automation. In contrast, the operation of the vacuum suction mode is fast, and the suction time duration can be controlled easily by a valve with controlled opening for a specified duration. Thus, vacuum suction mode is commonly used in commercial CE equipment for sample introduction. Moreover, it is easier for automation, and a larger and variable sample size can be introduced using vacuum suction compared to gravity feed mode.

A syringe pump operated by a high-precision stepper motor provides flexibility for flowrate control and ease of use for procedure automation. It has been commonly used for moving nanofluids from one position to another on a microfluidic chip (MC). However, when sequential movement at different durations is required in an MC fabricated with a complex microchannel pattern with numerous openings to the external environment, control of nanofluid movement is complicated, as numerous two- and three-way on-chip valves are required to avoid the conflict in pressure among different parts of the nanofluidic channel pattern. Thus, a syringe pump is used in MCs with simple microchannel patterns to move nanofluid to a limited number of vials in a closed system. It is not suitable for handling complex operations, such as those used for sample cleanup and analyte enrichment, or in multidimensional separation as described in Section 3.4.

3.2.2 Magnetic-Activated Mode

The manipulation of nanofluids by magnetic-activated mode provides an attractive means for nanofluid control as the control action is performed outside the MC, thereby simplifying the design for the MC-CE device for a given task. There are currently two major schemes used in the magnetic-activated mode. The first scheme uses a micrometer-sized, needle-shaped, Teflon-coated magnet added to the MC vial to mix reagents and samples prior to the separation and determination of the target analytes. This scheme can extend the working range by an order of magnitude in the determination of highly variable metabolites in urine [8–11]. Details on the operation are discussed in Chapter 4.

The second scheme uses ferrofluid added to the microchannel to perform the required operations upon magnetic activation. Ferrofluid is a stable colloidal suspension of subdomain magnetic particles dispersed in a liquid carrier immiscible with an aqueous solution. Each of the 10-nm particles present in the ferrofluid is coated with a surfactant to prevent its agglomeration and to maintain thermal stability. The water-immiscible ferrofluid (APG S12n) available from Ferrofluidic Inc. (Nashua, NH, USA) contains carrier (synthetic ester oil, 85% by volume), surfactant (10% by volume), and magnetic solids (5% by volume). In the absence of a magnetic field, the randomly distributed ferroparticles exhibit no net magnetization. In the presence of a magnetic field, ferroparticles act as a homogeneous magnetic liquid aligning along the applied magnetic field and moving toward areas with the highest flux. Thus, with careful manipulation of an external magnetic field, ferrofluid can be used to direct the movement of nanofluids to the desired location in the MC and to act as a stop valve for nanofluids, with a holding force proportional to the gradient of the applied field and the degree of magnetization of the ferrofluid.

In addition to pushing the aqueous nanofluid toward the desired position, the magnetized ferrofluid can also act as a stop valve to block the return movement of the nanofluid after it has reached the mixing coil. At the mixing coil, a rotation magnet is used to activate the ferrofluid to move several segments of nanofluid in a circular channel, with the aim to produce a homogenized segment of nanofluid for subsequent operations [12–18]. Details on the operation are given in Section 3.4.2.1 and on its application in binding assay in Chapter 13. With concurrent on-and-off operations of two stop valves acting in tandem, it can be used as an on-chip three-way valve for nanofluid control in mixing application [8,19,20].

The advantages for using ferrofluid to manipulate nanofluids are its low cost of operation and flexibility to integrate with other operations. A limitation is the slow movement of the magnetic-activated ferrofluid to avoid its breakup during operation. As mechanical movement is required for the operation of magnetic-activated control, it is too clumsy to automate for multiple operations required for complex operations.

3.2.3 Electric Field–Driven Mode

Use of an electric field for nanofluid control has been well established in electrophoresis for the separation of both charged and neutral analytes based on the differences in their charge, mobility, and differential migration driven by an electric field under high voltage (HV) applied to the device. There are two modes for transporting nanofluids in MC devices: under an electric field gradient and by electro osmotic flow (EOF), with the former for charged substances and the latter for both charged and neutral substances. Thus, for moving nanofluid to a desired position in the MC-CE device, EOF should be used for sample introduction because it is nonbiased for both charged and neutral substances. Thus, a low electric field gradient should be used for a longer period during sample introduction to avoid differential movement of analytes and to ensure the fill up of the sample channel segment by the sample introduced by EOF. In contrast, HV should be applied for separation during a CE run for analyte separation in the separation capillary.

The advantages for using an electric field to move nanofluids are as follows:

1. The nanofluid plug can be directed to the desired position depending on the sign of voltage imposed at the start and terminal vials of the MC.
2. In a suitably designed voltage-switching program, nanofluid can go directly to the terminal vial within a complex microfludid channel network in an open system without the need for using any three-way valve or two-way stop to direct or block its movement.

3. The operation does not involve any mechanical movement, and only HV switching at suitable vials for a specified time is required.

4. Integration with CE for analyte separation is easy and direct because both MC and CE are under voltage control. The limitations are the hazards associated with the use of HV and the need to control humidity during operation to avoid self-discharge through humid air. Thus, hermetically sealed power supplies and humidity control for the power supply are required for electric field–driven manipulation.

3.2.4 Applicability of Different Manipulation Modes for MC-CE Operations

For the pressure-driven mode, the gravity feed method is suitable for addition of reagent and sample for systems that do not require a high volume throughput. For vacuum suction, a faster and a larger sample size can be introduced, and it is easy to apply for automation implementation. A syringe pump to manipulate nanofluids offers a flexible and controllable means to deliver samples and reagents for MC. However, for a complex microchannel system, the instrument cost increases rapidly, as many pumps, valves, and stops are needed to direct the flow of nanofluids to different locations of the MC for desired operation.

The magnetic-activated mode using ferrofluid to drive nanofluids to desired positions provides a highly flexible means for manipulating nanofluids in MC-CE devices, because it can also act as an on-chip stop valve to direct or block the movement of nanofluids. A limitation is the slow operation needed to avoid the breakup of ferrofluids during operation. Thus, the magnetic-activated mode poses difficulties for automation of the analytical procedure in MC-CE devices, in particular, for those devices requiring numerous operations in sequence in MCs with complex microchannel patterns.

For the electric field–driven mode, nanofluid flow directed by electric field gradients provides a desirable means to move nanofluids to designated positions; in particular, it is desirable for handling multiple samples for multichannel operation. It provides a high throughput platform for microscale analysis. For a complex operation system using several techniques in sequence, such as multidimensional analysis, the electric field–driven mode is much preferred compared to the other control modes, because nanofluids can be manipulated without the need for pressure control in a complex microchannel network with connectivity to numerous vials under ambient conditions. The instrumental requirement is simple, requiring only an HV power supply and an associated switching system to perform the desired voltage–time program for controlling microelectrodes placed inside desired vials of the MC.

Based on a carefully designed microfluidic channel pattern with an associated control program for applying HVs at desired vials in sequence, no on-chip valve or on-chip stop is necessary for intended application as shown in Sections II–IV of this book. Operation and procedure automation are much simpler using MC-CE devices with online continuous flow and on-chip integration of MC with CE separation to prevent contamination, reduce analysis time, and enhance analyte detection. The applicability of the pressure-driven, magnetic-activated, and electric field–driven modes for nanofluid manipulation is summarized in Table 3.1. For performing complex operations with on-chip integration of sample pretreatment and analyte enrichment, the voltage-driven mode is preferred for the manipulation of nanofluids to achieve the targeted lab-on-a-chip operations for handling complex real samples.

3.3 Instrumentation and Facilities for Electric Field–Driven Nanofluid Manipulation

To meet the need for the development of portable CE equipment and MC research, multiple HV supplies are available commercially (see Appendix I for information on suppliers of portable CEs, HV power supplies, and associated cables) with the capability to apply multiple HVs to designated vials for a specified sequence under user-friendly software (CorelDRAW 10, Corel Corporation, Ottawa, Ontario) control to define the control level, time duration, and sequence for an eight-channel HV power supply. Two commercial systems have been investigated, with results discussed for their integration with MC-CE devices. Both systems offer eight-channel HVs with different ways to control voltage. The special features and limitations of these two systems are given in the following sections.

3.3.1 Power Supply with Eight Switchable HV Channels Operated by One HV Source

The first system investigated is a power supply with eight switchable channels operated by one HV source. Switching between channels can be controlled at eight switchable positions across an HV resistor for different durations to provide eight HV channels with voltages adjustable between 0 and +3500 V (model MP-3500-FP, Major Science, Taiwan). The HV system is shown in Figure 3.1. In this system, both the HV and ground voltage can be switched at the same time to a pair of microelectrodes placed at selected vials. With external connection to other microelectrodes, it can control more than two microelectrodes, with one group of electrodes at HV and the other group at ground voltage at a given switchable position. The control of only one HV at a time during each switch has limited the applicability of the device. In addition, the adjustable voltages between 0 and +3500 V are not sufficient for many analytical CE separations. The system has been used for assessing the binding assay of bilirubin by albumin, with details given in Chapter 13.

TABLE 3.1

Applicability of Magnetic-Activated and Pressure-Driven and Electric Field–Driven Modes for On-Chip Operations by MC-CE Devices

Microfluid Manipulation Mode	Valves (Operation)	Microfluid Flowrate	Easy for Automation	Control Flexibility	Operation Time	Sample Pretreatment	Remarks
Pressure-Driven Mode							
Gravity feed	No (sample intake); yes (operation)	Low	Yes	Limited	Slow	No	Sample intake and solution introduction
Syringe pump	No (sample intake); yes (operation)	Fast and adjustable	Yes	Good	Moderate	Limited	Sample intake and reagent introduction Multioperation on-chip for simple tasks
Vacuum suction	No (sample intake); yes (operation)	Moderate	Yes	Limited	Fast	No	Sample intake and solution introduction
Magnetic-activated mode	No (sample intake); yes (operation)	Low	Moderate	Good	Slow	Limited	Flexible but slow on-chip operation
Electric field–driven mode	Valveless	Fast and adjustable	Yes	Good	Fast	Yes	Flexible multioperation on-chip for complex tasks

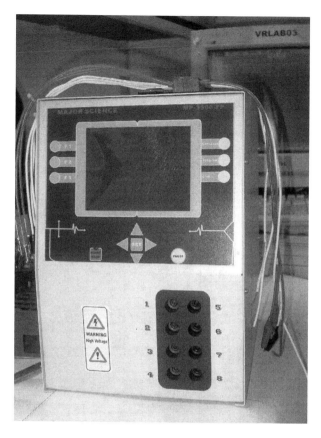

FIGURE 3.1
An eight high-voltage (HV) microfluidic power supply with one adjustable HV between 0 and +3500 V and eight switching positions across the resistors.

3.3.2 Power Supply with Eight Programmable HV Channels Independently Operated by Eight HV Sources

To broaden the scope of application, an Octo-Channel Multi-Output High Voltage System (~0–4000 V, EMCO High Voltage Corporation, Sutter Creek, CA, USA) was acquired and tested for HV control in MC-CE devices under a control software compiled by the Technology Support Centre (University of Hong Kong) for independent control of eight HV channels (Figure 3.2a). One of the eight channels has to be designated as ground and cannot be changed by software control, making many essential operations impossible to perform on-chip because the ground electrode has to be changed in different sequences to direct the flow of nanofluid to the desired vials.

The upper voltage at 4000 V is sufficient for nanofluid manipulation within the MC, but not sufficient for analytical separation within the capillary. To overcome the above-mentioned obstacles, an HV switching system

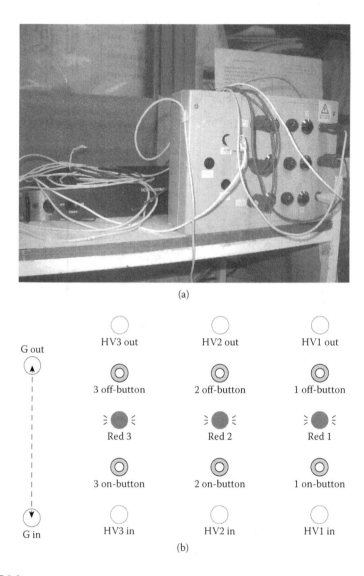

FIGURE 3.2
(a) Photograph of the front control panel for voltage switching by the eight-channel high-voltage power supply. (b) Schematic diagram showing the high- and low-voltage buttons and control.

was constructed to incorporate two HVs: one HV for nanofluid manipulation within the MC and the other HV for running the CE separation procedure. It can switch the ground voltage to different vials during the various operations performed in sequence for the implementation of the analytical procedure using the MC-CE device. The operation of the HV switching system is given in the next section.

3.3.3 Electronic Circuit for Multi-HV Switching to Achieve Desired Operations of the MC-CE Devices

For switching HVs, safety considerations for workers are required during operation. For operation of the electronic control circuit (Figure 3.3), low-voltage circuits are normally used by the worker with output to control HV switching operation inside a grounded box. Thus, there are two zones: the first zone for interface with the control circuit operated at low voltages (those filled with grey color and placed under low voltage control circuit (between two 0Vs marked above) as shown in Figure 3.3) and the second zone for HV switching (those filled up with black color and placed under high voltage control circuit (HV1, HV2 and HV3) as shown in Figure 3.3). Cables with matching colors are used to connect between the MC-CE device and the HV power to avoid wrong connection (Figure 3.2). Red cables are used for HV and white cables are designated as ground for connection to the HV supply, whereas yellow cables are used to connect to MC-CE devices.

Special HV cables are used, which can handle up to 20 kV or more, depending on different operations. Normal HV switches purchased from suppliers of electronic accessories, such as RS Components Ltd. or LabSmith, Inc. (see Appendix II), can handle up to 10 kV, which is sufficient for most analytical applications. Companies specializing in HV switches and cables, such as GIGAVAC Advanced Switching Solutions and ELAND Cables, can supply HV switches and cables that can handle much higher than 10 kV, although at a much higher cost.

For safe operation of the switching system, as shown in Figure 3.3, the following procedure should be performed to switch off all connecting buttons to the ground prior to starting a new HV control sequence with selected HV buttons. 1) Push all Off buttons (1, 2, and 3) first to disconnect all HV supplies at the start. All red lights should be off. 2) Switch all microelectrodes in vials by pushing the HV-Out and Ground-Out buttons before starting the HV sequence program. 3) Push 1 On-button, 2 On-button, or 3 On-button in suitable sequence to connect predetermined HV to the MC-CE devices following the control sequence.

3.4 Nanofluid Manipulation to Achieve Desired Operations for MC-CE Devices

Of the major operations performed by MC-CE devices, three basic operations required for multidimensional analysis—injecting the sample, mixing of microfluids, and performing complex operations—are described with results in this section. Other operations required for specific use are given in detail in Sections II–IV of this book as required under specific applications.

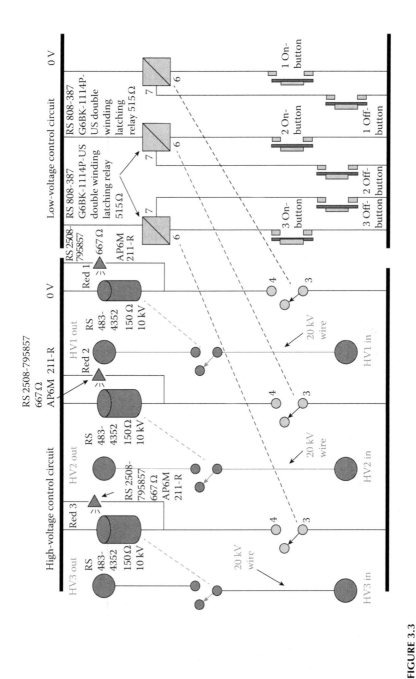

FIGURE 3.3
Circuit diagram for control of switching high voltage at microelectrodes placed at designated vials.

3.4.1 Operations to Introduce Samples and Nanofluids to MC-CE Devices

Sample introduction techniques, such as vacuum suction and hydrodynamic and electrokinetic injection, commonly used in CE are not suitable for MC-CE devices due to different requirements, such as a larger injected sample size on-chip for CE separation and no requirement for other laboratory facilities for sample injection for on-site operation of MC-CE devices. To search for a suitable mode for sample introduction for MC-CE devices, the sample injection methods for MC are reviewed here and shown in Figure 3.4. Their applicability for sample introduction for MC-CE devices is discussed in the following sections.

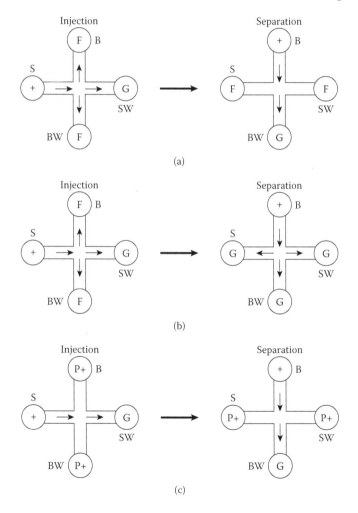

FIGURE 3.4
Various modes of sample injection for MC-CE devices. (a) Floated injection/floated separation. (b) Floated injection/grounded separation. (c) Pinched injection/pinched separation.
(Continued)

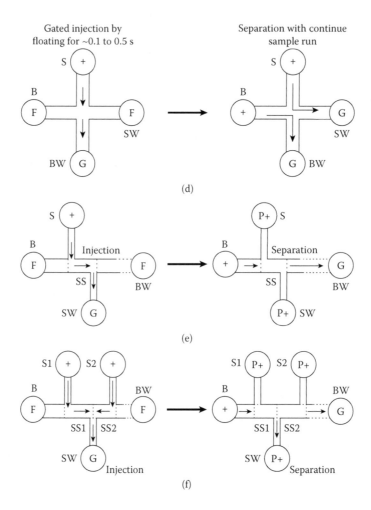

FIGURE 3.4 (Continued)
Various modes of sample injection for MC-CE devices. (d) Gated injection under continue buffer and sample run. (e) Double-T injection/pinched separation. (f) Triple-T injection/pinched separation. S, sample vial; S1, sample 1/standard 1 vial; S2, sample 2/standard 2 vial; SS1, sample segment 1; SS2, sample segment 2; B, buffer vial; BW, buffer waste vial; SW, sample waste vial.

3.4.1.1 Floated Injection/Floated Separation

The simplest sample introduction mode, using floated injection and floated separation, has been employed in the first MC design with a cross-channel configuration. The operation is easy to realize because the required HV applied across the sample to grounded sample waste vial during sample injection can be readily switched over to across the buffer to buffer waste vials during the separation stage, whereas the other vials are kept floating (Figure 3.4a). Thus, only one HV is used in couple with an HV switch in a double poles/double throws format. Although the design and operation are simple, it has suffered the following two problems. The first problem is the diffusion of sample from the sample channels to the separation channel during the separation phase, giving rise to a tailing peak profile. The second problem is the diffusion of the sample into the two floated separation channels during sample injection. The extra amount of sample entering the floated separation channel during injection is variable, with a larger amount over a longer injection time, giving rise to peak broadening and lowering of the separation efficiency.

3.4.1.2 Floated Injection/Grounded Separation

To prevent the diffusion of sample from the sample channels into the separation channel during separation, an improved design has been developed. In the injection phase of the floated injection, the same arrangement is used as described in the first design. However, during the separation phase, both the inlet and outlet of the sample vial are connected to the ground (Figure 3.4b). Compared to the simplest injection, the diffusion of sample from the sample channels to the separation channel is reduced as the buffer stream is pushing the sample in the sample channel back into the grounded vials during the separation phase. In this design, one HV and two HV switches are required for the operation of the grounded separation. However, the second problem for the diffusion of the sample into the two floated separation channels during sample injection is still unresolved, and the extra sample is higher with the use of a longer injection time.

3.4.1.3 Pinched Injection/Pinched Separation

To address the problem for diffusion of the sample into the floated separation channel during sample injection, the application of pinched voltage to the microelectrodes placed in the buffer waste and buffer vials is required. The applied potential should be the same as that at the channel cross-section to counteract the migration of analytes into the separation channel and to keep the sample stream within the sample channel. Under the control of a suitably adjusted pinched potential, a well-defined sample plug can be kept during sample injection.

In the subsequent separation phase, pinched voltages at the same potential as at the cross-section junction are applied at the sample and sample waste vials to keep the sample stream in the sample channel stationary, thereby preventing their leakage into the separation channel (Figure 3.4c). The use of pinched injection and pinched separation has become a common practice at present for introducing samples to MC. It requires two HVs and two HV switches for its operation.

3.4.1.4 Gated Injection/Continue Buffer and Sample Run

To enable fast sample injection for continue MC operation, a gated injection has been developed as shown in Figure 3.4d. The MC is designed for process control under a continuous separation and sample run with a fast and gated injection of new samples as needed. Under continuous background run before sample injection, both the buffer and the sample streams meet each other at the cross junction of the T channel configuration and make a 90° turn toward the buffer and sample waste vials, respectively. During sample injection, the buffer and buffer waste vials are floated for a short time, typically 0.1–0.5 s, to allow the sample stream to enter the separation channel for a specific time. The background run is then reassumed to conduct CE separation of the amount of sample previously entering the separation channel. The advantages of a gated injection are a fast sample injection and capability to change sample for continue injection. The limitation is poor detection sensitivity as a result of very limited volume of sample to be injected into the system. In addition to the requirement of two HVs and two HV switches, the scheme requires an accurately controlled voltage program for leakage control and fast HV switches for its operation.

3.4.1.5 Double-T Injection/Pinched Separation

To improve the repeatability for sample introduction, a double-T injector has been developed using a back-to-back arrangement of the microchannel in a T configuration (Figure 3.4e). The volume for the channel segment between the two vials in a double-T configuration is constant; thus, the volume of the microfluid trapped in the channel segment is repeatable and independent of the flowrate as far as the flushing out of the original fluid is completed during the injection operation. Pinched separation is needed during separation to prevent the diffusion of samples back into the separation channel.

The inner channel surface of the MC is required to exhibit sufficient EOF for the operation of the double-T injector to sustain sufficient flowrate for the microfluid during the sample introduction operation. For poly(methyl methacrylate) (PMMA) plastic with the channel pattern fabricated by laser ablation, the vaporization of PMMA as described in Chapter 2 produces an oxidized surface with sufficient hydroxyl functional groups to sustain EOF for the operation of the double-T injector and for subsequent movement of the injected nanofluid segment. The negative surface charge of the MC is the same as the inner surface of the separation capillary wall made up of fused silica; thus, the operation is simplified because the same buffer constituents can be used to move nanofluids in both the MC and CE.

3.4.1.6 Triple-T Injection/Pinched Separation

Additional inverted T configuration can be added easily using a software-aided laser ablation method for the fabrication of the required micro-channel pattern. A triple-T injection with pinched separation is shown in Figure 3.4f. Two different nanofluid segments can be introduced into a channel segment next to each other for subsequent operations, such as standard addition, dilution, or performance of a binding assay between the two different substances injected. The two sample vials (S1 and S2) are pinched to prevent back diffusion during the transfer of the microfluids for subsequent operations.

In addition to producing a microchannel surface with an oxidized surface to sustain EOF, the use of laser for fabrication provides an easy way for the design of a channel pattern with different channel lengths or depths for fabricating individual T injectors on demand. Thus, different desired volume ratios of two or more substances to be mixed can be made easily on a specific MC to satisfy the requirements for different analytical tasks. MC-CE devices with suitable injection volumes and a desirable number of substances to be investigated can be prepared at a specific channel segment for injection into a mixer for a mixing operation alone or in combination with other subsequent on-chip operations. The mixing operations that are important for applications such as on-chip titration for binding assay, on-chip dilution, and on-chip standard addition are discussed in detail in the following sections.

3.4.2 Mixing Operations

There are two major modes of mixing operation: active and passive. Active mixing requires external sources of activation, whereas passive mixing does not require input from external sources. Passive mixing is more simple to perform than active mixing, but active mixing is quicker. Selected cases using active and passive mixing are given in Sections 3.4.2.1 and 3.4.2.2, respectively.

3.4.2.1 Active Mixing

For mixing within microchannels, magnetic activation mode is commonly used, with internally placed magnets activated by an external rotating magnetic field. The simplest is the use of an extremely small magnetic needle with Teflon coating to mix substances in MC vials until a uniform solution is produced before injection and separation [8,21,22].

The availability of a magnetic ferrofluid containing magnetizable, nano-sized particles enables on-channel magnetic activation, making it possible to produce magnetic valves and microfluid stops. Thus, a complex operation can be performed on MC-CE devices. For the mixing operation, two types of magnets are used: a driving magnet and a rotation magnet. The four-step operation for on-chip mixing of different nanofluids generated at given microchannel segments for binding assay is shown in Figure 3.5, and the details are given in Chapter 13 [12].

As shown in Figure 3.5a, the driving magnet (M1) is used to drive different nanofluid segments with suitable length to produce a desirable volume ratio by the activated ferrofluid segment to the mixing coil for mixing. M1 is then acting as a stop valve to block the return of the sample segments to the sample microchannel during mixing. The mixing operation is performed by four cylindrical rotating magnets (M2), each 15 mm in thickness and 10 mm in diameter, coupled to a motor mounted at the bottom of the circular mixing channel. After a suitable number of rotation cycles are performed to ensure satisfactory mixing of the nanofluids from the target microchannel segments, the testing solution is directed by a linearly moving driving magnet to the separation capillary to conduct the analytical CE separation for analyte determination.

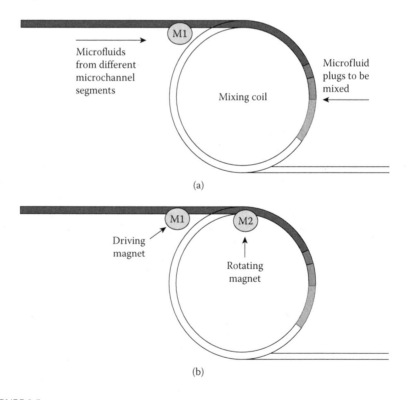

(a)

(b)

FIGURE 3.5
Magnetic-activated mode for on-chip mixing of nanofluids. (a) Introducing different nanofluids contained in microchannel segments by the driving magnet to the mixing coil. (b) Coupling of rotating magnet to activate the ferrofluid plug at the mixing coil.

(Continued)

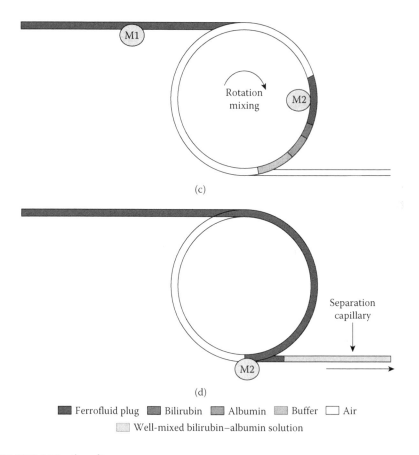

(c)

(d)

■ Ferrofluid plug ▨ Bilirubin ▦ Albumin ▨ Buffer □ Air
▨ Well-mixed bilirubin–albumin solution

FIGURE 3.5 (Continued)
Magnetic-activated mode for on-chip mixing of nanofluids. (c) Mixing different nanofluid segments by the ferrofluid plug coupled with the rotating magnet to produce a well-mixed solution. (d) Introducing the well-mixed solution by the driving magnet to the separation capillary. (From Nie, Z. and Fung, Y.S. *Electrophoresis* 2008, 29, 1924–1931.)

3.4.2.2 Passive Mixing

The double-T sample segment in MC is typically 5 mm in length, 150 μm in width, and 100 μm in depth, with an estimated volume of 80 nL, about the same volume as used for a CE hydrodynamic injection of a sample. Compared to a typical capillary (50 μm inside diameter and 40 cm long) for an MC-CE device, the sample volume from a double-T injector is <0.1% of the total inner capillary volume. Thus, direct in-capillary mixing is possible prior to separation and determination of analytes. In general, passive in-column mixing is preferred for small charged analytes, in particular, after on-column stacking because the compression of analyte ions during transient isotachophoresis enhances the mixing process.

This enhancement has been shown in the determination of anionic metabolites in urine [11,23,24], with details given in Chapter 5.

The success of passive mixing driven by electric field depends on the characteristic of the inner microchannel surface; this surface is very rough after ablation by the CO_2 laser operated in a pulse mode. Charged analytes such as metabolites are well mixed under zone electrophoresis. For large molecules such as proteins, mixing by diffusion is slow. However, isoelectric focusing (IEF) can be used to isolate and mix proteins with similar IEF values. The isolation of similar proteins with specific IEF values is discussed in the following sections.

3.4.3 Complex Operations for Multidimensional Separation

To determine analytes present at low levels in a sample containing a lot of similar substances, multidimensional separation is required using different techniques applied in sequence to tackle a difficult analytical task. Complex online and on-chip operations are often used to perform multidimensional separation for MC-CE devices. The manipulation of nanofluids in MC-CE devices for the separation of minor proteins in a sample with high protein content, such as the determination of functional proteins in milk, is described below to illustrate how the MC-CE device can be used to carry out the required operations for multidimensional separation.

To be able to separate minor proteins from samples with major proteins, the procedures for sample cleanup and analyte enrichment should be performed prior to capillary zone electrophoresis (CZE) separation and determination of the analyte proteins [25–31]. The schematic diagram showing the MC-CE device developed for the above-mentioned task is given in Figure 3.6;

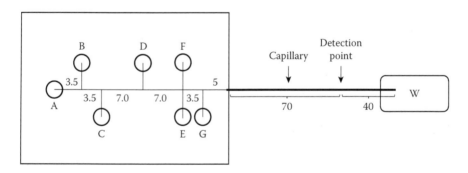

FIGURE 3.6
MC-CE device for determination of minor proteins in protein-rich samples. All dimensions are shown in millimeters. Vials: analyte (A), sample (B), terminating electrolyte (C), background electrolyte (D), sample waste (E), catholyte (F), leading electrolyte (G), CE waste (W). (From Wu, R.G., Wang, Z.P., Zhao, W.F., Yeung, W.S.B. and Fung, Y.S. *J Chromatogr A* 2013, 1304, 220–226.)

this device can carry out IEF, desalting, stacking, and CZE in a suitable sequence [26]. The minor analyte proteins are first isolated from the protein-rich sample by IEF and kept at the channels between vials A and F. With the anode set at D, C, and A in sequence, isolated proteins from different fractions with different IEF values are moving out in a different sequence. With vial F set at grounded potential and vials D, C, or B as anode for an initial short period, excess salts are removed via vial F from the different isolated protein fractions. Afterward, the cleaned proteins are stacked up by switching the grounded potential to vial G. Finally, with anode set at vial G and ground potential at vial W, the CZE separation is carried out to separate the analyte proteins from similar proteins from the protein-rich sample. The settings for HV and time duration at different vials during the MC-CE operation are shown in Table 3.2 [26]. Two HV switches are required for the two CZE runs, which have been shown sufficient to isolate, separate, and determine all analyte proteins present in infant formula samples.

With the removal of salts from the sample, the symmetry of the migration peaks has improved, showing sharp peaks in the electropherogram (Figure 3.7) [26]. A total time of 18 min is required for both sample pretreatment (on-chip) and separation, a time considerably shorter than the hours required to remove major protein prior to conducting CE separation as currently practiced in laboratory procedures. With the capability to perform on-chip manipulation of operations required to clean up samples and enrich analytes, the scope of application of the MC-CE devices has been broadened to on-site analysis with significant reduction in operating time and the capability to deliver urgent results on-site to assist in timely decisions.

TABLE 3.2

HV Settings for Isolation and Separation of Functional Proteins

Separation Mode	Time (min)	HV Setting (V)							
		A	B	C	D	E	F	G	W
IEF	2	980	840	700	420	140	0	140	Float
ITP_CZE1	1	3500	3500	3500	3600	3310	3310	3210	0
	7	3490	3490	3490	3490	3490	3490	3600	0
ITP_CZE2	1	3530	3530	3600	3330	3140	3140	3050	0
	Manually stopped	3490	3490	3490	3490	3490	3490	3600	0

From Wu, R.G., Wang, Z.P., Zhao, W.F., Yeung, W.S.B. and Fung, Y.S. *J Chromatogr A* 2013, 1304, 220–226.

Note: HV, high voltage; IEF, isoelectric focusing; ITP, isotachophoresis, CZE, capillary zone electrophoresis.

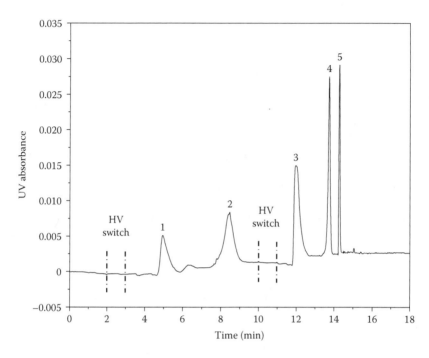

FIGURE 3.7
Electropherogram for the determination of functional proteins LF, IgG, and α-LA in infant formula by MC-CE device with multidimensional separation. β-LgA (5), β-LgB (4), α-LA (3), IgG (2), and LF (1). (From Wu, R.G., Wang, Z.P., Zhao, W.F., Yeung, W.S.B. and Fung, Y.S. *J Chromatogr A* 2013, 1304, 220–226.)

3.5 Summary and Remarks

This chapter introduces the various desired operations manipulated on nanofluids at MC-CE devices to achieve intended application. The required instrumentation and facilities are presented and discussed for their operations to manipulate nanofluids. The active and passive modes for on-chip mixing are discussed and compared for their application for standard addition, dilution, and binding assay via on-chip titration.

The manipulated operations on samples such as sample injection to MC-CE devices, sample cleanup, and dilution are described, together with the manipulated operations for analyte isolation and enrichment. The various sample injection modes are presented and compared for their applicability in MC-CE devices. With double-T injector and suitable on-chip mixing, various applications are given, such as binding assay for bilirubin and albumin, on-chip dilution, and standard addition for the determination of metabolites with highly variable concentrations in urine. Details on their application in different areas as described in Sections II–IV of this book are summarized in Table 3.3.

TABLE 3.3

Microfluid-Manipulated Procedures for MC-CE Operations in Different Application Areas

Manipulated Procedure	Operation	Application Area	Chapter	Reference
Sample-Manipulated Operation				
Sample injection	Double-T injector	On-site assay for portable devices	4, 12	[19–22,24–27,32,33]
Sample dilution	Magnetic-activated mixing	Organic and inorganic ions in urine	4	[8–9,11]
Sample cleanup	Imprinted polymer, salt elution	Carbonyl in air, minor protein in milk	6, 7	[2,34,35]
Analyte-Manipulated Operation				
Analyte isolation	IEF	Minor proteins in infant formula	7	[26,28,30,36,37]
Analyte enrichment	Stacking, imprinted polymer	Metabolites in urine, carbonyl in air	4, 6	[2,9–11,21,35,38]
Standard addition	Dual-channel mixing	Nephrolithiasis biomarkers in urine	5	[24,39]
Binding assay	Magnetic-activated ferrofluid for on-chip titration	Binding assay for free bilirubin in serum	13	[15–18,39–41]
Mixing Operation				
Active mixing	Postcolumn reagent mixing	Detect photochemically or electrochemically inactive analytes	9, 10, 11	[5–7,42–48]
Passive mixing	Dual-channel mixing	Nephrolithiasis biomarkers in urine	5	[24,39]
Enhanced Scope of Detection				
Manipulate microfluid flow	Dual-electrode detection	Enhance detection selectivity, detect electrochemically inactive analytes	8, 9	[5,7,44,49,50]
Integrate particle counting	GEMMA, LIF	Characterize protein and organelles by highly sensitive particle counting	14, 15, 16	[51–57]

Note: IEF, isoelectric focusing; GEMMA, gas-phase electrophoretic mobility molecular analyzer; LIF, laser-induced fluorescence.

In addition to integrating the MC-CE device with various sample preparation operations, the scope of detection can also be enhanced by nanofluid manipulation, such as the addition and mixing of reagents after separation to enable the detection of optically or electrochemically inactive analytes (see Chapters 9–11), as well as the integration of the device with highly sensitive detection modes, such as postcolumn particle counter, gas-phase electrophoretic mobility molecular analyzer (GEMMA), and laser-induced fluorescence (LIF) detection. Details are given in Chapters 14–16. Based on improvement of various sample preparation procedures as well as enhancement of the detection selectivity and sensitivity, the MC-CE devices have shown the capability to expand their scope for application to explore uncharted areas.

References

1. Wood, C.G. Immunoassays & co.: Past, present, future?—A review and outlook from personal experience and involvement over the past 35 years. *Clin Lab* 2008, 54, 423–438.
2. Sun, H., Chan, K.Y. and Fung, Y.S. Determination of gaseous and particulate carbonyls in air by gradient-elution micellar electrokinetic capillary chromatography. *Electrophoresis* 2008, 29, 3971–3979.
3. Guo, W.P., Lau, K.M., Sun, H. and Fung, Y.S. Microfluidic chip-capillary electrophoresis for emergency onsite biomedical analysis. *Sep Sci* 2010, 2, 3–10.
4. Qin, J., Fung, Y.S. and Lin, B.C. DNA diagnosis by capillary electrophoresis and microfabricated electrophoretic devices. *Expert Rev Mol Diagn* 2003, 3, 387–394.
5. Fung, Y.S. and Du, F.Y. Microfluidic chip-capillary electrophoresis devices—Dual electrode detectors for direct and indirect determination of analytes in complicated sample matrixes. *Abstract IL2, 11th Asian-Pacific International Symposium on Microscale Separations and Analysis (APCE 2011)*, Hobart, Australia, November 27–30, 2011, p. 21.
6. Du, F.Y., Chen, Q.D., Zhao, W.F. and Fung, Y.S. A new microfluidic-chip capillary electrophoresis device with a serial dual-electrode detector for biomedical analysis. *Abstract T21, 4th International Symposium on Microchemistry and Microsystems (ISMM 2012)*, Zhubei City, Taiwan, June 10–13, 2012, pp. 326–327.
7. Du, F.Y., Wu, R.G. and Fung, Y.S. Microfluidic chip-capillary electrophoresis devices for determination of proteins and amino acids in biofluids. *Abstract, the 13th Asian-Pacific International Symposium on Microscale Separations and Analysis (APCE 2013)*, Jeju, November 3–6, 2013, Korea Paper LOC-WK02, p. 1.
8. Guo, W.P., Lau, K.M. and Fung, Y.S. Microfluidic chip-capillary electrophoresis for two orders extension of adjustable upper working range for profiling of inorganic and organic anions in urine. *Electrophoresis* 2010, 31, 3044–3052.
9. Guo, W.P. and Fung, Y.S. Profiling of organic anions in urine by microfluidic chip-capillary electrophoresis. *Abstract P4-03, 9th Asia-Pacific International Symposium on Microscale Separation and Analysis (APCE 2009) and 1st Asian-Pacific International Symposium on Lab-on-Chip (APLOC2009)*, Shanghai, China, October 28–31, 2009, p. 228.

10. Guo, W.P. and Fung, Y.S. Microfluidic chip-capillary electrophoresis with adjustable on-chip sample dilution for profiling of urinary markers. *Proceedings, 14th International Conference on Miniaturized Systems for Chemistry and Life Sciences (μTAS 2010)*, Groningen, The Netherlands, October 3–7, 2010, Paper W11A, pp. 1481–1483.

11. Guo, W.P. and Fung, Y.S. Microfluidic chip-capillary electrophoresis with adjustable on-chip sample dilution for profiling of urinary markers. *Abstract PB11, 10th Asian-Pacific International Symposium on Microscale Separations and Analysis (APCE 2010)*, Hong Kong, China, December 10–13, 2010, p. 102.

12. Nie, Z. and Fung, Y.S. Microchip capillary electrophoresis for frontal analysis of free bilirubin and study of its interaction with human serum albumin. *Electrophoresis* 2008, 29, 1924–1931.

13. Fung, Y.S. and Nie, Z. Microfluidic chip capillary electrophoresis for biomedical analysis. *Sep Sci* 2009, 1, 21–26.

14. Mo, S.L. and Fung, Y.S. Quantum-dots mediated LIF detection of free bilirubin by frontal analysis using microfluidic chip-capillary electrophoresis device. *Abstract P32, 11th Asian-Pacific International Symposium on Microscale Separations and Analysis (APCE 2011)*, Hobart, Australia, November 27–30, 2011, p. 95.

15. Nie, Z. and Fung, Y.S. Multi-segment incremental sample injection for the determination of bilirubin binding capacity of albumin and study of drug interaction by microchip capillary electrophoresis. *Abstract, 22nd International Symposium on Microscale Bioseparations & Methods for System Biology (MSB 2008)*, Berlin, Germany, March 9–13, 2008, Paper P.284-Tu, p. 417.

16. Nie, Z. and Fung, Y.S. Determination of binding capacity of albumin for bilirubin by in situ titration at a circular ferrofluid driven micromixer in microfluidic chip-capillary electrophoresis. *Abstract P01-05, 8th Asia-Pacific International Symposium on Microscale Separation and Analysis (APCE 2008)*, Kaohsiung, Taiwan, November 2–5, 2008, p. 99.

17. Fung, Y.S. Microfluidic chip-capillary electrophoresis for assessing bilirubin-protein interaction and quality control of Chinese medicine. *Abstract, the 8th Chinese Symposium on Microscale Bioseparations & Methods for System Biology (CMSB 2008)*, Guilin, China, November 21–24, 2008, p. 4.

18. Nie, Z., Sun, H. and Fung, Y.S. Determination of binding capacity of albumin for bilirubin by microfluidic chip-capillary electrophoresis based on in situ titration at a circular ferrofluid driven micromixer. *Abstract P2-02, the 9th Asia-Pacific International Symposium on Microscale Separation and Analysis (APCE 2009) and the 1st Asian-Pacific International Symposium on Lab-on-Chip (APLOC2009)*, Shanghai, China, October 28–31, 2009, p. 155.

19. Guo, W.P., Lau, K.M. and Fung, Y.S. Microfluidic chip-capillary electrophoresis for emergent biomedical analysis in Hong Kong. *Abstract K27, the 9th Asia-Pacific International Symposium on Microscale Separation and Analysis (APCE 2009) and the 1st Asian-Pacific International Symposium on Lab-on-Chip (APLOC2009)*, Shanghai, China, October 28–31, 2009, p. 36.

20. Guo, W.P., Lau, K.M. and Fung, Y.S. Microfluidic chip-capillary electrophoresis for point-of-care analysis. *Proceedings, International Symposium on Microchemistry and Microsystems (ISMM 2010)*, Hong Kong, May 28–30, 2010, P31, pp. 144–145.

21. Guo, W.P. and Fung, Y.S. Microfluidic chip-capillary electrophoresis for the determination of organic acids in biofluids. *Abstract, the 8th Chinese Symposium on Microscale Bioseparations & Methods for System Biology (CMSB 2008)*, Guilin, China, November 21–24, 2008, p. 197.

22. Fung, Y.S. Microfluidic chip-capillary electrophoresis for biomedical applications. *Abstract, Separation Science Singapore*, Biopolis Park, Singapore, August 26–28, 2009, p. 36.
23. Fung, Y.S. Microfluidic chip-capillary electrophoresis devices—From sample pretreatment to analyte detection for clinical diagnosis and monitoring of biomarkers and metabolites. *Abstract KL21, 10th Asian-Pacific International Symposium on Microscale Separations and Analysis (APCE 2010)*, Hong Kong, China, December 10–13, 2010, p. 48.
24. Guo, W.P. and Fung, Y.S. Microfluidic chip-capillary electrophoresis with dynamic multi-segment standard addition for rapidly identifying nephrolithiasis markers in urine. *Electrophoresis* 2011, 32, 3437–3445.
25. Wu, R.G., Yeung, W.S.B. and Fung, Y.S. 2-D t-ITP/CZE determination of clinical urinary proteins using a microfluidic-chip capillary electrophoresis device. *Electrophoresis* 2011, 32, 3406–3414.
26. Wu, R.G., Wang, Z.P., Zhao, W.F., Yeung, W.S.B. and Fung, Y.S. Multi-dimension microchip-capillary electrophoresis device for determination of functional proteins in infant milk formula. *J Chromatogr A* 2013, 1304, 220–226.
27. Wu, R.G., Wang, Z., Fung, Y.S., Seah, D.Y.P. and Yeung, W.S.B. Assessment of adulteration of soybean proteins in dairy products by 2D microchip-CE device. *Electrophoresis* 2014, 35, 1728–1734.
28. Wu, R.G., Fung, Y.S. and Yeung, W.S.B. Determination of lactoferrin and β-lactoglobulin in dairy products by microfluidic-chip capillary electrophoresis. *Abstract P4-10, the 9th Asia-Pacific International Symposium on Microscale Separation and Analysis (APCE 2009) and the 1st Asian-Pacific International Symposium on Lab-on-Chip (APLOC2009)*, Shanghai, China, October 28–31, 2009, p. 238.
29. Wu, R.G., Fung, Y.S. and Yeung, W.S.B. Microfluidic-chip capillary electrophoresis for analysis of clinical urinary proteins. *Abstract P136, the 25th International Symposium on Microscale Bioseparations (MSB 2010)*, Prague, Czech Republic, March 21–25, 2010, p. 129.
30. Wu, R.G., Fung, Y.S. and Yeung, W.S.B. Determination of lactoferrin and immunoglobulin G in baby formula by microfluidic-chip capillary electrophoresis. *Abstract PB8, 10th Asian-Pacific International Symposium on Microscale Separations and Analysis (APCE 2010)*, Hong Kong, China, December 10–13, 2010, p. 99.
31. Wu, R.G., Fung, Y.S. and Yeung, W.S.B. Microfluidic-chip capillary electrophoresis for determination of clinical urinary proteins. *Abstract PB9, 10th Asian-Pacific International Symposium on Microscale Separations and Analysis (APCE 2010)*, Hong Kong, China, December 10–13, 2010, p. 100.
32. Mo, S.L. and Fung, Y.S. Development of microchip—Capillary electrophoresis for quality control of herbal medicine. *Abstract, 8th Chinese Symposium on Microscale Bioseparations & Methods for System Biology (CMSB 2008)*, Guilin, China, November 21–24, 2008, p. 187.
33. Nie, Z. and Fung, Y.S. Microfluidic chip-capillary electrophoresis for quality assessment of a complex herbal preparation. *Abstract, the 7th Asia-Pacific International Symposium on Microscale Separation and Analysis (APCE 2007)*, Singapore, December 16–19, 2007, Paper 5:48, pp. 5:141.
34. Fung, Y.S. Enhancing analytical capability of capillary electrophoresis by integrated microfluidic chip-capillary devices. *Abstract, the 12th Asian-Pacific International Symposium on Microscale Separations and Analysis (APCE 2012)*, Singapore, December 16–19, 2012, SICC7-A-0238, p. 1.

35. Sun, H. and Fung, Y.S. Integrating micellar electrokinetic capillary chromatography with molecular imprinting polymer- solid phase extraction for ambient carbonyl determination. *Abstract, the 12th Asian-Pacific International Symposium on Microscale Separations and Analysis (APCE 2012)*, Singapore, December 16–19, 2012, SICC7-A-0240, p. 1.

36. Wu, R.G., Fung, Y.S. and Yeung, W.S.B. Microfluidic chip-capillary electrophoresis for separation and determination of low abundance proteins. *Abstract, Microfluidic and Nanofluidic Devices for Chemical and Biochemical Experimentation Symposium, 2010 International Chemical Congress of Pacific Basin Societies (Pacifichem 2010)*, Honolulu, HI, December 15–20, 2010, TECH Paper 547.

37. Wu, R.G., Fung, Y.S. and Yeung, W.S.B. Microfluidic chip-capillary electrophoresis for separation and determination of low abundance proteins. *Abstract P01-02, the 8th Asia-Pacific International Symposium on Microscale Separation and Analysis (APCE 2008)*, Kaohsiung, Taiwan, November 2–5, 2008, p. 96.

38. Zhao, W.F., Du, F.Y., Chen, Q.D. and Fung, Y.S. Microchip-capillary electrophoresis devices incorporating magnetically activated microbeads with molecularly imprinted polymer for determination of antibiotics in milk. *Abstract M22, 4th International Symposium on Microchemistry and Microsystems (ISMM 2012)*, Zhubei City, Taiwan, June 10–13, 2012, pp. 138–139.

39. Fung, Y.S. Microfluidic chip—Capillary electrophoresis for investigation of binding interaction between free bilirubin, trace metals and essential minerals with biological significant proteins. *Abstract I-03, the 8th Asia-Pacific International Symposium on Microscale Separation and Analysis (APCE 2008)*, Kaohsiung, Taiwan, November 2–5, 2008, p. 65.

40. Fung, Y.S. and Nie, Z. Microfluidic chip capillary electrophoresis for biomedical analysis. *Sep Sci China* 2009, 1, 18–23.

41. Sun, H., Nie, Z. and Fung, Y.S. Determination of free bilirubin and its binding capacity by HSA using a microfluidic chip-capillary electrophoresis device with a multi-segment circular-ferrofluid-driven micromixing injection. *Electrophoresis* 2010, 31, 3061–3069.

42. Du, F.Y. and Fung, Y.S. Development of CE-dual opposite carbon-fiber microdisk electrode detection for peak purity assessment of polyphenols in red wine. *Electrophoresis* 2010, 31, 2192–2199.

43. Du, F.Y. and Fung, Y.S. Differential amperometric dual electrode detection for peak purity assessment of polyphenols in red wine after separation by microfluidic chip-capillary electrophoresis. *Abstract P3-01, the 9th Asia-Pacific International Symposium on Microscale Separation and Analysis (APCE 2009) and 1st Asian-Pacific International Symposium on Lab-on-Chip (APLOC2009)*, Shanghai, China, October 28–31, 2009, p. 212.

44. Du, F.Y. and Fung, Y.S. Microfluidic-chip capillary electrophoresis device incorporating a new serial dual-microelectrode detector for the determination of amino acids and proteins in food. *Abstract PB12, 10th Asian-Pacific International Symposium on Microscale Separations and Analysis (APCE 2010)*, Hong Kong, China, December 10–13, 2010, p. 103.

45. Chen, Q.D. and Fung, Y.S. Capillary electrophoresis with immobilized quantum dot fluorescence detection for rapid determination of organophosphorus pesticides in vegetables. *Electrophoresis* 2010, 31, 3107–3114.

46. Chen, Q.D., Zhao, W.F. and Fung, Y.S. Determination of acrylamide in potato crisps by capillary electrophoresis with quantum dot-mediated LIF detection. *Electrophoresis* 2011, 32, 1252–1257.

47. Chen, Q.D. and Fung, Y.S. Microfluidic-chip capillary electrophoresis with immobilized quantum dots detection for the separation and determination of organophosphorus pesticides in contaminated vegetables. *Abstract P01-07, the 8th Asia-Pacific International Symposium on Microscale Separation and Analysis (APCE 2008),* Kaohsiung, Taiwan, November 2–5, 2008, p. 101.

48. Chen, Q.D., Zhao, W.F., Du, F.Y. and Fung, Y.S. Determination of pesticides by microchip-capillary electrophoresis device with magnetically immobilized QD for LIF detection. *Abstract T22, 4th International Symposium on Microchemistry and Microsystems (ISMM 2012),* Zhubei City, Taiwan, June 10–13, 2012, pp. 328–329.

49. Fung, Y.S., Kwok, W.P. and Du, F.Y. Microfluidic chip-capillary electrophoresis devices for metabolomics applications. *Abstract L-121, 19th International Symposium on Capillary Electroseparation Techniques (ITP 2012),* Baltimore, MD, September 30–October 3, 2012, p. 9.

50. Du, F.Y., Mo, S.Y. and Fung, Y.S. Dual microelectrodes as CE and HPLC post column detector for peak purity assessment and protein determination. *Extended Abstract, the 13rd International Symposium on Electroanalytical Chemistry (13th ISEAC),* Changchun, China, August 19–22, 2011, pp. 63–64.

51. Zhao, W.F., Chen, Q.D., Wu, R.G., Wu, H., Fung, Y.S. and O, W.S. Capillary electrophoresis with LIF detection for assessment of mitochondrial number based on the cardiolipin content. *Electrophoresis* 2011, 32, 3025–3033.

52. Zhao, W.F., O, W.S. and Fung, Y.S. Microfluidic chip-capillary electrophoresis with laser induced fluorescence detection for assessing changes in mitochondria from HepG2 cells. *Abstract P01-06, the 8th Asia-Pacific International Symposium on Microscale Separation and Analysis (APCE 2008),* Kaohsiung, Taiwan, November 2–5, 2008, p. 100.

53. Zhao, W.F., Fung, Y.S. and O, W.S. Assessment of the mitochondria number based on cardiolipin by capillary electrophoresis-LIF detection. *Abstract PA17, 10th Asian-Pacific International Symposium on Microscale Separations and Analysis (APCE 2010),* Hong Kong, China, December 10–13, 2010, p. 89.

54. Ma, T.M., Wu, R.G. and Fung, Y.S. Hyphenation of microfluidic-chip capillary electrophoresis with gas-phase electrophoretic mobility molecular analyzer (GEMMA) for protein characterization. *Abstract PB10, 10th Asian-Pacific International Symposium on Microscale Separations and Analysis (APCE 2010),* Hong Kong, China, December 10–13, 2010, p. 101

55. Ma, T.M. and Fung, Y.S. Microchip CE device for coupling gas-phase electrophoretic mobility molecular analyzer with capillary electrophoresis for study of metal-protein binding in milk. *Abstract M20, 4th International Symposium on Microchemistry and Microsystems (ISMM 2012),* Zhubei City, Taiwan, June 10–13, 2012, pp. 134–135.

56. Fung, Y.S. Microchip-capillary devices for metabolomics application. *Abstract, the 1st International Conference on Urine Omics (URINOMICS),* Caparica-Almada, Portugal, September 9–11, 2013, Paper O54, pp. 141–142.

57. Fung, Y.S. Microfluidic chip—Capillary electrophoresis for biomedical analysis. *Abstract APCAS 2, 2012 Asia Pacific Conference on Analytical Science (APCAS),* Manila, Philippines, April 11–13, 2012, pp. 26–27.

Section II

Integration to Improve Sample Preparation and Cleanup

4

Single-Channel Mixing: Extending Analytical Range to Handle Samples with High Analyte Variation for Profiling Urinary Organic and Inorganic Anions

Wenpeng Guo

First Affiliated Hospital of Shenzhen University
Shenzhen, People's Republic of China

CONTENTS

4.1 Introduction

The increase in levels of some organic anions, which is known as acidemia, can result in a decrease in diabetic ketoacidosis and lactic acidosis [1–4]. The concentration of 2-keto-glutarate, for instance, is an important biomarker for diagnosis of hyperammonemic coma. The concentration ratios between organic anions are sometimes more valuable than a single anion, such as the concentration ratios of acetoacetate to 3-hydroxybutyrate and pyruvate to lactate [5–10]. It is important to profile short-chain organic and inorganic acids in urine to make informed medical decisions for metabolite disorders and acidosis diagnosis.

Current analytical methods for organic and inorganic acid determination include high-performance liquid chromatography and gas chromatography-mass spectrometry for profiling. Complicated instrumentation with high operational and capital costs are used in both techniques. Capillary electrophoresis (CE), as used for determining organic and inorganic acids in urine and serum [8,11–18], provides an economical and efficient method for anion determination. However, large variation in analyte concentrations necessitates considerable dilution effort for accurate measurements to be obtained.

Microfluidic chip (MC) technology has developed rapidly in the past few decades [19–22]. Its scope of application has further expanded with the recent development of microfluidic chip-capillary electrophoresis (MC-CE) devices used for on-site determination of urine biomarkers and metabolites [23,24]. The major problem in monitoring such metabolites is their high analyte variability, requiring frequent on-site dilution to meet a working range requirement for an individual analyte [8,16,25]. Thus, an MC-CE device has been developed to enable on-chip dilution prior to metabolite separation and quantitation. Herein, an MC-CE device is fabricated for on-site dilution to meet the demand for urgent assay of metabolites [26].

4.2 MC-CE Device for Working Range Extension

4.2.1 Design of the MC-CE Device

The extended working range required to determine the anions commonly found in urine samples is given in Table 4.1. The enhancement factors ranged from 2 to 40 for commonly found urinary anions. Thus, the MC-CE device should be able to show flexible enhancement factors for 40 times dilution. The MC-CE device (Figure 4.1) includes two functional units: one unit for on-sample dilution and the other unit for on-chip separation. The on-chip magnetic fluid–activated valve was used to dilute samples to the desired concentration [26]. Thereby, urine samples with highly variable inorganic and organic anions could be handled and diluted for MC-CE separation and quantitation in the second unit.

The on-chip valves [26–29] include thermopneumatic and screwed stop valves, and precision machining is needed to avoid leakage [27]. Hartshorne et al. [27] successfully made use of ferrofluid to make valves with a dead volume as small as 25 nL [26]. A circular valve with an extremely low dead volume (<10 nL) has been developed here (Figure 4.1), with a ferrofluid-activated circular valve, mixing vial (MV), and two capillaries to introduce buffer and sample to the MC-CE device.

TABLE 4.1

Least Enhancement Factors Required for the Upper Range of
Commonly Found Urinary Anions

Anion	Required Range (mM)	Working Range[a] (mM)	Minimum Enhancement Factor[b]
Acetate	0.1–2.0	0.005–0.5	4
Aspartate	1.0–4.5	0.015–1.0	5
Butyrate	2.3–4.0	0.010–0.5	8
Carbonate	2.0–8.5	0.010–1.5	6
Chloride	10.0–200	0.010–5.0	40
Citrate	0.5–7.0	0.005–0.5	14
Glucuronate	0.1–4.0	0.010–0.5	8
Glutamate	1.0–10.0	0.010–1.0	10
Lactate	2.0–5.0	0.010–0.5	10
L-Ascorbate	0.5–4.0	0.020–2.0	2
Oxalate	0.3–2.0	0.010–1.0	2
Phosphate	1.5–10	0.020–2.0	5
Pyruvate	0.1–2.0	0.020–0.5	4
Succinate	0.1–7.8	0.005–0.5	16
Sulfate	1.5–18	0.005–2.0	9

From Guo, W.P., Lau, K.M. and Fung, Y.S. *Electrophoresis* 2010, 31, 3044–3052.
[a] CE working range by hydrodynamic injection.
[b] Minimum enhancement factor = (upper required range/upper CE range).

The dilution unit shown in Figure 4.1 consists of an MV and a sampling capillary (capillary 1). Considering the volume of the MV with a micromagnet (10.5 μL) to the minimum volume of sample or standard to be added to the MV (0.1 μL), the highest dilution factor or the maximum enhancement factor is estimated as 100. Five seconds is sufficient for complete mixing at a rotation speed of 50 rpm. Another additional dilution factor of 5 can be added through adjustment of the height and time for sample injection. Thus, the dilution ratio ranges from 1 to 500, much larger than the estimated factor from 2 to 40 shown in Table 4.1.

4.2.2 Operation of the Fabricated MC-CE Device

Details on the operation of the MC-CE device are given in Guo et al. [26], with four major operations in sequence shown in Figure 4.2: (1) washing/cleaning, (2) dilution/mixing, (3) sample injection, and (4) separation and quantitation. The ferrofluid was guided magnetically into valve position A, raising MV to the buffer vial (BV) level, and applying high voltage across BV (–25 kV) to Waste Vial (WV) at (0 V) for CE separation and ultraviolet (UV) detection at 240 nm. The time required was 20 min, including 2 min for cleanup; 1–2 min for mixing and dilution; and a final 15 min for injection, separation, and detection.

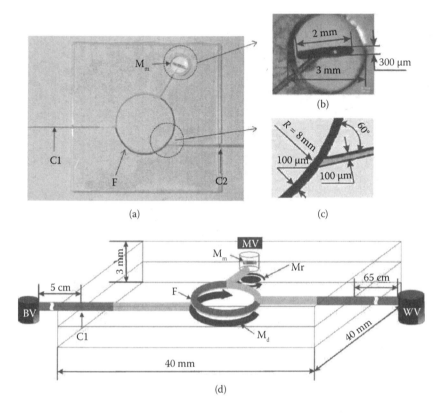

FIGURE 4.1
Layout of the MC-CE device. Top view (a), enlarged view (b, mixing vial), enlarged view (c, circular valve), and perspective view (d, device in operation). C1 and C2, capillary 1 and 2, respectively; F, ferrofluid; M_m, micromagnet; R, radius of circular valve; M_d, driving magnet, Mr, rotating magnet; BV, vials for buffer; MV, vials for mixing; WV, vials for waste. (From Guo, W.P., Lau, K.M. and Fung, Y.S. *Electrophoresis* 2010, 31, 3044–3052.)

4.3 Fabrication and Validation of the MC-CE Device for Working Range Extension

4.3.1 Design and Fabrication of the MC-CE Device

Details on the fabrication of the MC-CE device are given in Guo et al. [26], and the general layout is shown in Figure 4.1. A flexible microchannel design was ablated by the CO_2 laser onto polymethyl methacrylate polymer base plates to create the desired pattern. Two capillaries were used, with one capillary for separation and the other for sample introduction. Guided by an external magnet, the ferrofluid was introduced into the circular valve (Figure 4.1a). A micromagnet (2 mm × 0.3 mm × 0.1 mm) was then

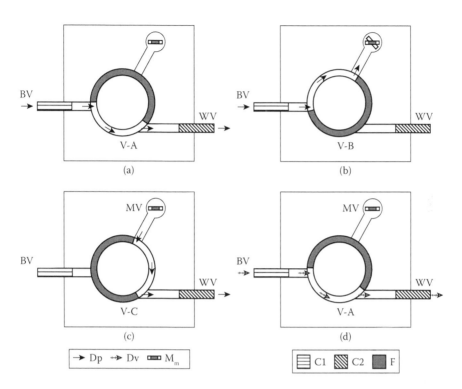

FIGURE 4.2
MC-CE operation procedure. Cleaning and washing (a), dilution and mixing (b), sample injection (c), and on-chip CE separation (d). Dv, high-voltage driving; Dp, pressure driving. Other abbreviations are as provided in Figure 4.1. (From Guo, W.P., Lau, K.M. and Fung, Y.S. *Electrophoresis* 2010, 31, 3044–3052.)

added to the MV as shown in Figure 4.1b. Capillary 2 was then blocked by the magnetically activated magnetic fluid (Figure 4.1c), subsequently by adding buffer and sample to the MV for mixing prior to the initiation of the MC-CE operation procedure (Figure 4.1d).

4.3.2 Validation of the Fabricated MC-CE Device

Under expected enhancement factors (10–50), the MC-CE device was verified for the expected enhancement factor by injecting 2 mM oxalate, with experimental and expected results shown in Table 4.2 [26]. The differences in results are assessed by a paired *t*-test, with no significant statistical difference observed at the 95% significant level. The use of an additional step in the MC-CE procedure for dilution is found to increase variability to an acceptable level, as shown by a moderate increase from 4.0–6.3% to 4.3–7.9% of the peak area repeatability (relative standard deviation [RSD], $n = 3$) for oxalate with 10–50 dilution factors.

TABLE 4.2

Verification of Enhancement Factors by MC-CE Device

Injected Standard[a] V_{ox} (μL)	Injected Buffer[b] V_{bf} (μL)	Expected Enhancement (V_{bf}/V_{ox})	Expected Concn.[c] C_E (μM)	Measured Concn.[d] C_M (μM)	Difference[e] $D = C_E - C_M$
0.20	2.0 (20)	10	200	201.8	−1.80
0.20	4.0 (40)	20	100	98.9	1.10
0.20	6.0 (60)	30	67	66.0	1.00
0.20	8.0 (80)	40	50	51.6	−1.60
0.20	10 (100)	50	40	37.7	2.30
				Mean ($n = 5$)	0.20
				SD ($n = 5$)	1.79

From Guo, W.P., Lau, K.M. and Fung, Y.S. *Electrophoresis* 2010, 31, 3044–3052.

[a] 2 mM oxalate (C_0), injection (0.5 psi, 4 s).

[b] Buffer, injection (1 psi, time in seconds as stated in parentheses).

[c] $C_E = C_0 \times (V_{ox}/V_{bf})$.

[d] Measured concentration (C_M) determined by oxalate standard solutions.

[e] Paired *t*-test, $t_{cal} = 0.25$, $t_{3,95} = 3.18$.

4.4 Optimization of Operations and Procedures of the MC-CE Device for Working Range Extension

For on-chip CE separation, parameters affecting the run are optimized. Indirect photometric detection mode was used because most anions exhibit low UV absorption. Buffer at pH 8.4 was found to give the best overall separation of the target anions. To optimize detection sensitivity for indirect UV detection, the alkylamines tetraethylenepentamine (TEPA), triethylenediamine (TETA), ethylenediamine (EDA), hexamethonium bromide (HEMB), and diethylenediamine (DETA) were investigated because they provided strong UV-absorbing chromophores. The results shown in Figure 4.3 indicate suppression of electroosmotic flow (EOF) by all amines, with sharp initial reduction (0–0.5 mM) followed by a gradual decrease at higher concentrations [26]. TEPA is used as the EOF modifier due to its effectiveness to achieve lower EOF. The best compromised separation for the anions is obtained using 1.5 mM TEPA.

The effect of 1,3,5-benzenetricarboxylic acid (BTA) on signal-to-noise (S/N) ratio is studied. With BTA <0.01 mM, no analyte peak is observed and broad peaks appear for BTA <1 mM. With BTA within the range of 2–3 mM, the S/N ratios are doubled, as attributed to sharper peaks and better resolutions. With BTA >3 mM, the S/N ratio is decreased gradually, as attributed to a larger noise at high UV-absorbing anions concentration. Although a better separation can be achieved at a higher BTA, undesirable Joule heating is also increased. Hence, 3 mM is chosen as the optimum BTA concentration.

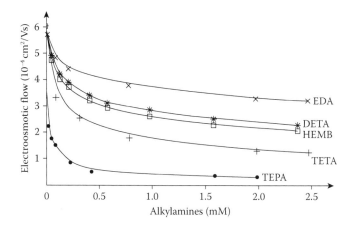

FIGURE 4.3
Effect of alkylamines on the electroosmotic flow. EDA, ethylenediamine; DETA, diethylene-diamine; HEMB, hexamethonium bromide; TETA, triethylenediamine; TEPA, tetraethylene-pentamine. (From Guo, W.P., Lau, K.M. and Fung, Y.S. *Electrophoresis* 2010, 31, 3044–3052.)

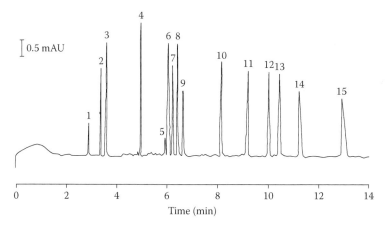

FIGURE 4.4
MC-CE separation of a standard anion mixture. Buffer (pH 8.4) composed of Tris (15 mM), TEPA (1.5 mM), and BTA (3 mM). Capillary = 50 μm inside diameter, 45/65 cm for effective/total length, respectively. Sample injection = 40 s × 0.5 psi. Buffer injection = 40 s × 1 psi. Magnet rotation = 5 s, 50 rpm. Separation: BV = –25 kV, WV = ground, 15 min. Anion migration order (15 to 1, 1.0 mM each) = glucuronate, L-ascorbate, glutamate, aspartate, n-butyrate, lactate, pyruvate, acetate, citrate, carbonate, phosphate, succinate, oxalate, sulfate, chloride. (From Guo, W.P., Lau, K.M. and Fung, Y.S. *Electrophoresis* 2010, 31, 3044–3052.)

The electropherogram for 15 simple urinary anions is given in Figure 4.4: oxalate, sulfate, chloride, phosphate, succinate, citrate, carbonate, pyruvate, acetate, n-butyrate, lactate, L-ascorbate, glutamate, glucuronate, and aspartate [26]. Hydrodynamic sampling at 8-cm difference in height is used to inject diluted samples. Satisfactory precision is shown for both peak height

TABLE 4.3

Working Parameters for Simple Urinary Inorganic and Organic Anions Using Developed MC-CE Device

Anion	Migration Time[a] (RSD, %)	Peak Area (RSD, %)	Peak Height (RSD, %)	Detection Limit[b] (µM)	Working Range[c] (mM)	Correlation Coefficient (R)
Acetate	0.62	5.1	5.2	1.0	0.005–50	0.997
Aspartate	0.57	5.5	5.6	2.5	0.015–100	0.997
Butyrate	0.53	5.4	5.3	2.0	0.010–50	0.995
Carbonate	0.65	4.8	4.9	3.0	0.010–150	0.996
Chloride	0.67	4.4	4.7	2.0	0.010–500	0.999
Citrate	0.87	5.5	5.5	1.0	0.005–50	0.997
Glucuronate	0.75	6.5	6.6	6.5	0.010–50	0.992
Glutamate	0.61	5.2	5.1	2	0.010–100	0.997
Lactate	0.65	4.8	4.9	1.5	0.010–50	0.997
L-Ascorbate	0.58	4.9	5.0	3	0.020–200	0.996
Oxalate	0.52	4.3	4.4	2.0	0.010–100	0.998
Phosphate	0.62	4.9	4.8	4.0	0.020–200	0.997
Pyruvate	0.64	4.9	5.0	5.0	0.020–50	0.995
Succinate	0.56	4.9	5.0	0.5	0.005–50	0.999
Sulfate	0.64	4.1	4.2	0.8	0.005–200	0.998

From Guo, W.P., Lau, K.M. and Fung, Y.S. *Electrophoresis* 2010, 31, 3044–3052.
[a] RSD (n = 5).
[b] Limit of detection (S/N = 2).
[c] Working ranges (100-fold on-chip dilution, n = 5).

and area measurement (Table 4.3) [26]. Corrected peak area is found to be better than peak height for quantitative measurement. Use of a longer injecting time increases the method's sensitivity but reduces the separation efficiency and resolution.

4.5 MC-CE Device for Urinary Anion Profiling

The electropherogram is shown in Figure 4.5, with peaks identified by spiking standard anions to the urinary sample [26]. Ten anions (sulfate, chloride, phosphate, carbonate, succinate, pyruvate, oxalate, acetate, glucuronate, and citrate) were baseline separated for identification and quantitation. Peaks 9 and 10 were not identified because they are outside the confidence interval for migration time at the 95% level. The results indicate low detection limits (0.5–6.5 µM, S/N = 2), wide working ranges (0.005–500 mM), and satisfactory repeatability (RSD, n = 5) of 4.1–6.5% and 0.52–0.87% for peak area and migration time, respectively.

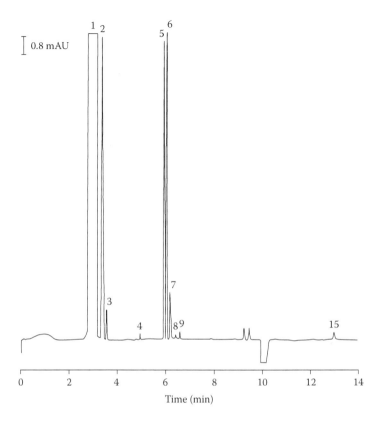

FIGURE 4.5
MC-CE electropherogram from a typical urinary sample. Designated peaks and all operation same as Figure 4.4, except 20-fold sample dilution. (From Guo, W.P., Lau, K.M. and Fung, Y.S. *Electrophoresis* 2010, 31, 3044–3052.)

Urinary samples are injected into the MC-CE device after filtration. To remove interference due to inner capillary surface adsorption, a rinsing cycle is used between consecutive runs with improved repeatability for both migration time and electroosmotic flow. Under optimized conditions, 15 anions in four samples can be analyzed within 1 h. The method reliability was established by recovery test and parallel measurement with ion chromatography (IC) method. The recovery results are shown in Table 4.4; satisfactory recoveries were achieved, ranging from 99.7 to 94.2%, for the anions investigated [26].

The IC method for urinary anions determination is used as a parallel method for comparison. The results using statistical matched pair *t*-test are given in Table 4.5, indicating no significant difference in results with the IC and MC-CE methods [26]. The MC-CE device is thus shown to provide a quick, economical, sensitive, and selective procedure for routine assay of urinary anions.

TABLE 4.4

Recovery Test of Spiked Anions by MC-CE Device

Anion	Present (μM)	Added (μM)	Found (μM)	Recovery (%)
Acetate	4.0	100	98	94.2
Carbonate	410	100	501	98.2
Chloride	7100	100	7180	99.7
Citrate	30	100	125	96.1
Glucuronate	3.0	100	97	94.2
Oxalate	16.5	100	112	96.1
Phosphate	405	100	491	97.2
Pyruvate	3.0	100	98	95.1
Succinate	4.0	100	98	94.2
Sulfate	625	100	712	98.2

From Guo, W.P., Lau, K.M. and Fung, Y.S. *Electrophoresis* 2010, 31, 3044–3052.
Note: Replicate analyses ($n = 5$) of on-chip 20-fold dilutions urine samples.

TABLE 4.5

Comparison of MC-CE and IC Methods for Assay of Urinary Inorganic and Organic Anions

Analyte	MC-CE (M)	IC (mM)	Difference[a]
Acetate	0.08	0.07	0.01
Carbonate	8.2	7.8	0.4
Chloride	142	139	3
Citrate	0.60	0.57	0.03
Glucuronate	0.06	0.05	0.01
Oxalate	0.33	0.31	0.02
Phosphate	8.1	8.3	0.2
Pyruvate	0.06	0.05	0.01
Succinate	0.08	0.07	0.01
Sulfate	12.5	12.1	0.4
		Mean ($n = 10$)	0.41
		SD ($n = 10$)	0.92

From Guo, W.P., Lau, K.M. and Fung, Y.S. *Electrophoresis* 2010, 31, 3044–3052.
[a] $t_{cal} = 1.40$, $t_{8,95} = 2.31$.

4.6 Summary

An MC-CE device is fabricated for profiling urinary short-chain inorganic and organic acids. Two functional units for dilution and separation are incorporated in the device. The dilution unit delivers enhancement factors up to 500-fold, sufficient to meet the demands from samples with highly variable analytes. The calculated enhancement factors have been verified by experimental results. The large injected sample volume (10–50 nL) enables a 1000 times injection for checking doubtful results, extending the working range for urine samples with highly variable analytes and determining the measurement precision. A complete assay of 15 commonly found urinary anions in four samples can be achieved within 1 h. The method has been shown to exhibit wide working ranges (0.005–500 mM), low detection limits (0.5–6.5 µM, S/N = 2), and satisfactory repeatability (RSD, n = 5) of 4.1–6.5% and 0.52–0.87% for peak area and migration time, respectively.

References

1. Charles, Y.C.P., John, R.P., Beverley, A.H. and Margaret, S.P. Predictive value of kidney stone composition in the detection of metabolic abnormalities. *Am J Med* 2003, 115, 26–29.
2. Olkhov, R.V., Fowke J.D. and Shaw, A.M. Whole serum BSA antibody screening using a label-free biophotonic nanoparticle array. *Anal. Biochem.* 2009, 385, 234–241.
3. Dolnik, V. and Dolnikova, J. Capillary electrophoretic determination of oxalate in amniotic fluid. *J Chromatogr A* 1995, 716, 269–272.
4. Addis, T. and Watanabe, C.K. The volume of urine in young healthy adults on a constant diet. *J Biol Chem* 1916, 27, 267–272.
5. Kveim, M. and Bredesen, J.E. A gas chromatographic method for determination of acetate levels in body fluids. *Clin Chim Acta* 1979, 92, 27–32.
6. Auliffe, J., Lind, L.J. and Leith, D.E. Hypoproteinemic alkalosis. *Am J Med* 1986, 81, 86–90.
7. Fencl, V. and Rossing, T.H. Acid-Base Disorders in Critical Care Medicine. *Annu Rev Med* 1989, 40, 17–27.
8. Hagen, L., Walker, V. and Sutton, R. Plasma and urinary oxalate and glycolate in healthy subjects. *Clin Chem* 1993, 39, 134–138.
9. Lv, Y., Zhang, Z. and Chen, F. Chemiluminescence biosensor chip based on a microreactor using carrier air flow for determination of uric acid in human serum. *Analyst* 2002, 127, 1176–1179.
10. Nelson, B.C., Rockwell, G.F. and Campfield, T. Capillary electrophoretic determination of oxalate in amniotic fluid. *Anal Chim Acta* 2000, 410, 1–3.
11. Shi, H. and Ma, Y. A simple and fast method to determine and quantify urinary creatinine. *Anal Chim Acta* 1995, 312, 79–83.
12. Fry, S.B. and Ian, D.R. The Determination of Oxalate in Urine and Plasma by High Performance Liquid Chromatography. *Ann Clin Biochem* 1991, 28, 581–584.
13. Pacenti, S.D. and Villanelli, M.F. Determination of organic acids in urine by solid-phase microextraction and gas chromatography–ion trap tandem mass spectrometry previous 'in sample' derivatization with trimethyloxonium tetrafluoroborate. *Biomed Chromatogr* 2008, 22, 1155–1158.

14. Tsukasa, N., Norio, Y.F., Mikiko, S.M., Kazunobu, S. and Tomonori, T. Determination of rabeprazole and its active metabolite, rabeprazole thioether in human plasma by column-switching high-performance liquid chromatography and its application to pharmacokinetic study. *J. Chromatogr B.* 2005, 824, 238–243.

15. Holmes, R. Measurement of urinary oxalate and citrate by capillary electrophoresis and indirect ultraviolet absorbance. *Clin Chem* 1995, 41, 1297–1301.

16. Ashby, R.A. and Sleet, R.J. The role of citrate complexes in preventing urolithiasis. *Clin Chim Acta* 1992, 210, 157–160.

17. Baena, A.C. and Beatriz, C.B. Analysis of carboxylic acids in biological fluids by capillary electrophoresis. *Electrophoresis* 2005, 26, 2622–2630.

18. Chen, L.Z., Yun, Z. and Chen, G. Fabrication, modification, and application of poly(methyl methacrylate) microfluidic chips. *Electrophoresis* 2008, 29, 1801–1808.

19. Harrison, D.J., Fluri, K.S., Fan, Z. and Manz, A. Micromachining a Miniaturized Capillary Electrophoresis-Based Chemical Analysis System on a Chip. *Science* 1993, 261, 895–898.

20. Park, S.G., Lee, S.K., Moon, J.H. and Yang, S.M. Holographic fabrication of three-dimensional nanostructures for microfluidic passive mixing. *Lab Chip* 2009, 9, 3144–3150.

21. Liu, C., Luo, Y., Maxwell, E.J., Fang, N. and Chen, D.Y. Reverse of Mixing Process with a Two-Dimensional Electro-Fluid-Dynamic Device. *Anal Chem* 2010, 82, 2182–2185.

22. Tang, Y. and Wu, M. The simultaneous separation and determination of five organic acids in food by capillary electrophoresis. *Food Chem* 2007, 103, 243–247.

23. Laugere, F., Guijt, R.M., Bastemeijer, J., Van der Steen, G., Berthold, A., Baltussen, E., Sarro, P., et al. On-Chip Contactless Four-Electrode Conductivity Detection for Capillary Electrophoresis Devices. *Anal Chem* 2003, 75, 306–309.

24. Wang, J. and Chatrathi, M.P. Microfabricated Electrophoresis Chip for Bioassay of Renal Markers. *Anal Chem* 2003, 75, 525–529.

25. Miyake, O., Yoshimura, K., Takahara, S. and Okuyama, A. Possible causes for the low prevalence of pediatric urolithiasis. *Urology* 1999, 53, 1229–1233.

26. Guo, W.P., Lau, K.M. and Fung, Y.S. Microfluidic chip-Capillary Electrophoresis for two orders extension of adjustable upper working range for profiling of inorganic and organic anions in urine. *Electrophoresis* 2010, 31, 3044–3052.

27. Hartshorne, H., Backhouse, C.J. and Lee, W.E. Ferrofluid-based microchip pump and valve. *Sens Actuators B* 2004, 99, 592–600.

28. Kim, J.H., Na, K.H., Kang, C.J., Jeon, D. and Kim, Y.S. A disposable thermopneumatic-actuated microvalve stacked with PDMS layers and ITO-coated glass. *Microelectron Eng* 2004, 73, 864–869.

29. Weibel, D.B., Kruithof, M., Potenta, S., Sia, S.K., Lee, A. and Whitesides, G.M. Torque-Actuated Valves for Microfluidics. *Anal Chem* 2005, 77, 4726–4733.

5

Dual-Channel Mixing: Standard Addition by Dual-Channel Continuous Flow Mixing for Assay of Complex Nephrolithiasis Biomarkers in Urine

Wenpeng Guo

First Affiliated Hospital of Shenzhen University
Shenzhen, People's Republic of China

CONTENTS

5.1 Introduction

Nephrolithiasis (kidney or renal stones) [1] is a common disorder. The risk for formation of these stones can be predicted from a urine assay, indicating abnormally low levels of inhibitors (citrate [CA] and creatinine [Cr]) and high levels of crystallization promoters (oxalate [OA] and uric acid [UA]) [2,3]. The incidents of milk adulteration with melamine lead to the demand for a rapid method to assess the risk for formation of kidney stones in babies for which milk was the only diet [4].

Current methods for nephrolithiasis markers include enzyme assay [5,6], ion chromatography [7,8], high-performance liquid chromatography (HPLC) [9–11], gas chromatography-mass spectrometry [12], and capillary electrophoresis [13–18]. All of these methods use laboratory equipment for assay after tedious and lengthy pretreatment procedures; thus, they are ultimately too slow for urgently needed results. For monitoring promoters or inhibitors, two separate procedures are generally required to provide the results needed for risk assessment; a single method capable of delivering timely results for assay of both inhibitors and promoters in kidney stones risk assessment is needed.

The recent development of microfluidic chip-capillary electrophoresis (MC-CE) devices [19–21] provides a promising combined technique for on-site assay of urinary nephrolithiasis markers due to its portability, efficient separation, and rapid and field-operable procedures. The MC-CE device has been shown to perform 100 times on-chip dilution before MC-CE determination of urinary anions [22–24]. However, the device requires extra operation time for off-line active mixing and washing, and the need to activate stirring and valve control by applying an external magnetic field. To simplify the operation, an MC-CE device integrated with a continuous passive mixing mode is designed, fabricated, and tested.

5.2 Design and Operation of the MC-CE Device

All continuous mixing [25] operations are active, except one [26], with a mixed active and passive mixing mode. Although a continuous passive mixing mode is simple in operation, there are few found in the literature, except one using several electrodes for mixing [26]. The reason is attributed to the use of zigzag channels [27], which give rise to undesirable band broadening associated with a large dead volume. Thus, a compact mixer was designed and fabricated by laser ablation with a small mixing volume.

Design of a Y-merging flow (Figure 5.1) with electro-osmotic flow (EOF)-driven passive mixing mode was fabricated with channel 1 (Ch1) to allow electrokinetic injection from sample vial (SV) (−350 V) to one or two sample segments while waste vials (W)1 or W1 + W2 are grounded [24]. The volume of the each microchannel segment is calculated to be 9.6 nL, which is adequate for on-chip mixing sample and standards for standard addition.

The MC-CE device fabricated (Figure 5.1) was washed with 0.1 M NaOH (5 min), deionized water (10 min), and running buffer (5 min) before use. Urine samples after 50-fold dilution and 0.45 µm syringe filter filtration are directly used for injection. Sample and standard solutions were loaded by electrokinetic injection onto a double-T channel segment for 10 s. For electrophoretic run, high voltages were applied to SV (−3 kV), buffer vial (B)1 (−6 kV), common waste vial (CW) (0 V), W2 (0 V), and W1 (0 V) for MC-CE separation and detection at 210 nm.

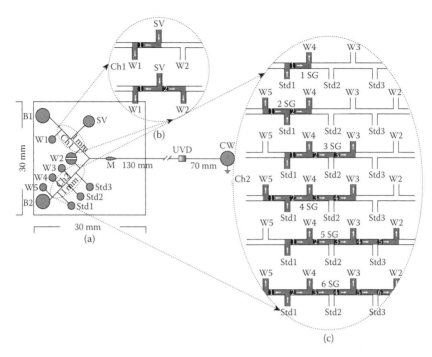

FIGURE 5.1
Schematic diagrams showing dual-channel multisegment Y-merging flow (a); sample and standard loading channel 1 (Ch1) (b); standard markers loading segments (SG) Ch2 (c). Std(1–3), standard vials; SV, sample vial; W(1–5), waste vials; B(1 and 2), buffer vials; UVD, ultraviolet detector; M, mixer; CW, common waste vial. (From Guo, W.P. and Fung, Y.S. *Electrophoresis* 2011, 32, 3437–3445.)

5.3 Fabrication and Verification of the MC-CE Device to Determine Nephrolithiasis Markers in Complex Samples

The Y-emerging dual-channel pattern was ablated by the CO_2 laser onto a poly(methyl methacrylate) (PMMA) plate for injection of standard and sample via channel 2 (Ch2) and Ch1, respectively. A separation capillary was placed inside a designated channel between PMMA plates for 15 min of bonding under 0.6 MPa at 92°C to fabricate the MC-CE device (Figure 5.1). After cooling, the MC-CE device was washed with 0.1 M HNO_3 (2 min) and 0.1 M NaOH (5 min) before cleaning in an ultrasonic bath and drying before storage.

To verify accuracy of on-chip standard addition, standard solutions (0.5 mM UA) were injected into the MC-CE device for standard addition to a given sample. The analyte enhancement factors upon standard addition by experiments based on peak areas and calculations are shown in Table 5.1. No significant differences in results were found between the expected and measured data [24].

TABLE 5.1

Verification of Enhancement Factor by MC-CE in Standard Addition Mode

Injected Volume in Ch1, V_1 nL (Segments)	Injected Volume in Ch2, V_2 nL (Segments)	Expected Enhancement Factor (E_E)	Measured Peak Area	Measured Enhancement Factor (E_M)	Difference ($D = E_M - E_E$)
9.6 (1)	0 (0)	1.0	0.712	1.00	0.000
9.6 (1)	4.8 (1)	1.5	1.09	1.53	0.030
9.6 (1)	14.4 (3)	2.5	1.89	2.65	0.150
9.6 (1)	28.8 (6)	4.0	3.08	4.33	0.330
				Mean ($n = 4$)	0.128
				SD ($n = 4$)	0.130

From Guo, W.P. and Fung, Y.S. *Electrophoresis* 2011, 32, 3437–3445.

5.4 Optimization of the MC-CE Operation Procedures

Two promoters, OA and UA, and two inhibitors, CA and Cr, were monitored as urinary biomarkers [3]. The current capillary electrophoresis (CE) methods require two separate procedures for separating anions and cations. The reasons are attributed to the large difference in the concentration of the biomarkers as well as to conditions for their separation and quantitation. Herein, the strategy for assay of these four biomarkers in a single MC-CE run is tested.

The pK_a values of inhibitors and promoters are given in Table 5.2 [24]. Cr is found in a different chemical form compared to the other three markers, as it exists in cationic or neutral form in buffers (pH 4–10); OA, UA, and CA exist as anions in alkaline buffer. Thus, it is possible to separate three markers in a single run using an alkaline buffer. Although Cr is omitted in the assessment using an alkaline buffer, results from the other three markers are sufficient for assessing risk for kidney stone formation.

An alkaline buffer previous developed for anion separation at pH 8.4 [22] was modified to pH 10.3 to shorten the run time and to enhance the stacking efficiency for the low-conductivity sample. Detection at 210 nm was used as it gave a compromised performance for the three biomarkers.

A borate buffer adjusted to pH 10.3 (cetyltrimethylammonium bromide [CTAB] [0.5 mM] + borates [20–80 mM]) was investigated for its effect on background noise and migration time. The best performance with adequate analyte separation and background noise reduction was shown using 20 mM borates; further reduction gives rise to peak overlap and noisy baseline. Thus, the optimized conditions for the running buffer are pH 10.3 with 20 mM borate and 0.5 mM CTAB as buffer constituents. The effect of cationic (CTAB) and anionic (sodium dodecyl sulfate [SDS]) surfactants on the separation efficiency is shown in Table 5.3 [24]. With addition

TABLE 5.2

pK_a Values from Nephrolithiasis Markers
Investigated

pK_a Value	CA	OA	UA	Cr
		Biomarker		
pK_{a1}	3.09	1.25	5.8	12.3
pK_{a2}	4.75	4.14	—	19.3
pK_{a3}	6.41	—	—	—

From Guo, W.P. and Fung, Y.S. *Electrophoresis*
2011, 32, 3437–3445.
Note: CA, citrate; OA, oxalate; UA, uric acid;
Cr, creatine.

TABLE 5.3

Effect of Surfactant on MC-CE Separation

		SDS[a]		CTAB[b]	
			Surfactant Additive		
Na$_2$BO$_4$ (mM)		20	80	40	20
Migration time	CA	2.21	6.03	2.63	1.49
(min)	OA	3.4	ND[c]	ND[c]	1.72
	UA	4.81	14.3	10.1	2.19
Sensitivity	CA	1.3	1.3	1.1	2.2
(mAU/mM)	OA	0.8	ND[c]	ND[c]	1.6
	UA	16	1.4	1.2	23

From Guo, W.P. and Fung, Y.S. *Electrophoresis* 2011, 32,
3437–3445.
Note: CTAB, cetyltrimethylammonium bromide; SDS,
sodium dodecyl sulfate; CA, citrate; OA, oxalate;
UA, uric acid.
[a] SDS buffer (pH 10.3) contains SDS (5 mM) + Na$_2$BO$_4$
(20 mM).
[b] CTAB buffer (pH 10.3) contains CTAB (0.5 mM) +
Na$_2$BO$_4$ (20–80 mM).
[c] ND = not detected.
Mixed standards: 1 mM CA + 1 mM OA + 0.1 mM UA.

of 5 mM SDS, both migration time and background noise were increased, whereas with addition of 0.5 mM CTAB, reduction in both migration time and background noise were found and attributed to the combined effect of EOF and electric field. Thus, the cationic surfactant CTAB was subsequently used.

From the best results obtained [3,10,11,16] using HPLC with ultraviolet detection (UVD), UA, CA, and OA can be determined in a single run by MC-CE device, whereas two procedures are required by HPLC, first for OA and CA [10] and then for UA [11]. The limits of detection (LODs) for

OA and CA by MC-CE device are lower than those via HPLC, whereas UA shows similar LODs for both techniques. The lengthy protein removal procedures used in HPLC are not required by MC-CE device; hence, MC-CE is much faster and uses a smaller sample size.

Compared to CE alone, similar LODs are obtained for OA and lower LODs for CA and UA using the MC-CE device. This difference is attributed to a higher stacking efficiency for MC-CE using buffer with a much higher conductivity compared to sample. For CE alone, diluted samples (five fold) and nearly neural buffer (pH 6.5) were used, conditions not favorable for stacking. UA absorption at 195 nm is better than at 210 nm, thereby compensating for the lower stacking efficiency and giving rise to similar LODs for both methods. With the use of standard addition with matching sample matrix, a much wider working range was observed for MC-CE compared to CE. In addition, the analyte migration time using MC-CE device is constant at both micromolar and millimolar ranges (Figure 5.2) [24]. As only one procedure is used in MC-CE, the analysis time is half that of CE.

In summary, three major nephrolithiasis markers—UA, OA, and CA—exhibited migration times under optimized conditions at 2.17, 1.76, and 1.47 min, respectively, at both micromolar (Figure 5.2a) and millimolar (Figure 5.2b) concentration ranges. The standard addition working ranges were found to vary for the different analytes: CA (5 µM–8 mM), OA (10 µM–8 mM), and UA (1 µM–8 mM). The repeatabilities (percent relative standard deviation [%RSD], $n = 5$) as shown in Table 5.4 for migration time varied from 0.27 to 1.33% and for peak area from 2.87 to 5.26% [24]. The LODs for the analytes were UA (0.8 µM), OA (7.3 µM), and CA (3.1 µM).

5.5 Application of the Dual-Channel MC-CE Device for Assay of Urinary Biomarkers

MC-CE assay results for urinary biomarkers are shown in Table 5.5 [24]. With extended working range using standard addition, the MC-CE device is capable of assaying concentration ranges for biomarkers UA, OA, and CA in a single run, after appropriate dilution and use of suitable microchannel segments for standard and sample injection. In addition, the device delivers LODs below 10% of the control levels (Table 5.5).

A typical MC-CE electropherogram using standard addition for the assay of urinary nephrolithiasis biomarkers is shown in Figure 5.3. One segment (9.6 nL from a diluted urine sample, 50-fold) is injected electrokinetically (350 V for 10 s) into Ch1. Standards (one segment, CA; two segments, OA and UA) were then loaded (–350 V for 5–10 s) in Ch2. With reload of urine sample at Ch1, the second run is then conducted for 4 min by holding 0 V at CW

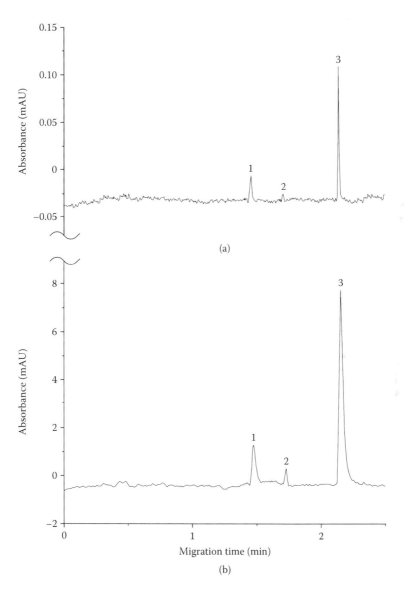

FIGURE 5.2
MC-CE electropherogram for separation of nephrolithiasis markers. Buffer (pH 10.3) composed of Na_2BO_4 (20 mM) + 0.5 mM CTAB (0.5 mM). Migration order: 3 = UA, 2 = OA, 1 = CA. (a) CA (20 μM) + OA (20 μM) + UA (2 μM). (b) CA (1 mM) + OA (1 mM) + UA (0.1 mM). (From Guo, W.P. and Fung, Y.S. *Electrophoresis* 2011, 32, 3437–3445.)

TABLE 5.4

Repeatability of Migration Time and Peak Measurement by Using MC-CE Device and Conventional CE

Biomarker		Migration Time[a]		Peak Area[a]	
		MC-CE	CE	MC-CE	CE
Standard[b]	CA	1.33	2.2	3.64	1.8
	OA	1.3	2	5.26	2.7
	UA	0.27	1.7	2.87	1.6
Sample	CA	1.4	2.8	4.46	2.4
	OA	1.43	2.3	6.1	3.2
	UA	0.47	2.4	1.98	2.7

From Guo, W.P. and Fung, Y.S. *Electrophoresis* 2011, 32, 3437–3445.
Note: MC-CE, microfluidic chip-capillary electrophoresis; CA, citrate; UA, uric acid; OA, oxalate.
[a] $n = 5$.
[b] Mixed standards: UA (0.1 mM) + CA (1 mM) + OA (1 mM).

TABLE 5.5

Working Range and LOD for Simultaneous Assay of Urinary Nephrolithiasis Markers by MC-CE Device

Urinary Biomarker		Promoter		Inhibitor
		UA	OA	CA
One sample segment	LOD[a] (µM)	1.5	10.3	4.7
	Working range[b] (mM)	0.05–40	0.50–40	0.25–40
Two sample segments	LOD[a] (µM)	0.7	6.1	2.6
	Working range[b] (mM)	0.025–30	0.25–30	0.13–30
Overall performance	LOD[a] (µM)	0.7	6.1	2.6
	Working range[b] (mM)	0.025–40	0.25–40	0.13–40
Concentration range reported[c] (mM)		0.5–9.0	0.3–2.0	0.5–7.0
Significant concentration[d] (mM)		>4.0	>2.0	<1.9

From Guo, W.P. and Fung, Y.S. *Electrophoresis* 2011, 32, 3437–3445.
Note: CA, citrate; OA, oxalate; UA, uric acid; Cr, creatine; LOD, limit of detection.
[a] LOD (signal to noise = 3, $n = 3$).
[b] Standard addition working range ($R > 0.99$).
[c] Range of nephrolithiasis biomarkers reported [28–33].
[d] Critical concentrations if exceeded [28–33].

and −6 kV at B1/B2 to execute mixing, standard addition, and separation. The start of the migration time is marked by the system peak. Samples with low OA concentrations are difficult to identify in run 1. However, it can be picked up in run 2. To improve the peak shape for OA identification, two sample segments can be used in run 1 to assist in OA quantitation.

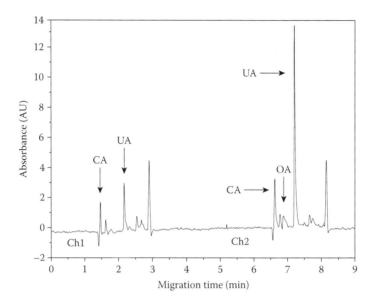

FIGURE 5.3
Electropherograms for separation and identification of different urinary markers using the dual-channel MC-CE device.

5.6 Summary

An MC-CE device is developed for field monitoring of important urinary nephrolithiasis markers. A passive online mixing mode is integrated with the MC-CE device to replace active mixing, which requires the use of external facilities and laboratory support. Under optimized conditions, only 10 min is required for a complete assay of the three major urinary markers UA, CA, and OA. Simple filtering and dilution are needed before sample injection. Satisfactory working range (0.025–40, 0.25–40, and 0.13–40 mM), low LOD (0.7, 6.1, and 2.6 µM), satisfactory repeatability (%RSD, $n = 5$) for peak area (1.98, 6.10, and 4.46%), and migration time (0.47, 1.43, and 1.40 min) are obtained for UA, OA, and CA, respectively.

The integration of on-chip standard addition to the dual-channel MC-CE device is shown to assist identification of urinary biomarkers in samples with a complex and variable matrix. It also improves the repeatability for measuring migration time and peak area and enhances quantitative determination of urinary biomarker concentrations. The analytical performances achieved have demonstrated the potential of MC-CE device for on-site monitoring of urinary biomarkers to produce results in time to assist in making decisions for applying intervention measures to patients under critical conditions.

References

1. Liang, B.A. Management and Prevention of Nephrolithiasis. *Hosp Physician* 1999, 2, 22–27.
2. Bihl, G. and Meyers, A. Recurrent renal stone disease-advances in pathogenesis and clinical management. *Lancet* 2001, 358, 651–656.
3. Barbas, C., Grac, A., Saavedra, L. and Muros, M. Urinary analysis of nephrolithiasis markers. *J Chrom B* 2002, 781, 433–455.
4. Li, G., Jiao, S., Yin, X., Deng, Y. and Wang, Y. The risk of melamine-induced nephrolithiasis in young children starts at a lower intake level than recommended by the WHO. *Pediatr Nephrol* 2010, 25, 135–141.
5. Li, M. and Madappally, M. Rapid enzymatic determination of urinary oxalate. *Clin Chem* 1989, 35, 2330–2333.
6. Petrarulo, M., Facchini, P., Cerelli, E., Marangella, M. and Linari, F. Citrate in urine determined with a new citrate lyase method. *Clin Chem* 1995, 41, 1518–1521.
7. Petrarulo, M., Marangella, M., Pellegrino, S., Linari, F. and Mentasti, E. Ion chromatographic determination of plasma oxalate in healthy subjects, in patients with chronic renal failure and in cases of hyperoxaluric syndromes. *J Chromatogr* 1990, 511, 223–231.
8. Hagen, L., Walker, V. and Sutton, R. Plasma and urinary oxalate and glycolate in healthy subjects. *Clin Chem* 1993, 39, 134–138.
9. Kataoka, K., Yakada, M., Kato, Y., Iguchi, M., Kohri, K. and Kurita, T. Determination of urinary oxalate by high-performance liquid chromatography monitoring with an ultraviolet detector. *Urol Res* 1990, 18, 25–28.
10. Khaskhali, M.H., Bhanger, M.I. and Khand, F.D. Simultaneous determination of oxalic and citric acids in urine by high-performance liquid chromatography. *J Chrom B* 1996, 675, 147–151.
11. Jen, J.F., Hsiao, S.L. and Liu, K.H. Simultaneous determination of uric acid and creatinine in urine by an eco-friendly solvent-free high performance liquid chromatographic method. *Talanta* 2002, 58, 711–717.
12. Pacenti, S.D. and Villanelli, M.F. Determination of organic acids in urine by solid-phase microextraction and gas chromatography–ion trap tandem mass spectrometry previous 'in sample' derivatization with trimethyloxonium tetrafluoroborate, *Biomed. Chromatogr.* 2008, 22, 1155–1158.
13. Verica, A.G., Luis, S. and Coral, B. Capillary electrophoresis for short-chain organic acids and inorganic anions in different samples. *Electrophoresis* 2003, 24, 1951–1981.
14. Beatriz, A.C. and Coral, B. Analysis of carboxylic acids in biological fluids by capillary electrophoresis. *Electrophoresis* 2005, 26, 2622–2632.
15. Christian, W.K. Determination of organic acids by CE and CEC methods. *Electrophoresis* 2007, 28, 3362–3378.
16. Munoz, J.A., Montserrat, L.M. and Valiente, M. Development and validation of a simple determination of urine metabolites (oxalate, citrate, uric acid and creatinine) by capillary zone electrophoresis. *Talanta* 2010, 81, 392–397.
17. Jennifer, L., Boughton, W. and Timothy, G. Determination of uric acid in human serum by capillary electrophoresis with polarity reversal and electrochemical detection. *Electrophoresis* 2002, 23, 3705–3710.

18. Tuma, P., Samcova, E. and Stulik, K. Determination of the spectrum of low molecular mass organic acids in urine by capillary electrophoresis with contactless conductivity and ultraviolet photometric detection—An efficient tool for monitoring of inborn metabolic disorders. *Anal Chim Acta* 2011, 685, 84–90.

19. Harrison, D.J., Manz, A., Fan, Z., Luedi, H. and Widmer, H.M. Capillary Electrophoresis and Sample Injection Systems Integrated on a Planar Glass Chip. *Anal Chem* 1992, 64, 1926–1932.

20. Guo, W.P., Lau, K.M. and Fung, Y.S. Microfluidic chi-capillary electrophoresis for emergency onsite analysis. *Sep Sci* 2009, 2, 3–10.

21. Guo, W.P. *Proceedings of the 2005 IEEE Engineering in Medicine and Biology 27th Annual Conference*, Clinical Laboratories on a Chip for Human Immunodeficiency Virus Assay, 2005, pp. 1274–1277.

22. Guo, W.P., Lau, K.M. and Fung, Y.S. Microfluidic chip-Capillary Electrophoresis for two orders extension of adjustable upper working range for profiling of inorganic and organic anions in urine. *Electrophoresis* 2010, 31, 3044–3052.

23. Guo, W.P. and Fung, Y.S. Microfluidic chip-capillary electrophoresis with adjustable on-chip sample dilution for profiling of urinary markers. *Proceedings of the 14th International Conference on Miniaturized Systems for Chemistry and Life Sciences*, 2010, pp. 1481–1483.

24. Guo, W.P. and Fung, Y.S. Microfluidic chip-capillary electrophoresis with dynamic multi-segment standards addition for rapidly identifying nephrolithiasis markers in urine. *Electrophoresis* 2011, 32, 3437–3445.

25. Ng, W., Goh, S., Lam, Y., Yang, C. and Rodriguez, I. DC-biased AC-electroosmotic and AC-electrothermal flow mixing in microchannels. *Lab Chip* 2009, 9, 802–809.

26. Yan, D., Yang, C., Miao, J., Lam, Y. and Huang, X. Enhancement of electrokinetically driven microfluidic T-mixer using frequency modulated electric field and channel geometry effects. *Electrophoresis* 2009, 30, 3144–3152.

27. Chen, J.K. and Yang, R.J. Electroosmotic flow mixing in zigzag microchannels. *Electrophoresis* 2007, 28, 975–983.

28. Coe, F.L., Parks, J.H. and Asplin, J.R. The pathogenesis and treatment of kidney stones. *N Engl J Med* 1992, 327, 1141–1152.

29. Curhan, G., Willett, W.C., Speizer, F.E. and Stampfer, M.J. Twenty-four-hour urine chemistries and the risk of kidney stones among women and men. *Kidney Int* 2001, 59, 2290–2298.

30. Addis, T. and Watanabe, C.K. The volume of urine in young healthy adults on a constant diet. *J Biol Chem* 1916, 27, 267–272.

31. Miyake, O., Yoshimura, K., Tsujihata, M. and Yoshioka, T. Possible causes for the low prevalence of pédiatrie urolithiasis. *Urology* 1999, 53, 1229–1234.

32. Garcia, A., Barbas, R. and Castro, M. Capillary electrophoresis for rapid profiling of organic acidurias. *Clin Chem* 1998, 44, 1905–1911.

33. Wishart, D.S., Knox, C. and Guo, A.C. HMDB: A knowledgebase for the human metabolome. *Nucleic Acids Res* 2009, 37, D603–D610.

6

Sample Cleanup and Analyte Enrichment: Integrating MIP/SPE with ME-MC-CE Device to Enhance Sample Cleanup and Multidimensional Analyte Enrichment for Hourly Determination of Atmospheric Carbonyl Compounds

Hui Sun

Guangzhou University
Guangzhou, People's Republic of China

Ying Sing Fung

The University of Hong Kong
Hong Kong SAR, People's Republic of China

CONTENTS

6.1 Introduction

The rising global concern on air pollution in particular for toxic pollutants give rise to the needs and requirements for their monitoring in ambient air and the associated new analytical methods developed for their monitoring and environmental impact assessment. As the control limit for toxic pollutants are extremely low, sample cleanup and analyte enrichment are required prior to their determination.

To extend the scope of current methodologies, MIP/SPE is integrated with ME-MC-CE device for multidimensional enrichment of carbonyl-DNPH derivatives prior to electrophoretic determination of carbonyl derivatives. The optimized working conditions are presented and discussed. Case study for the application of the ME-MC-CE device with multidimensional enrichment of carbonyl-DNPH derivatives for determination of carbonyl compounds in ambient air is presented to illustrate the advantages and limitations for the use of MC-CE device for environmental monitoring.

6.1.1 Needs and Requirements for Determination
        of Atmospheric Carbonyl Compounds

The analysis of atmospheric carbonyls has received a great deal of scientific attention due to their important role in atmospheric chemistry and their adverse effects on human health. Carbonyls present in photochemical smog are important precursors for the generation of toxic compounds, such as hydroxyl radicals (OH·), carboxylic acids, ozone, and peroxy acetyl nitrate [1–3]. In addition, several carbonyls are recognized as carcinogenic and mutagenic [4,5]. The rapid increase in atmospheric carbonyls has led to a growing worldwide concern, such that there is an international agreement for their reduction and control [6,7], thereby giving rise to the development of analytical tools for effective monitoring of atmospheric carbonyl compounds and apportionment of their sources.

6.1.2 Current Methodologies for Determination of Atmospheric Carbonyl Compounds and Assessment of Pollution Sources

Current determination of atmospheric carbonyl compounds can be divided into two methodologies: (1) sampling in the field followed by (2) analysis in the laboratory. The results obtained are used to determine the relative contribution of different pollution sources to the atmospheric carbonyl compounds collected at a given sample site for environmental impact assessment.

Due to the limited sensitivity of the current analytical methods, a long collection time is needed for sampling sufficient atmospheric carbonyls in the field for their subsequent determination in the laboratory. A normal measuring procedure for atmospheric carbonyl compounds typically takes 1 day to complete. Thus, results on weekly or daily levels of atmospheric carbonyl compounds are normally issued by environmental monitoring stations. To enable an effective monitoring of atmospheric carbonyl compounds, hourly results are needed, as one of the major sources of atmospheric carbonyls is from automobiles, with highly variable numbers throughout the day. Thus, a significant increase in the detection sensitivity for the analytical procedures is needed to reduce the sampling period required to 1 h for quantitation of atmospheric carbonyl compounds.

6.1.2.1 Sampling Methods for Collection of Atmospheric Carbonyl Compounds

The existing standard sampling protocols [8,9] use a sorbent coated with an excess amount of 2,4-dinitrophenylhydrazine (2,4-DNPH) for trapping atmospheric carbonyls by the formation of carbonyl 2,4-dinitrophenylhydrozone, a highly colored carbonyl derivative with extremely high detection sensitivity by an ultraviolet–visible spectrophotometer. The chemical reaction is shown in Equation 6.1.

$$\begin{array}{ccc} \text{Carbonyl} & \text{2,4-DNPH} & \text{2,4-Dinitrophenylhydrozone} \end{array}$$

The problem with the above-mentioned method is that it collects both the gaseous and particulate forms of atmospheric carbonyls. Carbonyls are semivolatile organic compounds (SVOCs) present in both gaseous and particulate forms. As different forms of SVOCs exhibit highly variable photochemical reaction rate, differential deposition velocities, and migration paths, it is import to differentiate gaseous and particulate carbonyls to determine the relative contribution of different pollution sources. A newly introduced sampling method based on denuder sampling [10,11] is used to collect gaseous and particulate carbonyl compounds from the air sampled at a specific site. The demand for hourly results and the capability to differentiate gaseous carbonyls from particulate carbonyls lead to the development, in the present work, of integrating a microfluidic chip-capillary electrophoresis (MC-CE) device with multidimensional enrichment of the analyte in the various analytical procedures. This MC-CE device can be used to achieve a significant increase in carbonyl detection sensitivity.

6.1.2.2 Laboratory Methods for Determination of Atmospheric Carbonyl Compounds

Various methods have been used for the determination of carbonyls in the atmosphere by using different separation procedures. Separation procedures such as high-performance liquid chromatography (HPLC) [12,13], gas chromatography (GC) [14–16], capillary liquid chromatography [17], capillary electrochromatography [18], and micellar electrokinetic capillary chromatography (MEKC) [10] have been reported in the literature.

Due to the requirement for analyte volatility in operation, GC methods are, in general, not suitable for the determination of nonvolatile and SVOC compounds, HPLC procedures are adopted based on the determination of the derivatives from carbonyl and DNPH for assessing atmospheric carbonyls [8,9]. However, several hours or even days are needed to collect enough ambient carbonyls using conventional HPLC methods due to their low mass detection sensitivity [19,20]. This long collection time is unfavorable for timely detection of sudden pollution events and for monitoring the migration of carbonyl pollutants.

To reduce the analysis time and improve the detection sensitivity, methods based on capillary chromatography have been adopted recently [13,10,17,18]. The major issues encountered in capillary electrophoresis (CE) determination of environmental pollutants are insufficient method sensitivity to handle pollutants present at low levels or in complex matrices. Thus, pretreatment methods for sample cleanup and analyte enrichment are needed prior to the scope of CE separation is broadened to allow the determination of trace pollutants in complex sample matrices.

6.1.3 Problems and Issues

A challenge for sampling ambient carbonyls based on the formation of DNPH derivatives is that the addition of excess 2,4-DNPH to enable a complete collection of ambient carbonyls leads to a high level of unreacted 2,4-DNPH residual. This residual can co-elute with carbonyl-DNPH derivatives and cause interference problems during chromatographic separation. In addition, the maximum preconcentration factor of the target analyte is limited by the precipitation of excess DNPH residual during the reduction of the sample volume by evaporation. Due to the poor preconcentration factor, only formaldehyde, acetaldehyde, and acetone can be observed in the CE electropherogram, whereas low abundant ambient carbonyls, such as benzaldehyde, acrolein, isobutyraldehyde, crotonaldehyde/butyraldehyde, isovaleraldehyde, hexaldehyde, methyl ethyl ketone, *p*-tolualdehyde, *o*-tolualdehyde, valeraldehyde, 2,5-dimethylbenzaldehyde, and propionaldehyde, are not detected [18,21,22].

Thus, it is important to get rid of the unreacted DNPH from the sample solution to improve the preconcentration factor to achieve hourly determination of ambient carbonyls, especially for the low abundant carbonyls.

6.2 Methodologies for Sample Cleanup and Analyte Enrichment

6.2.1 Preparation and Test of Molecularly Imprinted Polymers for Sample Cleanup

To produce the special recognition binding sites for template molecules in a polymer matrix, a molecularly imprinted polymer (MIP) technology has been used that involves copolymerization of functional monomers and cross-linkers in the presence of the template molecules. MIPs have been shown to possess advantages such as predetermined selectivity and high chemical stability suitable for use in various analytical areas. The application of MIPs for solid phase extraction (SPE) has exhibited great potential for cleanup of complex samples, even for differentiating analyte and impurities with similar properties [23–27].

Molecular imprinting technology is derived from the concept of creating designed recognition sites in macromolecular matrices by means of template polymerization. This technique is based on in situ copolymerization of cross-linkers and functional monomers to produce complexes with template (imprinted) molecules prior to polymerization. After removal of the template molecules from the imprinted material, binding sites are left behind, showing complementarity to the template in the subsequent rebinding experiment.

During carbonyl sampling, excess 2,4-DNPH is always used to ensure efficient collection of ambient carbonyls. However, the presence of unreacted 2,4-DNPH residuals, together with 2,4-dinitroaniline (2,4-DNAN) (a degradation product of 2,4-DNPH) and 2,4-dinitrophenylazine (2,4-DNPA) (produced by the reaction between 2,4-DNPH and NO_2), could affect quantitative determination of carbonyls by their co-elution together with carbonyl-DNPH derivatives during CE separation [28,29]. In addition, the maximum preconcentration factor of the target analyte is limited by the precipitation of excess DNPH residuals during volume reduction of sample solution during evaporation.

To get rid of the above-mentioned interfering compounds, 2,4-DNAN, instead of 2,4-DNPH, was used as the template for preparing MIPs based on the following considerations. First, due to the close similarity of the structures of 2,4-DNAN, 2,4-DNPH, and 2,4-DNPA (Figure 6.1), the MIPs produced using 2,4-DNAN as the template are expected to provide selective affinity to all three analytes. Second, if 2,4-DNPH is used as the template, a stable Schiff base (C=N) [17] will be formed in the MIPs' matrices due to a strong interaction between 2,4-DNPH and the carbonyl (C=O) moieties of the monomer methylacrylic acid (MAA) and cross-linker ethylene glycol dimethacrylate (EDMA) during polymerization. If a Schiff base has been formed, the template 2,4-DNPH would be difficult to remove from the resulted polymer matrix using a normal washing procedure and possibly leak out gradually during the subsequent SPE application procedure to cause interference. Based on the above-mentioned considerations, 2,4-DNAN was applied in the present work as a dummy template to prepare MIPs for cleanup of 2,4-DNPH and its analogs from the sample solutions. The schematic diagram showing the process for the preparation of MIP is shown in Figure 6.2.

A wet method is used to prepare MIP by a chemical reaction in a round-bottomed flask with the addition of 0.25 mol of DNAN, 1 mmol of MAA, 5 mmol of EDMA, and 30 mg of 2,2'-Azobisisobutyronitrile (AIBN) for dissolution in 16.5 mL of CH_3CN prior to polymerization at 60°C for 24 h. The 2- to 3-μm polymer particles (Figure 6.3) were collected and successively extracted with a mixture of methanol and acetic acid (9:1, v/v) to remove the templates from the imprinted polymer. A nonimprinted polymer (NIP) was prepared in the same way without the addition of template molecules.

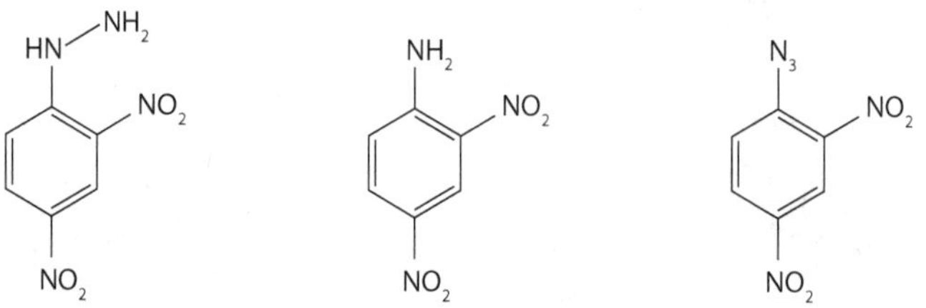

2,4-Dinitrophenylhydrazine
(DNPH)

2,4-Dinitroaniline
(DNAN)

2,4-Dinitrophenylazine
(DNPA)

FIGURE 6.1
Structures of compounds with similar structure as DNPH.

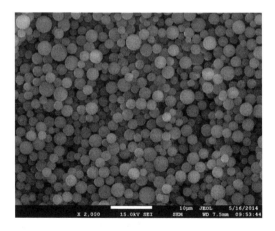

FIGURE 6.2
Schematic diagram showing the MIP preparative procedures.

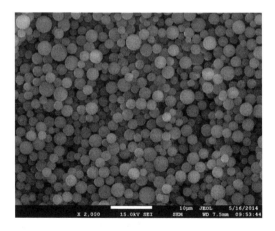

FIGURE 6.3
Scanning electron microscope image showing the morphology of MIP particles.

6.2.2 MIP/SPE for Selective Analyte Enrichment by Incorporation of MIPs in SPE

SPE provides a suitable means to clean up and enrich samples, because it is more rapid, simple, economical, and environmental friendly compared to traditional liquid–liquid extraction. However, the main problem associated with the use of commercially available C18 solid phase, such as C18 SPE column, is its low selectivity toward the target analyte. The use of MIPs prepared with a suitable template for a specific target analyte as a solid phase (MIP/SPE) could greatly improve the selectivity of the solid phase, with results shown in the following sections.

For SPE cartridge preparation, a slurry containing 100 mg of MIPs in 1.0 mL of methanol is packed into an empty SPE column. Polytetrafluoroethylene frits are then placed above and below the sorbent bed. For sample purification, the cartridge is subjected to washing and eluting treatment in sequence. Through the washing procedure, all carbonyl-DNPH derivatives are expected to be eluted out from the MIP/SPE cartridge for subsequent CE analysis, whereas DNPH and its analogs are expected to be entrapped in the MIP/SPE cartridge. In the next eluting (regeneration) step, the entrapped substances are eluted out and the SPE cartridge is regenerated. The selection of washing/regeneration solvent is important, because the properties of washing/regeneration solvent would affect the analyte adsorption–desorption kinetics and the selectivity of the MIP/SPE cartridge.

To optimize the operation conditions, 100 μL of a standard mixture of 0.125 mmol L^{-1} carbonyl-DNPH derivatives and 50 mmol L^{-1} 2,4-DNPH were loaded onto the SPE cartridge, followed by the addition of 4 mL of either toluene, chloroform, dichloromethane, methanol, or acetonitrile to wash the cartridges. All the extraction fractions were collected and evaporated to dryness before redissolution in acetonitrile/2 mmol L^{-1} sodium tetraborate (1:1, v/v) for MEKC separation. The results showed that when toluene was used as the washing solvent, the MIP/SPE cartridge exhibited high selectivity for DNPH, with 98% of DNPH held by the SPE column, whereas carbonyl-DNPH derivatives were eluted off, with an average recovery of $97 \pm 5.3\%$ ($n = 5$).

The selectivity is attributed to the specific affinity of MIP/SPE cartridge to DNPH. According to the molecular recognition principle of MIP, the interaction between the MIP/SPE cartridge and DNPH can be stabilized in the aprotic or low polar organic solvents, such as toluene [30,31]. In contrast, with the increase of hydrophilicity or polarity of the washing phase, the selectivity of the MIP/SPE cartridge could be deteriorated due to the disruption of the hydrogen bonds. Thus, when acetonitrile was used as the washing solvent, both DNPH and various carbonyl derivatives were all eluted out at the same time from the SPE cartridge. Therefore, in the present work, toluene was employed in the washing step to collect the purified carbonyl-DNPH derivatives, with an average recovery of $97 \pm 5.3\%$ ($n = 5$), and acetonitrile was used in the next regeneration step to elute out 2,4-DNPH from the SPE cartridge to regenerate the MIP/SPE cartridge.

To prepare the blank cartridge, NIP was synthesized and tested as SPE material for comparison. Due to the lack of specific recognition binding sites for 2,4-DNPH in the NIP cartridge, 2,4-DNPH was found to co-elute with the carbonyl-DNPH derivatives from the SPE cartridge. To assess the specific affinities of the MIP/SPE cartridge, the performance of the commercial

C18 SPE cartridge (Alltech, Milwaukee, WT) was compared to MIP/SPE on the effect for cleaning 2,4-DNPH from carbonyl-DNPH derivatives solution. The results showed that the commercial C18 cartridge cannot differentiate 2,4-DNPH from carbonyl-DNPH derivatives at all. The 2,4-DNPH and carbonyl-DNPH derivatives were eluted together from the C18 SPE cartridge under the same conditions. Thus, the ability of the MIP/SPE cartridge to remove 2,4-DNPH from the sample solution is attributed to the intrinsic recognition sites present in the MIP network.

6.2.3 Evaporation for Increasing Analyte Concentration

Evaporation of the extracting organic solvent is a simple and effective method to increase analyte concentration. However, the cleanup efficiency must be high. Ideally, only analytes are completed isolated from the sample matrix and left behind in the extraction solvent. However, there are always some other constituents extracted from the sample matrix, albeit an extremely small amount. To achieve a high preconcentration factor by solvent evaporation, the solubility of constituents extracted from the sample matrix would not be exceeded to interfere with the subsequent analytical procedure. The concentration of analyte present in the sample is expected to be extremely low to justify the use of an additional enrichment procedure. Thus, the precipitation of analyte by evaporative concentration due to exceeding solubility limit is not of concern, and the focus of a high evaporation ratio is the close-to-complete removal of other constituents extracted from the sample matrix.

Evaporative preconcentration (EC) after MIP/SPE extraction provides an ideal solution to increase the analyte concentration. The first step is for selective adsorption of analytes by MIP while other sample constituents are washing out, giving rise to effective cleanup, and the second step is for eluting analyte out from MIP by a high-purity solvent, producing a solution with a low level of analyte in almost pure solvent which is particularly suitable for evaporation enrichment of the analytes. Thus, coupling MIP/SPE with almost complete removal of other sample matrices to EC with high concentration factor for eluted analyte leads to the achievement of an extremely high analyte enrichment factor.

The highest enrichment achievable is evaporation to dryness at the smallest possible area for the solvent containing analytes eluted from MIP/SPE. A minimal volume of a second solvent can then be added to redissolve the deposited analytes back into solution. The procedure has another advantage for solvent switching, which is often needed to meet different requirements for analyte extraction and separation. For extraction, the solvent is normally a nonpolar solvent for enriching organic analytes from the aqueous samples. For separation such as CE, running buffer with a specific composition is needed and the redissolution solvent should be matched to the buffer for the subsequent CE run.

6.3 Optimization of Electrophoretic Working Conditions for Determination of Carbonyl Derivatives

6.3.1 Buffer Manipulation to Improve Resolution for Separation of Carbonyl Derivatives

A previously identified MEKC buffer solution containing 20% (v/v) methanol, 20 mM borate, and 50 mM sodium dodecyl sulfate (SDS) is selected for the separation of carbonyl derivatives [10]. As carbonyl-DNPH derivatives are highly soluble in acetonitrile, airborne carbonyl-DNPH derivatives collected from the sampler are usually dissolved in acetonitrile for HPLC or CE separation [20–22]. However, when carbonyl-DNPH derivatives were dissolved in pure acetonitrile, the peaks in the MEKC electropherograms were broad and poorly resolved, and the late-eluting carbonyl hydrozones, such as hexaldehyde, tolualdehyde, and dimethyl benzaldehyde derivatives, showed serious peak tailing, with peaks merged together as shown in Figure 6.4a.

An interesting phenomenon was observed when a mixture of acetonitrile and sodium tetraborate was used as the running buffer for MEKC

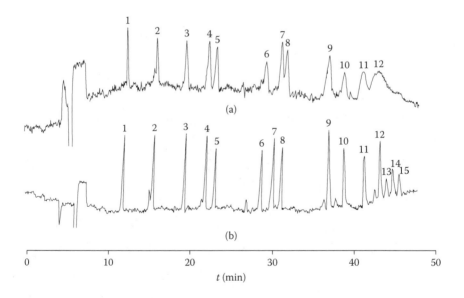

FIGURE 6.4
MEKC electropherogram of a standard mixture of DNPH and carbonyl-DNPH derivatives. (a) Mixture treated with MISPE cartridge and resolved in acetonitrile. (b) Mixture treated with MISPE cartridge and resolved in acetonitrile/2 mmol L^{-1} sodium tetraborate (1:1, v/v). Peaks: 15, 2,5-dimethylbenzaldehyde; 14, *p*-tolualdehyde; 13, *o*-tolualdehyde; 12, hexaldehyde; 11, benzaldehyde; 10, valeraldehyde; 9, isovaleraldehyde; 8, butyraldehyde; 7, crotonaldehyde; 6, methyl ethyl ketone; 5, propionaldehyde; 4, acrolein; 3, acetone; 2, acetaldehyde; and 1, formaldehyde.

separation of a carbonyl sample. When the salt concentration was increased from 0 to 2 mmol L^{-1} in the sample matrix, an obvious gain in the peak shape was observed. When the salt concentration was increased to higher than 2 mmol L^{-1}, little change in the signal intensity could be observed. Thus, the mixture of acetonitrile/2 mmol L^{-1} sodium tetraborate (1:1, v/v) was optimized in the present work for improving carbonyl resolution. A substantial gain in the peak height and a significant narrowing of the analyte peak for all carbonyl compounds investigated were observed upon optimization of the salt content of the MEKC solvent compared to the use of pure acetonitrile as the MEKC solvent.

Due to the improved separation efficiency, hexaldehyde, *o*- or *p*-tolualdehyde, and 2,5-dimethylbenzaldehyde could be baseline separated (Figure 6.4b), and the peak enhancement factor ranged from 1.01 to 9.13 (Table 6.1). The enhancement factor has been calculated as the ratio of the number of theoretical plates when the sample was dissolved in acetonitrile with 2 mmol L^{-1} sodium tetraborate added to that of a pure acetonitrile. The theoretical plate count was calculated using the following formula:

$$N = 5.54 \times (t_R/w_{1/2})^2$$

where N is the number of theoretical plates, t_R is the retention time of analytes, and $w_{1/2}$ is the peak width measured at one-half of the peak height.

TABLE 6.1

Comparison of MEKC Separation Efficiency for Carbonyl-DNPH Derivatives in Pure Acetonitrile and Acetinitrile/2 mmol L^{-1} Sodium Tetraborate (1:1, v/v) Solution

Carbonyl-DNPH Derivative	No. of Theoretical Plates ($\times 10^5$)		Enhancement Factor[c]
	Method I[a]	Method II[b]	
Hexaldehyde	2.008	0.219	9.13
Benzaldehyde	2.265	0.283	8.00
Valeraldehyde	1.687	0.273	6.18
Isovaleraldehyde	1.859	0.235	7.90
Butyraldehyde/isobutyraldehyde	1.430	0.268	5.33
Crotonaldehyde	1.250	0.273	4.58
Methyl ethyl ketone	1.380	0.289	4.78
Propionaldehyde	1.751	0.330	5.31
Acrolein	1.632	0.336	4.86
Acetone	1.346	0.352	3.82
Acetaldehyde	1.425	0.519	3.95
Formaldehyde	1.246	1.230	1.01

[a] Method I, sample was dissolved in acetonitrile/2 mmol L^{-1} sodium tetraborate.
[b] Method II, sample was dissolved in acetonitrile.
[c] The enhancement factor was calculated as the ratio of the number of theoretical plates when the sample was dissolved in acetonitrile/2 mmol L^{-1} sodium tetraborate (method I) solution to that using pure acetonitrile (method II).

6.3.2 Manipulation of Micelle to Enhance Separation Efficiency and Stacking of Carbonyl-DNPH Derivatives

As the use of stacking prior to separation could enhance separation efficiency and improve detection sensitivity attributed to sharper peaks, micelle manipulation is used in the present work to stack samples. Due to the low conductivity of the sample plug, the electric field strength in the sample plug is much higher than that in the background solution (BGS) zone. Anionic micelles would migrate quickly into the sample plug from the cathodic end but were retarded at the boundary between BGS and the anodic end of the sample solution zone. Thus, the carbonyl-DNPH derivatives in the sample solution could be carried by the micelle and focused at the boundary between BGS and the anodic end of the sample solution zone, realizing sample stacking. Then, the SDS-analytes group is separated by the MEKC mode and brought to the detector by electroosmotic flow [32]. However, if the neutral analytes (carbonyl-DNPH derivatives) were dissolved in pure acetonitrile, the micelle entering the sample plug could be corrupted to some extent by the highly concentrated acetonitrile. In contrast, the rapid distribution of the analytes to micelle should be inhibited as the solubility of the hydrophobic analytes is much higher in acetonitrile than micelle; hence, poor separation efficiency has been observed in pure acetonitrile.

From the results given in Table 6.1, the peak enhancement factors for the late-eluting species is higher than those of the early-eluting species; thus, the stacking effect increases with the retention factor (k) of the analytes [33]. Due to the improved separation efficiency, the extremely hydrophobic carbonyl-DNPH derivatives, such as benzaldehyde, hexaldehyde, o- and p-tolualdehyde, and 2,5-dimethyl benzaldehyde, could be baseline separated in the buffer with the optimized composition.

6.3.3 Optimized Analytical Performance

The analytical parameters under optimized working conditions are summarized in Table 6.2. The results showed repeatability, with relative standard deviations (RSDs) ($n = 5$) ranging from 2.3 to 8.5% for the peak area and from 0.36 to 536 mg L^{-1} for the working range. Detection limits based on a signal-to-noise ratio (S/N) = 2 ranged from 0.14 to 1.32 mg L^{-1}.

6.4 Multidimensional Enrichment of Carbonyl-DNPH Derivatives by Integrating MIP/SPE with ME-MC-CE Device

6.4.1 Fabrication of MC-CE Device Integrated with ME

Fabrication of a microevaporator-microfluidic chip-capillary electrophoresis (ME-MC-CE) device involved a CO_2 laser with a wavelength of 10.6 μm

TABLE 6.2

Analytical Parameters for MEKC Determination of Carbonyl-DNPH Derivatives in Acetonitrile/2 mmol L^{-1} Sodium Tetraborate (1:1, v/v) Solution

Carbonyl-DNPH Derivative	Detection Limit[a] (mg L^{-1})	Working Range (mg L^{-1})	Relative Standard Deviation* (%, n=5)
2,5-Dimethylbenzaldehyde	1.32	3.30–536	8.5
p-Tolualdehyde	1.15	2.88–480	7.6
o-Tolualdehyde	1.15	2.88–480	7.1
Hexaldehyde	0.44	1.11–320	4.1
Benzaldehyde	0.96	2.40–508	6.5
Valeraldehyde	0.86	2.16–412	3.3
Isovaleraldehyde	0.43	1.08–344	3.8
Butyraldehyde/isobutyraldehyde	0.32	0.81–259	6.9
Crotonaldehyde	0.32	0.81–224	6.3
Methyl ethyl ketone	0.36	0.90–288	5.4
Propionaldehyde	0.30	0.75–232	3.6
Acrolein	0.20	0.51–156	4.9
Acetone	0.23	0.57–139	2.3
Acetaldehyde	0.25	0.63–158	2.7
Formaldehyde	0.14	0.36–96	3.1

[a] Detection limit based on S/N = 2.

being used to ablate the poly(methyl methacrylate) (PMMA) chip (30 mm × 40 mm × 0.15 mm). The structure of the chip is shown in Figure 6.5. All ablated channels were 150 μm in width and 100 μm in depth. The distance between the double-T injector to each buffer vial was 3 mm. The length of the double-T injector was designed to be 1 mm to inject a suitable amount of sample introduced by suction from the microevaporation. There are two fused silica capillaries (50 μm inside diameter) embedded in the microfluidic chip (MC), one capillary for sample introduction and the other capillary for separation, with a physical length of 15 cm.

Before analysis, all MC channels were preconditioned successively for 5 min using 0.1 mol/L NaOH, H$_2$O, and the running buffer. When the MC-CE device was ready for operation, sample was introduced via a microdropper to the microevaporator (ME) for volume reduction from microliters to nanoliters volume. The evaporation rate was controlled by passing nitrogen slowly through the open end of the capillary from the waste vial at such a flowrate to allow the liquid to retreat toward the capillary slowly without leaving a deposit at the wall surface (ME) until its retraction to the capillary. The length of the capillary with liquid loading was then measured to calculate the final sample volume after evaporation. For the present work, exchange of solvent is required for the subsequent CE run. Evaporation to dryness at the inlet of the capillary was performed. Details on the operation of the ME for evaporation to dryness is given in the next section.

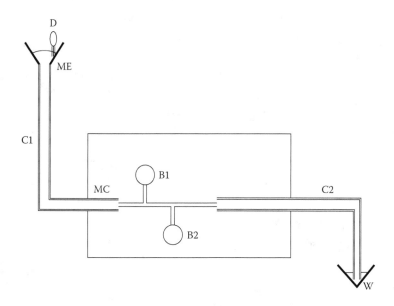

FIGURE 6.5
Schematic diagram showing the MC-CE device integrated with an ME. B1 and B2, buffer vials; C1, sampling capillary; C2, separation capillary; D, microdropper; MC, sampling capillary with metal coating; ME, microevaporator; W, waste vial.

6.4.2 Design and Operation of ME

The design of the ME in the present work is to evaporate the extraction solvent to dryness at the capillary inner surface. A given volume of the CE buffer at nanoliter size is then added to redissolve the analyte to achieve the highest possible enrichment. This is made possible for extraction solvent with near complete removal of other sample constituents and a high retention of analytes that are normally present at trace level in the original sample. The main problem lies with the introduction of nanoliter sample to the MC-CE device for conducting CE separation for analyte quantitation.

The embedding of the capillary to the double-T injector has been shown in our previous work to have a good fit with a slight distortion at the capillary head due to the flow of PMMA during thermal bonding [34]. Sample is sucked via capillary 1 (C1) into the double-T injector by a syringe operated at the open end of the separation capillary (C2), as shown in Figure 6.5. To enable the application of pinched voltage to avoid sample leakage during the CE run, the outer surface of the sampling capillary is sputtered with Pt with metalized surface extended to outside the MC for applying pinched voltage. Buffer vial B2 is used to remove excess amount of sample sucked in through the double-T injector. Buffer vial B1 is used to conduct the CE run with the waste vial (W) held at grounded potential.

For the multidimensional enrichment operation for analyte, the MIP/SPE operation offers the first enrichment as the volume of the eluting solvent from MIP/SPE is much less than the extracting solvent from the carbonyl-DNPH cartridge. The eluted organic solvent was then evaporated to ~0.5 mL of reduced solvent. Using a microdropper, the reduced solvent was added dropwise to the ME under nitrogen flow to evaporate to dryness at the inlet of the connecting capillary (Figure 6.5); 500 nL of the CE buffer was then added to redissolve the carbonyl-DNPH derivatives deposited at the inlet of the sampling capillary (C1). The volume of the redissolved sample solution was measured by the length of C1 occupied by the liquid sample plug prior to injection. Part of the sample plug was then sucked into the MC-CE device, and the injected volume is determined by the double-T injector channel segment after the removal of the excess sample via high voltage applied across B2 and W vials.

6.5 Application of ME-MC-CE Device with Multidimensional Enrichment of Carbonyl-DNPH Derivatives for Determination of Carbonyl Compounds in Ambient Air

6.5.1 Sampling and Laboratory Procedures

A DNPH-coated cartridge has been prepared as reported previously [10] for sampling total carbonyls near the University of Hong Kong at 0.25 L min^{-1} for 1 h. The versatile air pollutant sampler (VAPS) system equipped with filters and annular denuders [35–37] has been prepared for sampling gaseous and particulate carbonyls separately at 10 L min^{-1} for 1 h [10]. After sampling, the filters were extracted twice under ultrasonication by 5 mL of acetonitrile (containing 50 μL of 0.1% DNPH in acetonitrile) for 30 min, and the cartridge or annular denuders were washed twice with 5 mL of acetonitrile (HPLC grade). The volume of all extracts was then reduced to 0.1 mL prior to ME enrichment.

6.5.2 Results and Discussion

The electropherograms of the field samples are shown in Figures 6.6 and 6.7. Note that without SPE treatment, only three carbonyls—acetone, acetaldehyde, and formaldehyde—are detectable either in particulate phase or gaseous phase. However, after getting rid of DNPH and its analogs by MIP/SPE, the peaks for formaldehyde, acetaldehyde, and acetone were shown to exhibit a significant increase (Figures 6.6 and 6.7). Furthermore, owing to the high enrichment factor, many low abundant carbonyls could also be detected and quantified from the electropherograms.

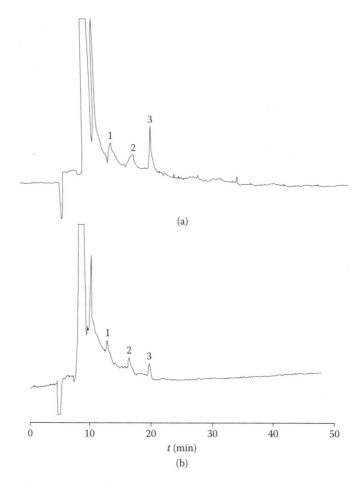

FIGURE 6.6
MEKC electropherograms of the ambient carbonyls collected using the VAPS method without treatment by the MIP/SPE cartridge. (a) Gaseous carbonyls. (b) Particulate carbonyls. Peaks: ③, acetone; ②, acetaldehyde; and ①, formaldehyde.

In contrast, Figure 6.7 also shows that 8 of the 15 carbonyls studied (formaldehyde, acetaldehyde, acetone, propionaldehyde, valeraldehyde, benzaldehyde, hexaldehyde, and *p*-tolualdehyde) were distributed both in the gaseous phase and particulate phase in the collected samples, whereas 3 of the 15 carbonyls studied (methyl ethyl ketone, butyraldehyde/ isobutyraldehyde, and isovaleraldehyde) existed only in gaseous phase. The other four carbonyls (acrolein, crotonaldehyde, *o*-tolualdehyde, and 2,5-dimethylbenzaldehyde) cannot be found either in gaseous phase or particulate phase in the collected samples.

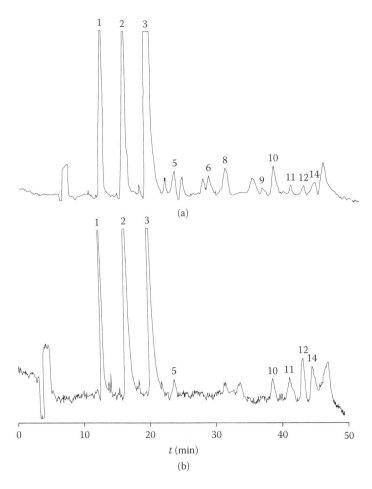

FIGURE 6.7
MEKC electropherograms of the ambient carbonyls collected using the VAPS method and treated with MIP/SPE cartridge. (a) Gaseous carbonyls. (b) Particulate carbonyls. Peaks: 14, *p*-tolualdehyde' 12, hexaldehyde; 11, benzaldehyde; 10, valeraldehyde; 9, isovaleraldehyde; 8, butyraldehyde; 6, methyl ethyl ketone; 5, propionaldehyde; 3, acetone; 2, acetaldehyde; and 1, formaldehyde.

The distribution factors of the carbonyls between gaseous and particulate phases are listed in Table 6.3. The results showed that the ratios between particle and gas for C1–C5 carbonyls ranged from 0.15 to 0.32. For C6 carbonyls, such as benzaldehyde, hexaldehyde, and *p*-tolualdehyde, the particle/gas ratios were in the range 1.17–1.55. Thus, C1–C5 carbonyls existed mainly in gaseous phase, and carbonyls with the number of carbons above five existed mainly in particulate phase.

TABLE 6.3

Results for Determination of Atmospheric Carbonyls by VAPS

Carbonyl	Gaseous ($\mu g\ m^{-3}$)	Particulate ($\mu g\ m^{-3}$)	Particle/Gas Ratio	Total Content ($\mu g\ m^{-3}$)
2,5-Dimethylbenzaldehyde	—	—	—	
p-Tolualdehyde	0.30	0.35	1.17	0.65
o-Tolualdehyde	—	—	—	—
Hexaldehyde	0.22	0.34	1.55	0.56
Benzaldehyde	0.28	0.38	1.36	0.66
Valeraldehyde	0.62	0.20	0.32	0.82
Isovaleraldehyde	0.08	—	—	0.08
Butyraldehyde/ isobutyraldehyde	0.30	—	—	0.30
Crotonaldehyde	—	—		—
Methyl ethyl ketone	0.36	—	—	0.36
Propionaldehyde	0.45	0.09	0.20	0.54
Acrolein	—	—		—
Acetone	9.30	1.74	0.19	10.04
Acetaldehyde	6.38	1.61	0.16	7.99
Formaldehyde	7.82	1.21	0.15	9.03

To compare with the results obtained from the two sampling methods, the content of carbonyls in gaseous and particulate phases has been added together. Note that the two sampling methods have produced comparable results for most of the carbonyls, except p-tolualdehyde, hexaldehyde, and methyl ethyl ketone. The variation of the results is expected to be caused by the wind speed during sampling and the change of wind direction.

Under optimized conditions, the ambient carbonyls can be detected with the MIP/SPE coupled MEKC method, showing a working range of 0.0019–2.68 $\mu g\ m^{-3}$, and detection limit varying from 0.28 to 2.64 ng, or from 0.00047 to 0.0044 $\mu g\ m^{-3}$, for sampling 1 h at a sampling rate of 10 L min^{-1}. In ambient clean air, the aldehyde concentration is reported to be between 0.0005 and 0.002 ppm (0.6–2.5 $\mu g\ m^{-3}$) at the ground level [38]. As the detection limit of the methodology developed is 100 times below the concentration of aldehyde reported, the present method is sensitive enough to detect the low abundance carbonyls in atmosphere, monitor the hourly variation of carbonyl content, and predict the seasonal variation of a pollution source.

Furthermore, it should be noted that the method developed can be applied for atmospheric carbonyl determination in both urban and clean, unpolluted background areas. For a highly polluted environment, the sampling volume can be shortened or the solution diluted to meet the analytical requirement. Alternatively, the extremely low level of carbonyls in the background area can be determined by extending the sampling time to meet the quantitation requirement.

6.6 Conclusions

In this chapter, procedures for sampling, cleanup, and enrichment of atmospheric carbonyls are surveyed, and the limitations in the current analytical methodology are identified. To tackle the problem for insufficient sensitivity to detect minor carbonyls and to distinguish gaseous and particulate carbonyl for identification and apportionment of pollution sources, a novel procedure is developed based on multidimensional analyte enrichment by integrating MIP/SPE with ME-MC-CE device. A sampling cartridge coated with DNPH was used to collect separately the gaseous and particulate carbonyls. The unreacted DNPH in sample solution was removed effectively by using a highly selective MIP/SPE cartridge to achieve a high enrichment factor for the carbonyl derivatives. With manipulation of micelle stacking and acetonitrile/salt content in MEKC buffer, the separation efficiency for carbonyl-DNPH derivatives is significantly enhanced. Coupled with multidimensional sensitivity enhancement by integrating MIP/SPE with ME-MC-CE device, successful separation of 15 ambient carbonyls in gaseous or particulate phase is shown, with satisfactory detection limits and wide working range for hourly sampling. For field application using the device developed, the particle/gas ratios of a wide range of the ambient carbonyls are determined. In summary, the device provides a highly selective and sensitive tool for fast identification of atmospheric carbonyl compounds either in gaseous phase or in particulate phase, with information needed for identification and quantitation of pollution sources.

Acknowledgments

The support on the present work by the National Natural Science Foundation of China (21107018, 21477026), Science and Technology Key Project of Ministry of Education (212129), Natural Science Foundation of Guangdong Province (2014A030313525) is acknowledged.

References

1. Salthammer, T. Formaldehyde in the ambient atmosphere: From an indoor pollutant to an outdoor pollutant. *Angew Chem Int Ed* 2013, 52, 3320–3327.
2. Barro, R., Regueiro, J., Llompart, M. and Garcia-Jares, C. Analysis of industrial contaminants in indoor air: Part 1. Volatile organic compounds, carbonyl compounds, polycyclic aromatic hydrocarbons and polychlorinated biphenyls. *J Chromatogr A* 2009, 1216, 540–566.

3. Read, K.A., Carpenter, L.J. and Arnold, S.R. Multiannual observations of acetone, methanol, and acetaldehyde in remote tropical Atlantic air: Implications for atmospheric OVOC budgets and oxidative capacity. *Environ Sci Technol* 2012, 46, 11028–11039.
4. Sarigiannis, D.A., Karakitsios, S.P. and Gotti, A. Exposure to major volatile organic compounds and carbonyls in European indoor environments and associated health risk. *Environ Int* 2011, 37, 743–765.
5. Huang, Y., Ho, S.S.H. and Ho, K.F. Characteristics and health impacts of VOCs and carbonyls associated with residential cooking activities in Hong Kong. *J Hazard Mater* 2011, 186, 344–351.
6. World Health Organization. *WHO Guidelines for Indoor Air Quality—Selected Pollutants.* WHO Regional Office for Europe, Geneva, Switzerland, 2010.
7. U.S. Environmental Protection Agency. 2012. Integrated Risk Information System. http://www.epa.gov/iris
8. U.S. EPA and C. M. T.-A. *Office of Research and Development.* The Environmental Protection Agency, Cincinnati, OH, 1999.
9. ASTM. *Standard Test Method for Determination of Formaldehyde and Other Carbonyl Compounds in Air (Active Sampler Methodology).* Method D5197-97, ASTM International, West Conshohocken, PA, 1997.
10. Sun, H., Chan, K.Y. and Fung, Y.S. Determination of gaseous and particulate carbonyls in air by gradient-elution micellar electrokinetic capillary chromatography. *Electrophoresis* 2008, 29, 3971–3979.
11. Temime, B., Healy, R.M. and Wenger, J.C. A denuder-filter sampling technique for the detection of gas and particle phase carbonyl compounds. *Environ Sci Technol* 2007, 41, 6514–6520.
12. Prokai, L., Szarka, S., Wang, X. and Prokai-Tatrai, K. Capture of the volatile carbonyl metabolite of flecainide on 2,4-dinitrophenylhydrazine cartridge for quantitation by stable-isotope dilution mass spectrometry coupled with chromatography. *J Chromatogr A* 2012, 1232, 281–287.
13. Uchiyama, S., Tomizawa, T., Inaba, Y. and Kunugita, N. Simultaneous determination of volatile organic compounds and carbonyls in mainstream cigarette smoke using a sorbent cartridge followed by two-step elution. *J Chromatogr A* 2013, 1314, 31–37.
14. Schmarr, H.G., Potouridis, T., Ganss, S., Sang, W., Koepp, B., Bokuz, U. and Fischer, U. Analysis of carbonyl compounds via headspace solid-phase micro-extraction with on-fiber derivatization and gas chromatographic-ion trap tandem mass spectrometric determination of their O-(2,3,4,5,6-pentafluorobenzyl) oxime derivatives. *Anal Chim Acta* 2008, 617, 119–131.
15. Li, J., Feng, Y.L., Xie, C.J., Huang, J., Yu, J.Z., Feng, J.L., Sheng, G.Y., Fu, J.M. and Wua, M.H. Determination of gaseous carbonyl compounds by their pentafluorophenyl hydrazones with gas chromatography/mass spectrometry. *Anal Chim Acta* 2009, 635, 84–93.
16. Pang, X.B., Lewis, A.C. and Hamilton, J.F. Determination of airborne carbonyls via pentafluorophenylhydrazine derivatisation by GC-MS and its comparison with HPLC method. *Talanta* 2011, 85, 406–414.
17. Prieto-Blanco, M.C., Moliner-Martinez, Y., Lopez-Mahia, R. and Campins-Falco, P. Determination of carbonyl compounds in particulate matter PM2.5 by in-tube solid-phase microextraction coupled to capillary liquid chromatography/mass spectrometry. *Talanta* 2013, 115, 876–880.

18. Fung, Y.S. and Long, Y.H. Determination of carbonyl compounds in air by electrochromatography. *Electropheresis* 2001, 22, 2270–2277.

19. Herrington, J.S., Fan, Z.H., Lioy, P.J. and Zhang, J. Low acetaldehyde collection efficiencies for 24-hour sampling with 2,4-dinitrophenylhydrazine (DNPH)-coated solid sorbents. *Environ Sci Technol* 2007, 41, 580–585.

20. Ho, S.S.H., Ip, H.S.S., Ho, K.F., Ng, L.P.T., Chan, C.S., Dai, W.T. and Cao, J.J. Hazardous airborne carbonyls emissions in industrial workplaces in China. *J Air Waste Manag Assoc* 2013, 63, 864–877.

21. Dabek-Zlotorzynska, E. and Lai, E.P.C. Separation of carbonyl 2,4-dinitrophenylhydrazones by capillary electrochromatography with diode array detection. *J Chromatogr A* 1999, 853, 487–496.

22. Pereira, E.A., Cardoso, A.A. and Tavares, M.F.M. Determination of low-aliphatic aldehydes indoors by micellar electrokinetic chromatography using sample dissolution manipulation for signal enhancement. *Electrophoresis* 2003, 24, 700–706.

23. Karimian, N., Vagin, M., Zavar, M.H.A., Chamsaz, M., Turner, A.P.F. and Tiwari, A. An ultrasensitive molecularly-imprinted human cardiac troponin sensor. *Biosens Bioelectron* 2013, 50, 492–498.

24. Yan, H.Y., Liu, S.T., Gao, M.M. and Sun, N. Ionic liquids modified dummy molecularly imprinted microspheres as solid phase extraction materials for the determination of clenbuterol and clorprenaline in urine. *J Chromatogr A* 2013, 1294, 10–16.

25. Lai, J.P., Xie, L., Sun, H. and Chen, F. Synthesis and evaluation of molecularly imprinted polymeric microspheres for highly selective extraction of an anti-AIDS drug emtricitabine. *Anal Bioanal Chem* 2013, 405, 4269–4275.

26. Basozabal, I., Gomez-Caballero, A., Diaz-Diaz, G., Guerreiro, A., Gilby, S., Goicolea, M.A. and Barrio, R.J. Rational design and chromatographic evaluation of histamine imprinted polymers optimised for solid-phase extraction of wine samples. *J Chromatogr A* 2013, 1308, 45–51.

27. De Smet, D., Kodeck, V., Dubruel, P., Van Peteghem, C., Schacht, E. and De Saeger, S. Design of an imprinted clean-up method for mycophenolic acid in maize. *J Chromatogr A* 2011, 1218, 1122–1130.

28. van Leeuwen, S.M., Hendriksen, L. and Karst, U. Determination of aldehydes and ketones using derivatization with 2,4-dinitrophenylhydrazine and liquid chromatography-atmospheric pressure photoionization-mass spectrometry. *J Chromatgr A* 2004, 1058, 107–112.

29. Pötter, W. and Karst, U. Identification of chemical interferences in aldehyde and ketone determination using dual-wavelength detection. *Anal Chem* 1996, 68, 3354–3358.

30. Lai, J.P., Lu, X.Y., Lu, C.Y., Ju, H.F. and He, X.W. Preparation and evaluation of molecularly imprinted polymeric microspheres by aqueous suspension polymerization for use as a high-performance liquid chromatography stationary phase. *Anal Chim Acta* 2001, 442, 105–111.

31. Andersson, L., Hardenborg, E. and Sandberg-Stall, M. Development of a molecularly imprinted polymer based solid-phase extraction of local anaesthetics from human plasma. *Anal Chim Acta* 2004, 526, 147–154.

32. Kim, J.B. and Terabe, S. On-line sample preconcentration techniques in micellar electrokinetic chromatography. *J Pharm Biomed Anal* 2003, 30, 1625–1643.

33. Quirino, J.P. and Terabe, S. Sweeping of analyte zones in electrokinetic chromatography. *Anal Chem* 1999, 71, 1638–1644.

34. Sun, H., Nie, Z. and Fung, Y.S. Determination of free bilirubin and its binding capacity by HSA using a microfluidic chip-capillary electrophoresis device with a multi-segment circular-ferrofluid-driven micromixing injection. *Electrophoresis* 2010, 31, 3061–3069.
35. Delgado-Saborit, J.M., Stark, C. and Harrison, R.M. Use of a versatile high efficiency multiparallel denuder for the sampling of PAHs in ambient air: Gas and particle phase concentrations, particle size distribution and artifact formation. *Environ Sci Technol* 2014, 48, 499–507.
36. Bao, L.F., Matsumoto, M., Kubota, T., Sekiguchi, K., Wang, Q.Y. and Sakamoto, K. Gas/particle partitioning of low-molecular-weight dicarboxylic acids at a suburban site in Saitama, Japan. *Atmos Environ* 2012, 47, 546–553.
37. Ortiz, R., Enya, K., Sekiguchi, K. and Sakamoto, K. Experimental testing of an annular denuder and filter system to measure gas-particle partitioning of semivolatile bifunctional carbonyls in the atmosphere. *Atmos Environ* 2009, 43, 382–388.
38. National Research Council. *Formaldehyde and Other Aldehyde*. National Academy Press, Washington, DC, 1981.

7

Enhancing Separation Efficiency: On-Chip Multidimensional Separation and Determination of Urinary Proteins by MC-CE Device

Ruige Wu

Singapore Institute of Manufacturing Technology
Singapore

Ying Sing Fung

The University of Hong Kong
Hong Kong SAR, People's Republic of China

CONTENTS

7.1 Introduction

With the advances in protein biomarker and protein drug research related to organ function or other aspects of health, the demand has increased for assay of protein biomarkers and pharmaceutical proteins in body fluids. Brief reviews and discussions are given here on the needs and requirements for protein assay and current analytical methods for their determination in urine, followed by a discussion of the problems and issues for the analytical task to be carried out for protein determination in urine samples.

7.1.1 Needs and Requirements

Accurate determination of the excretion levels of various proteins in urine is very important for early diagnosis of diabetes. Traditionally, albumin has served as a key indicator of problems with normal functioning of the kidneys, via a urine analysis, with the albumin concentration usually <20 mg/L in urine of healthy persons [1,2]. But, assay of other urinary proteins, such as immunoglobulin G (IgG), albumin, transferrin (TRF), and β2-microglobulin (β2-MG), can provide useful guidance parameters for therapy treatment [3–7]. IgG, β2-MG, and TRF are important indicators for diagnosis of tubular and glomerular injuries [3–5,8]. Unlike albumin, the concentrations reported for β2-MG, TRF, and IgG show considerable variation. From accumulated medical information, it is strongly advised to include other biomarker proteins, such as β2-MG, TRF, and IgG, in addition to albumin in a urine protein test, as their presence in urine provides stronger confirmatory results than albumin alone for clinical diagnosis of diabetic nephropathy [5,9,10].

7.1.2 Current Analytical Methods

Various qualitative and quantitative methods have been used for determination of proteins in urine, such as high-performance liquid chromatography [11], capillary electrophoresis (CE) [11], liquid chromatography-tandem mass spectrometry [12], immunoassay [13], and two-dimensional electrophoresis [14,15]. These analytical methods can be classified into two major groups. The first group consists of instrumental methods such as chromatography and electrophoresis, which require sample cleanup prior to separation and determination of the analyte proteins. Urine is normally used as the sample for the determination of protein as it is readily available from patients. Due to the high concentration of salts and metabolic wastes present in urine, various time-consuming sample pretreatment procedures are needed, such as ultrafiltration, dialysis, and precipitation [16–18]. Moreover, protein loss may occur during the preparative steps, thereby reducing the sensitivity for detecting urinary proteins present in trace levels [19,20]. The various instrumental methods normally determine a target group of proteins for a specific application.

The second group of methods uses specific immunochemistry techniques for the determination of urinary proteins, such as radioimmunoassay (RIA), immunoturbidimetric assay, and enzyme-linked immunosorbent assay [4,17,21,22]. However, methods based on immunoassay can only determine a specific protein during each test. The total time for assay of a target set of proteins in urine may take several hours to complete; thus, these methods are not suitable for screening a set of protein biomarkers.

7.1.3 Problems and Issues

The instrumental methods are more sensitive than immunoassay and subject to less interference in urine samples with a complex sample matrix. The major problems hampering the detection of urinary proteins are the high concentration of salts and metabolic wastes in urine and the relatively low concentration of urinary proteins. Conventional sample pretreatment to solve these two problems usually takes a long time and requires the use of complicated and time-consuming procedures. However, due to the high instrumental cost and the need for supporting laboratory facilities during operation, instrumental methods are unlikely to be used for routine monitoring of protein biomarkers at clinics.

In this work, a microfluidic chip-capillary electrophoresis (MC-CE) device with integrated two-dimensional isotachophoresis (ITP)-capillary zone electrophoresis (CZE) is developed to provide an on-site lab-on-a-chip device for a rapid assay of urinary proteins. With incorporation of sample cleanup and analyte enrichment procedures on-chip, urine samples can be analyzed directly without using conventional time-consuming sample pretreatment procedures. Thus, supporting laboratory facilities during operation are not required, and the reduction to microscale lab-on-a-chip operation has been shown to reduce significantly the sample preparation time compared to conventional laboratory-based sample preparation procedures.

7.2 Integrating On-Chip Sample Cleanup and Analyte Enrichment to Online CE Procedure for Protein Assay by MC-CE Device

The development of microfluidic devices for protein separation has increased rapidly in the past 20 years. Micellar electrokinetic capillary chromatography, zone electrophoresis, gel electrophoresis, ITP, and isoelectric focusing alone, as well as a combination of two or more of these techniques, have been developed to separate complicated protein mixtures [23–27]. The use of ITP to increase sample loading has recently been combined with microfluidic devices after sample cleanup to achieve a high preconcentration factor [28–30]. This approach has considerable potential as a methodology to determine minor proteins in urine samples, and in the present work, ITP is integrated prior to high-efficiency CE separation. To handle the difficult task of determining minor proteins in urine that contain high levels of interfering substances, the developed MC-CE device incorporates the following procedures on-chip: on-chip desalting and sample cleanup followed by application of ITP to enrich analytes and then a high-efficient CE procedure to remove interfering substances present in urine prior to the determination of target minor proteins for clinical urine tests.

7.3 Configuration and Basic Operations of MC-CE Device for Assay of Urinary Proteins

The MC-CE device developed is shown in Figure 7.1 [31]. The separation capillary is embedded in a poly(methyl methacrylate) (PMMA) chip with the desired microchannel pattern. The PMMA chip is produced by thermal bonding of two PMMA plates (40 mm × 27 mm × 1.5 mm). On-chip desalting and preconcentration of proteins are performed in the microfluidic path from vial A to vial C/E. The capillary is thermally bonded between two PMMA plates with the capillary entrance aligned at the microchannel exit to act as the inlet for the separation of urinary proteins coming out from the microfluidic chip (MC) by CZE.

An eight-channel high-voltage (HV) system is used to impose HVs to the MC-CE device at different vials under electronic control. A microscope equipped with a charge-coupled device is used to record the progress of ITP at different areas of the microchannels fabricated on the MC-CE device. An ultraviolet-visible detector is used to detect analyte proteins at 280 nm separated by the capillary.

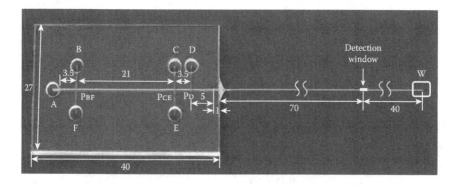

FIGURE 7.1

MC-CE device fabricated for urinary protein assay. Numbers are in millimeters. (From Wu, R.G., Yeung, W.S.B. and Fung, Y.S. *Electrophoresis* 2011, 32, 3406–3414.)

Samples are loaded to the microchannel segment (P_{BF} to P_{CF}) by HV applied across vial E to vial B. ITP is then used to stack protein zones from the sample vial to the double-T injector segment (P_{CE} and P_D) via HV applied across vial A and vial D. Once protein zones are stacked at the double-T injector segment, HV is switched across vial W to vial C to separate proteins by CZE.

7.4 Optimization of Analytical Procedures for MC-CE Device

7.4.1 On-Chip Desalting and Cleanup of Urine Sample Matrix

The presence of high concentrations of salts and metabolites in urine could interfere with the CE separation of proteins [21,32]. To verify this, 0.1 M NaCl was added to a mixture of standard proteins containing TRF, human serum albumin (HSA), β2-MG, and IgG [33]. The electropherogram is shown in Figure 7.2 [31], which indicates that all protein peaks were merged under a broad baseline. To solve this problem, HV was first applied across vial A and vial D for desalting via vial D and to achieve the ITP condition for analyte

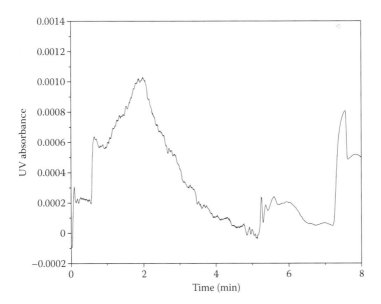

FIGURE 7.2
Electropherogram from a standard protein mixture under high-salt environment by CZE mode alone. Standard protein mixture: 0.5 mg/mL IgG, HSA, TF, and β2-MG with 0.1 M NaCl added to provide high-salt environment. BGE: 0.05% HEMC + 0.1 g/L PEO + 0.025 M Tris-HCl at pH 3.6. Injected sample channel length: 3.5 mm (P_{CE} to P_D). Peak identification: IgG (4), HSA (3), β2-MG (2), and TRF (1). HV: $W = 0$, $C = 4300$ V, and the rest are floating. (From Wu, R.G., Yeung, W.S.B. and Fung, Y.S. *Electrophoresis* 2011, 32, 3406–3414.)

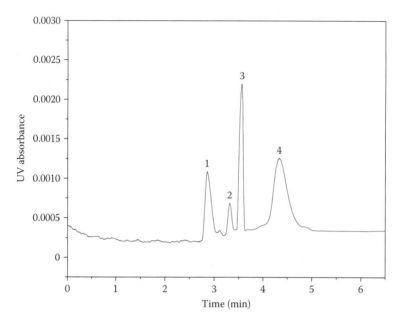

FIGURE 7.3

Electropherogram of a standard protein mixture under high-salt environment by MC-CE device. Standard protein mixture: 0.05 mg/mL IgG, HSA, TF, and β2-MG with 0.1 M NaCl added. Sample injection length: 21.0 mm. Peak identification: IgG (4), HSA (3), β2-MG (2), and TRF (1). HV (ITP) for 5 min: $D = 0$, $A = 940$ V, and the rest floating. HV (CZE) for 6.5 min: $W = 0$, $C = 4300$ V, and the rest are floating. (From Wu, R.G., Yeung, W.S.B. and Fung, Y.S. *Electrophoresis* 2011, 32, 3406–3414.)

protein stacking at the sample channel segment (between P_{CE} and P_{BF}). HV was then applied across vial W to vial C to inject the stacked protein zones (P_{CE} to P_D) to the separation capillary for their determination.

Desalting performance was investigated on this MC-CE device using a standard protein mixture under a high-salt environment. A baseline separation was obtained for the four analyte proteins on the developed MC-CE device, as shown in Figure 7.3 [31], indicating that the device can solve the interference of a high-salt environment for analysis of urinary proteins. The results indicate that the removal of excess salts from the urine sample is necessary to obtain a satisfactory peak profile for CE determination of urinary proteins.

7.4.2 On-Chip Enrichment of Analyte Proteins by ITP

A leading electrolyte (LE) and a background electrolyte (BGE) at pH 3.6 have been reported to be capable of achieving good ITP-CZE performance for acidic proteins (isoelectric point [pI] <7.0) [34]. As all target urinary proteins can be considered acidic proteins, such as IgG (pI 5.5~8.3), TRF (pI 5.5~6.6), β2-MG (pI 5.5), and HSA (pI 4.7), BGE and LE are adjusted to pH 3.6 for

TABLE 7.1

EFs for Four Urinary Proteins by MC-CE Device

Protein	CZE ($\times 10^{-5}$)[a]	ITP-CZE ($\times 10^{-5}$)[a]	EF[b]	E_{SS}[b]	E_{STACK}[b]
HSA	7.39	283	38.3	6	6.4
IgG	17.3	295	17.0	6	2.8
B2-MG	2.44	104	42.6	6	7.1
TRF	5.19	181	34.9	6	5.8

Note: Conditions were LE at pH 3.6 = 0.1 g/L PEO + 0.05% HEMC + 50 mM Tris-HCl. Terminating electrolyte (TE) at pH 3.0 = 0.1 g/L PEO + 0.05% HEMC + 50 mM HAc. BGE = LE. Standard protein mixture = β2-MG, TRF, IgG, and HSA (0.05 mg/mL each).

[a] Peak area for CZE and ITP-CZE.
[b] EF = overall enrichment factor = [peak area (MC-CE)]/[peak area (CZE alone)]. E_{SS} = enrichment by sample size effect = [injected length of MC-CE (P_B to P_{CE} = 21 mm)]/[injected length of CZE alone (P_D to P_{CE} = 3.5 mm)]. E_{STACK} = enrichment by protein stacking = EF/E_{SS}.

determination of these four urinary proteins by ITP and CZE in the present study. A BGE buffer (0.05% hydroxyethyl methylcellulose [HEMC], 0.1 g/L polyethylene oxide [PEO], and 0.025 M Tris-HCl) at pH 3.6 was found to deliver the best separation resolution [31].

The relationship between sample channel length and ITP stacking of protein zones was studied in previous experiments by using four similar MC-CE devices with four different sample injection lengths: 7.0, 14.0, 21.0, and 28.0 mm (from P_{BF} to P_{CE}). The MC-CE device with 21-mm sample injection length achieved the best ITP stacking effects [31].

Results for the stacking effect of analyte proteins by the MC-CE device (ITP-CZE) compared to CZE alone are shown in Table 7.1. The overall enrichment factors (EFs) were calculated by the following formula: EF = [ITP-CZE]/[CZE].

For calculations based on the measurement of peak area, the calculated EFs ranged from 17 to 42.6 for the four urinary proteins studied. Considering the different channel lengths used by ITP-CZE (21 mm) and CZE alone (3.5 mm), ITP enrichment is contributed by two factors: protein stacking effect and sample size. These two factors give a similar contribution weighting to the overall EF for β2-MG, TRF, and HSA, whereas the sample size effect is more important in sensitivity enrichment for IgG (Table 7.1).

7.4.3 Analytical Parameters for Assay of Urinary Proteins by MC-CE Device under Optimized Working Conditions

The analytical parameters for assay of a standard protein mixture by the MC-CE device are given in Table 7.2. Satisfactory least square lines (correlation coefficient >0.999) were achieved in the working range, with detection limits of 0.5, 0.05, 0.3, and 0.6 mg/L (signal-to-noise ratio [S/N] = 3) for IgG, β2-MG, TRF, and HSA, respectively. Linear ranges of 0.1 ~ 20 mg/L for β2-MG,

TABLE 7.2

Analytical Parameters for Assay of Urinary Proteins by MC-CE Device Using a Standard Protein Mixture

Analytical Parameter	β2-MG	TRF	HSA	IgG
Regression equation	$Y = 0.00153X +$ 0.00000183	$Y = 0.00643X -$ 0.00000139	$Y = 0.00833X -$ 0.00000480	$Y = 0.0124X +$ 0.0000291
Correlation coefficient	0.9998	0.9997	0.9997	0.9996
Linear range (mg/L)	0.1~20	1~450	1~450	1~400
Detection limit[a] (mg/L)	0.05	0.3	0.6	0.5
Repeatability[b] (RSD%)	2.7	3.1	2.8	3.6
Critical level[c] (mg/mL)	>0.0002	>0.01	>0.02	>0.015

[a] Detection limit calculated by $S/N = 3$ and $n = 3$.
[b] Relative standard deviation calculated by peak area ($n = 3$).
[c] Critical levels required intervention provided by clinical laboratory.

1–400 mg/L for IgG, and 1–450 mg/L for TRF and HSA were obtained by the device. To meet the demand to assess the critical levels of proteins in urine, the linear ranges from the developed MC-CE device are found to be sufficient to cover the concentration ranges expected from the levels of IgG, TRF, and HSA found in urine samples.

7.5 Applicability Study for Determination of Urinary Proteins by MC-CE Device Integrated with Multidimensional Separation

The electropherograms obtained from three typical urine samples for the assay of four urinary proteins by the MC-CE device are shown in Figure 7.4 [31]. The peaks were identified by matching both migration times and peak profiles with spiked standard proteins. Recovery tests indicated close to 100% recovery for all four analyte proteins (Table 7.3) [31].

A paired *t*-method was used to calculate the reliability of the developed method due to a large difference in concentrations among the four urinary proteins in clinical samples. The calculated *t* of 1.55 (versus theoretical *t* of 2.23) shows that there is no significant difference between the values determined by the MC-CE device and RIA in the clinical lab at 95% confidence limits (Table 7.4) [31].

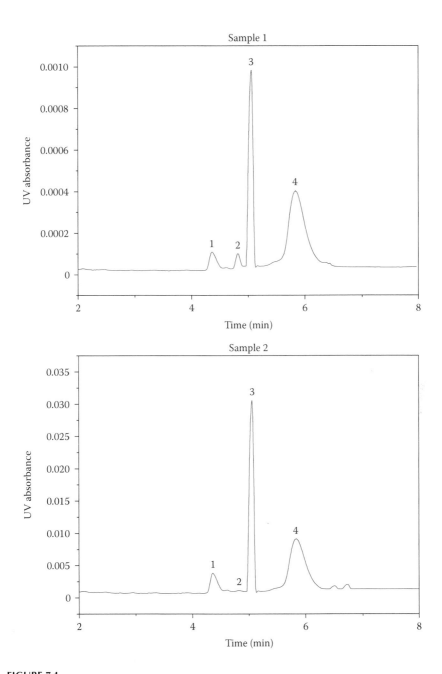

FIGURE 7.4
Electropherogram of typical clinical urine samples by MC-CE device. Injection length (P_{BF} to P_{CE}) = 21.0 mm. Peak identification: IgG (4), HSA (3), β2-MG (2), and TRF (1). HV (ITP) for 1.5 min: $D = 0$, $A = 940$ V, and the rest are floating. HV (CZE) for 6.5 min: $W = 0$, $C = 4300$ V, and the rest floating. (From Wu, R.G., Yeung, W.S.B. and Fung, Y.S. *Electrophoresis* 2011, 32, 3406–3414.)   *(Continued)*

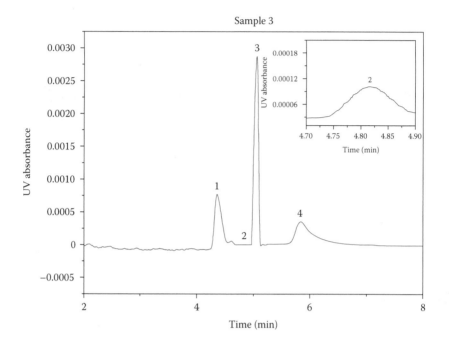

FIGURE 7.4 (Continued)
Electropherogram of typical clinical urine samples by MC-CE device. Injection length (P_{BF} to P_{CE}) = 21.0 mm. Peak identification: IgG (4), HSA (3), β2-MG (2), and TRF (1). HV (ITP) for 1.5 min: $D = 0$, $A = 940$ V, and the rest are floating. HV (CZE) for 6.5 min: $W = 0$, $C = 4300$ V, and the rest floating. (From Wu, R.G., Yeung, W.S.B. and Fung, Y.S. *Electrophoresis* 2011, 32, 3406–3414.)

TABLE 7.3

Recovery Test of HSA, TRF, β2-MG, and IgG from a Standard Protein Mixture Spiked to Clinical Urine Sample

Sample Tested[a]	HSA	TRF	β2-MG	IgG
Concn. determined by MC-CE (mg/mL)	0.011 ± 0.0005	0.0022 ± 0.0001	0.0034 ± 0.0001	0.0080 ± 0.0004
Concn. added (mg/mL)	0.01	0.01	0.01	0.01
Final concn. determined by MC-CE (mg/mL)	0.020 ± 0.0008	0.012 ± 0.0006	0.013 ± 0.0006	0.018 ± 0.0008
Recovery (%)	96	96	97	98

Source: Wu, R.G., Yeung, W.S.B. and Fung, Y.S. *Electrophoresis* 2011, 32, 3406–3414.

[a] Recovery test performed on clinical samples (from a clinical laboratory in Guangzhou, China). [Protein] expressed as mean ±1 SD ($n = 3$).

TABLE 7.4

Parallel Results from Determination of Four Urinary Proteins in Urine by RIA and MC-CE Device

Protein	Urine Sample	RIA (mg/L)[a]	MC-CE (mg/L)	Difference[b]
HSA	1	10.5	11.1	0.6
	2	310	327	17
	3	30.2	31.9	1.7
TRF	1	2.1	2.21	0.11
	2	69.3	73.6	4.3
	3	18.5	17.4	−1.1
IgG	1	7.65	7.98	0.33
	2	178	185	7
	3	8.61	8.89	0.28
β2-MG	1	3.62	3.45	−0.17
	2	6.81	6.76	−0.05
	3	0.150	0.163	0.013
			Mean	2.50
			SD	5.10
			t_{cal}	1.55

From Wu, R.G., Yeung, W.S.B. and Fung, Y.S. *Electrophoresis* 2011, 32, 3406–3414.

[a] Supplied by a clinical laboratory in hospital.

[b] $t_{10, 95} = 2.23$, $n = 3$.

7.6 Conclusions

The main problems hampering the detection of urinary proteins are the high concentrations of salts and metabolic wastes in urine and the relatively low concentrations of urinary proteins. Conventional sample pretreatment to solve these two problems usually takes a long time and requires the use of complicated and lengthy procedures. In the present work, an MC-CE device was developed for analysis of protein biomarkers in clinical urine samples. The device is integrated with on-chip sample pretreatment procedures such as sample desalting and cleanup via on-chip–controlled voltage switching as well as analyte enrichment via ITP prior to CZE separation of urinary proteins at trace levels in urine.

About 40-fold enrichment in detection sensitivity was observed for TRF, HSA, β2-MG, and IgG using the MC-CE device compared to CE alone using CZE separation mode. The detection limits were 0.3, 0.6, 0.5, and 0.05 mg/L for TRF, HSA, IgG, and β2-MG, respectively, which were sufficient for medical diagnosis. The enrichment is equally attributed to the use of a larger sample size and protein stacking for β2-MG, TRF, and HSA. Enrichment for IgG was mainly due to the use of a larger sample size.

The reliability of the MC-CE device is verified by close to 100% recoveries and excellent agreement of results by the device with those obtained by clinical laboratory using the established RIA method. Urine samples can be directly determined by the device without using time-consuming sample preparation procedures. The assay can be completed in <10 min, much faster than >7 h for the conventional RIA method. The developed MC-CE device provides a promising, easily automated method with fast sample throughput to assist timely medical intervention against disorders that influence kidney function.

Acknowledgments

We would like to acknowledge the contribution from Professor William Shu-Biu Yeung, the Department of Obstetrics and Gynaecology, The University of Hong Kong, Hong Kong SAR, People's Republic of China on the collaborated research reported in this chapter.

References

1. Pagana, K.D. and Pagana, T.J. *Mosby's Manual of Diagnostic and Laboratory Test-Urine Studies.* Mosby Elsevier, St. Louis, MO, 2010, pp. 946–1026.
2. Bessonova, E.A., Kartsova, L.A. and Shmukov, A.U. Electrophoretic determination of albumin in urine using on-line concentration techniques. *J Chromatogr A* 2007, 1150, 332–338.
3. Hiratsuka, N., Shiba, K., Nishida, K., Iizima, S., Kimura, M. and Kobayashi, S. Analysis of urinary albumin, transferrin, N-acetyl-β-D-glucosaminidase and β2-microglobulin in patients with impaired glucose tolerance. *J Clin Lab Anal* 1998, 12, 351–355.
4. Kanauchi, M., Akai, Y. and Hashimoto, T. Transferrinuria in type 2 diabetic patients with early nephropathy and tubulointerstitial injury. *Eur J Internal Med* 2002, 13, 190–193.
5. Hong, C.Y. and Chia, K.S. Markers of diabetic nephropathy. *J Diabetec Complications* 1998, 12, 43–60.
6. Branten, A.J.W., van den Born, J., Jansen, J.L.J., Assmann, K.J.M. and Wetzels, J.F.M. Familial nephropathy differing from minimal change nephropathy and focal glomerulosclerosis. *Kidney Int* 2001, 59, 693–701.
7. Michael, F.C. and Jonathan, L.T. Proteinuria in adults: A diagnostic approach. *Am Fam Physician* 2000, 62, 1333–1340.
8. Niwa, T. Biomarker discovery for kidney diseases by mass spectrometry. *J Chromatogr B* 2008, 870, 148–153.
9. Kazumi, T., Hozumi, T., Ishida, Y., Ikeda, Y., Kishi, K., Hayakawa, M. and Yoshino, G. Increased urinary transferrin excretion predicts microalbuminuria in patients with type 2 diabetes. *Diabetes Care* 1999, 22, 1176–1180.

10. D'Amico, G. and Bazzi, C. Urinary protein and enzyme excretion as markers of tubular damage. *Curr Opin Nephrol Hypertens* 2003, 12, 639–643.

11. Jellum, E., Dollekamp, H. and Blessum, C. Capillary electrophoresis for clinical problem solving: Analysis of urinary diagnostic metabolites and serum proteins. *J Chromatogr B* 1996, 683, 55–65.

12. Spahr, C.S., Davis, M.T., McGinley, M.D., Robinson, J.H., Bures, E.J., Beierle, J., Mort, J., et al. Towards defining the urinary proteome using liquid chromatography-tandem mass spectrometry. I. Profiling an unfractionated tryptic digest. *Proteomics* 2001, 1, 93–107.

13. Hansen, H.P., Hovind, P., Jensen, B.R. and Parving, H.H. Diurnal variations of glomerular filtration rate and albuminuria in diabetic nephropathy. *Kidney Int* 2002, 61, 163–168.

14. Pang, J.X., Ginanni, N., Dongre, A.R., Hefta, S.A. and Opitek, G.J. Biomarker discovery in urine by proteomics. *J Proteome Res* 2002, 1, 161–169.

15. Pieper, R., Gatlin, C.L., McGrath, A.M., Makusky, A.J., Mondal, M., Seonarain, M., Field, E., et al. Characterization of the human urinary proteome: A method for high-resolution display of urinary proteins on two-dimensional electrophoresis gels with a yield of nearly 1400 distinct protein spots. *Proteomics* 2004, 4, 1159–1174.

16. Anderson, N.G., Anderson, N.L., Tollaksen, S.L., Hahn, H., Giere, F. and Edwards, J. Analytical techniques for cell fractions. XXV. Concentration and two-dimensional electrophoretic analysis of human urinary proteins. *Anal Biochem* 1979, 95, 48–61.

17. Tantipaiboonwong, P., Sinchaikul, S., Sriyam, S., Phutrakul, S. and Chen, S.T. Different techniques for urinary protein analysis of normal and lung cancer patients. *Proteomics* 2005, 5, 1140–1149.

18. Thongboonkerd, V., McLeish, K.R., Arthur, J.M. and Klein, J.B. Proteomic analysis of normal human urinary proteins isolated by acetone precipitation or ultracentrifugation. *Kidney Int* 2002, 62, 1461–1469.

19. Lafitte, D., Dussol, B., Andersen, S., Vzzi, A., Dupy, P., Jensen, O.N., Berland, Y. and Verdier, J.-M. Optimized preparation of urine samples for two-dimensional electrophoresis and initial application to patient samples. *Clin Biochem* 2002, 35, 581–589.

20. Khan, A. and Packer, N.H. Simple urinary sample preparation for proteomic analysis. *J Proteome Res* 2006, 5, 2824–2838.

21. Kolios, G., Bairaktari, E., Tsolas, O. and Seferiadis, K. Routine differential diagnosis of proteinurias by capillary electrophoresis. *Clin Chem Lab Med* 2001, 39, 784–788.

22. Reichert, L.J.M. Koene, R.A.P. and Wetzels, J.F.M. Urinary IgG excretion as a prognostic factor in idiopathic membranous nephropathy. *Clin Nephrol* 1997, 48, 79–84.

23. Lin, C.-C., Hsu, B.-K. and Chen, S.-H. Integrated isotachophoretic stacking and gel electrophoresis on a plastic substrate and variations in detection dynamic range. *Electrophoresis* 2008, 29, 1228–1236.

24. Ramsey, J.D., Jacobson, S.C., Culbertson, C.T. and Ramsey, J.M. High-efficiency, two-dimensional separations of protein digests on microfluidic devices. *Anal Chem* 2003, 75, 3758–3764.

25. Li, Y., Buch, J.S., Rosenberger, R., Devoe, D.L. and Lee, C.S. Integration of isoelectric focusing with parallel sodium dodecyl sulfate gel electrophoresis for multidimensional protein separations in a plastic microfluidic network. *Anal Chem* 2004, 76, 742–748.

26. Griebel, A., Rund, S., Schönfeld, F., Dörner, W., Konrad, R. and Hardt, S. Integrated polymer chip for two-dimensional capillary gel electrophoresis. *Lab Chip* 2004, 4, 18–23.
27. Emrich, C.A., Medintz, I.L., Chu, W.K. and Mathies, R.A. Microfabricated two-dimensional electrophoresis device for differential protein expression profiling. *Anal Chem* 2007, 79, 7360–7366.
28. Wainright, A., Nguyen, U.T., Bjornson, T. and Boone, T.D. Preconcentration and separation of double-stranded DNA fragments by electrophoresis in plastic microfluidic devices. *Electrophoresis* 2003, 24, 3784–3792.
29. Huang, H., Xu, F., Dai, Z. and Lin, B. On-line isotachophoretic preconcentration and gel electrophoretic separation of sodium dodecyl sulfate-proteins on a microchip. *Electrophoresis* 2005, 26, 2254–2260.
30. Fang, X., Wang, W., Yang, L., Chandrasekaran, K., Kristian, T., Balgley, B.M. and Lee, C.S. Application of capillary isotachophoresis-based multidimensional separations coupled with electrospray ionization-tandem mass spectrometry for characterization of mouse brain mitochondrial proteome. *Electrophoresis* 2008, 29, 2215–2223.
31. Wu, R.G., Yeung, W.S.B. and Fung, Y.S. 2-D t-ITP/CZE determination of clinical urinary proteins using a microfluidic-chip capillary electrophoresis device. *Electrophoresis* 2011, 32, 3406–3414.
32. Friedberg, M.A. and Shihabi, Z.K. Urine protein analysis by capillary electrophoresis. *Electrophoresis* 1997, 18, 1836–1841.
33. Putnam, D.F. Composition and concentrative properties of human urine. NASA contractor report, No. NASA CR-1802, available at http://ntrs.nasa.gov/archive/nasa/casi.ntrs.nasa.gov/19710023044.pdf, 1971.
34. Wu, R.G., Wang, Z.P., Zhao, W.F., Yeung, W.S. and Fung, Y.S. Multi-dimension microchip-capillary electrophoresis device for determination of functional proteins in infant milk formula. *J Chromatogr A* 2013, 1304, 220–226.

Section III

Integration to Enhance
Analyte Detection

8

Improving Detection Selectivity: MC-CE Device Integrated with Dual Opposite Carbon-Fiber Microdisk Electrodes Detection for Determining Polyphenols in Wine

Fuying Du

Wuhan University
Wuhan, People's Republic of China

Ying Sing Fung

The University of Hong Kong
Hong Kong SAR, People's Republic of China

CONTENTS

8.1 Introduction

Electrochemical detection, with limits of detection (LODs) in the subnanomolar range, is one of the most sensitive detection methods for capillary electrophoresis (CE) and microfluidic chip (MC). Recent integration of MC and CE for the development of microfluidic chip-capillary electrophoresis (MC-CE) devices greatly enhances the power of either technique alone to provide the most promising detection mode for portable application. Normally, an amperometric detection mode by holding the detecting electrode at a constant potential is employed for detecting electrochemically active analytes after electrophoretic separation. The identification of an analyte depends on the migration time and repeatability of the operational conditions for CE separation. For determination of trace analytes in complex real samples, such as biofluid or food containing interfering substances, additional parameters are needed for peak confirmation to reduce false-positive results. The use of the two working electrodes operating at different potentials provides additional information based on the current ratio to assist in the identification of specific analytes eluted out from the separation capillary at the same time, hence reducing false-positive results due to comigration impurities. This enhancement is made possible by the integration of MC with separation capillary to enable repeatable alignment of the detection electrodes with the separation capillary to achieve identical environmental and operational conditions for each of the two detection electrodes. In this way, the current ratio obtained is mainly due to the difference in potential imposed on each of the two electrodes. As each substance has its specific voltage–current profile, it is possible to assess the peak purity for a specific analyte coming out from the capillary based on the ratio of the peak currents at two different potentials. Thus, detection selectivity of the dual-electrode detector is enhanced compared to single-electrode detection. The capability to detect interfering, comigrating impurities in real samples is illustrated by the detection of polyphenols in red wine by integrating dual-electrode detection with CE in the MC-CE devices fabricated.

8.2 Needs and Requirements for Polyphenol Determination

Polyphenols are secondary metabolites that exhibit various beneficial effects on human health, including antioxidant, antibacterial, antithrombotic, anti-allergic, and anti-inflammatory activities [1–3]. People consuming food and beverages high in polyphenols have been shown to have a lower risk of cataracts, osteoporosis, cancer, cardiovascular disease, immune dysfunction, and brain disorders [2–4]. Red wine contains a variety of polyphenols, such as gallic acid, quercetin, (–)-epicatechin, (+)-catechin, and *trans*-resveratrol (see molecular structures in Figure 8.1), which are strong antioxidants capable of protecting low-density lipoproteins against oxidation more effectively than α-tocopherol [4]. This protection may result in lowering of mortality rate attributed to coronary heart disorder for regular red wine drinkers compared to nondrinkers. Furthermore, polyphenols in red wine are related

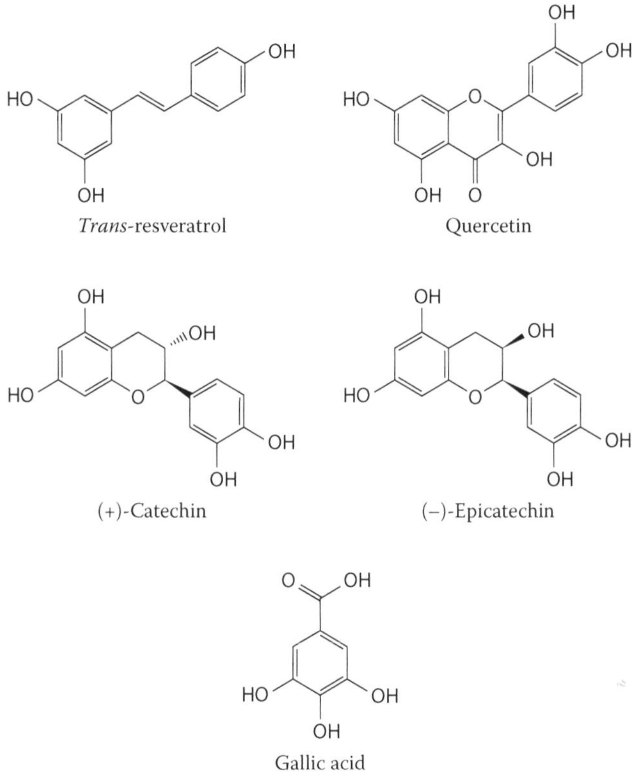

FIGURE 8.1
Molecular structure of gallic acid, (−)-epicatechin, (+)-catechin, quercetin, and *trans*-resveratrol.

with its taste and color [5,6]; therefore, the detection of polyphenols in red wine is of great importance for red wine quality assessment.

8.3 Analytical Techniques for Polyphenol Determination

The current analytical techniques for the detection of polyphenols are based on high-performance liquid chromatography (HPLC) [7–13], gas chromatography (GC) [14], thin-layer chromatography [15,16], CE [17–22], and MC devices [23–25]. Among the various methods developed, HPLC with ultraviolet-visible (UV-Vis) detection is the most popular analytical method due to the commercially available instrumentation at an affordable cost and the well-established HPLC procedures. However, the insufficient detection sensitivity for many minor polyphenols present in red wine and the limited resolution of HPLC columns for a complex sample matrix hinder its scope for application to important polyphenols, which are often present at trace levels in red wine. A sensitive method based on GC-mass spectrometry (MS) detection has been developed for the determination of *trans*-resveratrol in red wine, with a detection limit down to the nanogram per liter level [14]. However, tedious derivatization of polyphenols is required to convert them into suitable forms for GC-MS detection, giving rise to a more complex operation and the use of expensive instrumentation.

Owing to the high-separation efficiency, small sample size, short analysis time, and capability for on-site application, CE and MC devices are gaining considerable recognition. Both UV-Vis and electrochemical detections have been utilized after CE or MC separation to detect polyphenols. Although UV-Vis detection is the most widely used detection method in CE, its main disadvantage is its low detection sensitivity due to the short absorption pathway in the on-column optical detection cell. Compared to UV-Vis detection limits (10^{-7}–10^{-5} g/mL), electrochemical detection (10^{-8}–10^{-7} g/mL) gives a better sensitivity and improved selectivity as it does not respond to compounds such as proteins, lipids, and carbohydrates. Therefore, electrochemical detection has been widely used in tandem with CE separation to analyze phenolic constituents in beverage and food samples [17,26,27]. Despite the high resolution of CE separation, comigrating substances from a complex sample matrix such as red wine can interfere with polyphenol determination, giving rise to false-positive results. Therefore, an electrochemical detector with improved selectivity to detect the presence of comigrating substances after CE separation is needed to validate the eluted CE peaks.

8.4 Electrochemical Detection for CE and MC Devices

Due to high detection sensitivity, low cost, and portability, electrochemical detection is widely used for detection of drugs, metabolites, and analytes with electrochemically active functional groups in urine, serum, and biofluids. Microelectrodes fabricated from platinum or carbon as substrates are commonly used, and the position of the detection microelectrode has to be aligned accurately at the middle, just inside the separation capillary exit and away from the inner surface of the capillary wall. As the detection electrode can be fouled after depositing an insulation film at its surface during operation, replacement and realignment of the detection electrode are needed during the course of analysis. Technical skills are required for replacement of the working microelectrode, making electrochemical detection unpopular as a choice among the working analysts. In addition, the requirement of an electroactive analyte imposes a large restriction on the scope of application for electrochemical detection.

In contrast, the lack of a sensitive detection mode for MCs for on-site and personal analysis, and the low cost and portability of electrochemical instrumentation, make it a popular detection mode for MC devices in the form commonly known as microfluidic chip electrophoresis (MCE). Due to the lack of facilities for fabrication of a MC with the desired microchannel pattern in most research laboratories, most MCE devices reported in the literature are using MCs with microchannels prefabricated as a double-T configuration, which are available commercially. Most of the work is focusing on the assay of drugs and associated metabolites in urine. Although the devices reported are commonly called microchip-CE or MCE, there is no separation capillary embedded in the device and the separation of analyte depends on the very short microchannel fabricated at the surface of an MC in the form of a double-T configuration.

Zhang et al. [28] developed a microchip-CE device for assay of caffeine and associated major metabolite theophylline by electrochemical detection. The developed device can separate and detect caffeine and theophylline within 40 s. For both caffeine and theophylline, concentration limits of detection (cLODs) at 4 µM were achieved using this device. The same group also used this device for assay of morphine and codeine in urine. β-Cyclodextrin (β-CD) and acetonitrile were used as buffer additives to improve the separation efficiency [29]. Separation and determination of morphine and codeine in urine can be achieved within 140 s, with cLODs at 0.2 and 1 µM, respectively.

Ding et al. [30] reported a microchip-micellar electrokinetic chromatography device using pulsed electrochemical detection for the determination of β2-agonists, a prohibited growth promoter to increase muscle formation in animal feeding, in urine samples. The device can separate and detect four β2-agonists: ractopamine (RAC), cimaterol (CIM), terbutaline (TER), and salbutamol (SAL) within 200 s, with cLODs at 1.1, 0.91, 0.73, and 0.69 mM, respectively.

Electrochemiluminescence (ECL) detection had been used to determine the concentration of lincomycin, an antibiotic drug [31]. The working electrode used is indium tin oxide (ITO)–coated glass that is bonded to the polydimethyldisiloxane layer fabricated with microchannels. The microchip-CE device can complete the analysis for lincomycin within 40 s, with detection limits achieved at 3.1 and 9 µM for standard solution and urine sample, respectively.

Due to the very short separation pathway on-chip (1–2 cm), the separation efficiency of microchip-CE devices is limited. Thus, its current scope of application is restricted to the determination of analytes in samples with simple matrixes. The integration of MC with CE to produce MC-CE devices enables the use of a CE column with high-separation efficiency for analytes, thus greatly expanding the scope of electrochemical detection to complicated sample matrixes, such as red wine. Details on the development of dual opposite carbon-fiber microdisk electrode (DOCME) detection cell and its application for polyphenols determination are discussed in the following sections.

8.5 Recent Advancement in MC-CE Devices

MC-CE devices have been extensively investigated in recent years to extend the scope of application by integrating on-chip sample pretreatment procedures with high-efficiency CE separation. Sample pretreatments such as multi-injection [32], mixing [33], and conditioning and dilution [34], as well as preconcentration [35,36], have been demonstrated on MC-CE devices for the assessment of binding interaction between bilirubin and human serum albumin (HSA) by frontal analysis [32].

To enable on-site operation, a MC with a magnetic-activated micromixing injector has been shown to obviate the time-consuming steps for sample preparation. The device can detect the rising level of bilirubin before reaching the jaundice condition. Wu et al. [35] developed an MC-CE device to desalt and clean up sample on-chip and preconcentrate analyte by a two-dimensional isotachophoresis (ITP) procedure operating prior to a capillary zone electrophoresis (CZE) separation to determine urinary clinical protein samples. The method provided a rapid and sensitive detection without using the time-consuming conventional pretreatment steps by the use of a large sample size. Four important proteins (transferrin, immunoglobulin, β2-microglobulin, and HSA) were determined in urine, showing 40-fold enhancement in detection sensitivity.

An MC-CE device reported by Wu et al. [36] was used for determining functional proteins in infant formula based on a multidimensional separation strategy with protein isolation on-chip via isoelectric focusing (IEF) and analyte enrichment in capillary by IEF before CE separation. The transient ITP/CZE run was completed in 18 min. Enrichment up to 60-fold was realized for isolated protein fractions.

The next phase in development will focus on detection, after success has been demonstrated in various MC-CE devices with incorporation of different on-chip sample preparation procedures prior to CE separation for intended application. The design and assembly of the DOCME detection cell are discussed in the following sections, with results for determining polyphenols in red wine.

8.6 Design and Assembly of DOCME Detection Cell

A dual-electrode detector was integrated with a microfluidic device for detecting the comigrating interferents expected from the sample matrix. Although the concept for using dual-electrode detection to handle complex sample matrixes is well established, few reports are found in the literature. The major problem for using dual-electrode detection is attributed to the technical difficulties for precise alignment of the two working electrodes at the desired positions near the exit of the separation capillary [37]. In a CE-electrochemical detection system, end-column detection is the most frequently used configuration, as it is easy to construct. However, peak broadening is a serious issue to be addressed, because the electrode is located outside the separation path. To minimize peak broadening, the working electrode needs to be positioned close to the capillary outlet using a micromanipulator under a microscope. The precise positioning of the working electrode at the designated place requires skilled personnel and takes a lot of time. Fixing the working electrode permanently at the outlet of the capillary provides a simple solution. However, the whole detection system has to be discarded to replace fouled electrodes.

To enable a quick alignment and flexible exchange of microelectrodes and separation capillary, an exchangeable modular concept is designed in the DOCME system with precise positions fixed by screws under a constant pressure between two poly(methyl methacrylate) (PMMA) plates, which are fabricated with a microchannel pattern to guide the positioning of capillary to the microchannels. The construction of the electrochemical cell is illustrated in Figure 8.2. First, microchannels (500 µm in width and 360 µm in depth) are ablated by the CO_2 laser at the bottom plate. Second, two circular holes are drilled at the middle of both the top and bottom plates to allow electrolyte from the electrochemical cell access to the DOCME (Figure 8.2a). The positioning of the two carbon-fiber microdisk electrodes facing each other across the exit end of the separation capillary column is guided precisely by the microchannel pattern fabricated at the upper part of the bottom PMMA plate. This is possible due to the similar outer diameter (~360 µm) of both separation and detection capillaries (Figure 8.2b) and the high-precision, computer-aided microchannel fabrication.

The two carbon-fiber microelectrodes are then assembled at the top and bottom plates by fine horizontal adjustment under the microscope to make the two flat ends cut from the carbon fiber face each other at the capillary outlet (25 µm inside diameter) (Figure 8.2c). This design of the two working electrodes results in an ultrasmall detection volume (~0.02 nL by estimation, 25 µm × π × 15 µm × 15 µm). Furthermore, the arrangement of the two working electrodes in the opposite configuration across the capillary outlet makes them, under the same conditions, independent of the eluent flow pattern. This arrangement aims to solve the problem reported in the literature [38] for the end-on dual-detection mode that has exhibited different flow patterns under variable eluent conditions. To enable the measurement of the current ratio not affected by a change in flow conditions, the two working electrodes are positioned next to each other with both facing the outcoming eluent under the same conditions. Moreover, the chance of cross-contamination from dual electrodes placed sequentially next to each other is removed in the opposite arrangement of the two working electrodes.

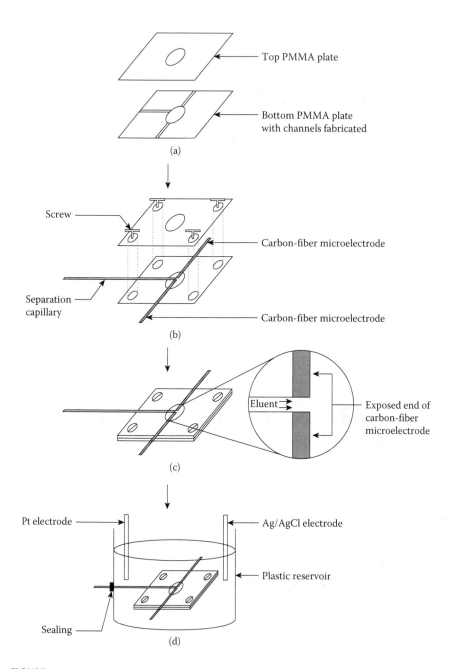

FIGURE 8.2
Preparation of electrochemical detection cell using DOCME. (a) Drill concentric holes at both top and bottom plates and channels fabricated at bottom only. (b) Screw at four corners to fix capillary and DOCME at required locations. (c) Align capillary with DOCME. (d) Immerse detection cell into buffer reservoir. (From Du, F.Y. and Fung, Y.S. *Electrophoresis* 2010, 31, 2192–2199.)

Last, the four corners of the top and bottom plates were fixed by screws, and the assembly was immersed in the buffer solution before the CE run (Figure 8.2d) [39]. As both the two working electrodes and the separation capillary are fixed in position by screws, plugged capillary or fouled working electrodes can be easily replaced. In this two-electrode detection system, a Ag/AgCl electrode acts as both reference and counter electrode, and two DOCME electrodes act as working electrodes. The working potentials of the two DOCME electrodes are controlled independently by two battery-operated potentiostats. A common ground is made by connecting the grounded Pt electrode and the Ag/AgCl electrode for short-circuit protection.

Trans-resveratrol at a concentration of 4 mg/L was injected into the separation capillary when both the working electrodes were held at +1.0 V to test the alignment of electrodes. The correct alignment of two electrodes is achieved when a current ratio close to 1.0 is obtained from the two working electrodes.

8.7 Integrating DOCME with MC-CE Device for Determining Polyphenols in Red Wine

8.7.1 Challenges for Determining Polyphenols in Red Wine

Red wine contains a complex sample matrix with a large number of polyphenols with similar molecular structures, giving rise to possible peak overlap in the electropherogram. Despite the high resolution of CE and the careful control of detection potential, overlap of migrating polyphenol peaks can occur, leading to undesirable false-positive errors. Therefore, assessing the purity of migration peaks attributed to a specific polyphenol is required for quality control in the production of red wine.

Matching migration time with that from the given standards is used for peak identification. However, erroneous peak identification can occur due to comigrating peaks. Therefore, the current ratio arising from two microelectrodes controlled at different potentials and placed at a position opposite facing each other at the exit of the separation capillary is used as a parameter for the identification of the migrated peak. As each substance has its specific voltage–current profile and the two microelectrodes are placed under identical conditions, the current ratio obtained by the working electrodes controlled at two different potentials reflects the specific voltage–current feature of target analytes. Hence, the current ratios can be used to assess peak purity of analytes exiting the separation capillary. Comigrating interferents are suspected of an unknown peak with significant mismatch of the current ratio. Matching both migration time and current ratio enhances the confidence for the identified peak and significantly increases the reliability for result quantitation. If a comigrating impurity is absent in the analyte peak, the analyte can be quantified with accurate results. Conversely, if a mismatch of current ratio is observed, it suggests the existence of interference, and thus inaccurate quantitative results for a specified analyte are avoided. Moreover, the observation of a comigration peak indicates that the CE separation conditions should be further improved. The current ratio can be used as a guide for the improvement of CE conditions until a pure peak is obtained.

8.7.2 Optimization of MC-CE/DOCME Devices for Determining Polyphenols in Red Wine

To establish the working conditions of DOCME for peak purity assessment, the voltage–current profile of five polyphenols are investigated. Peak current for an individual polyphenol was recorded by a carbon-fiber microelectrode held at a constant potential that varied from +0.1 to +1.1 V versus Ag/AgCl. With the increase of the detection potential beyond 0.5 V versus Ag/AgCl, peak currents are found to increase for all five polyphenols before reaching the limiting current plateau at +1.0 V, except for *trans*-resveratrol, which keeps a slight increase beyond +1.0 V. When a potential above +1.0 V is applied at the carbon-fiber microelectrode, the background current noise increases significantly. Therefore, to obtain a low detection limit for all five polyphenols, the detection potential for amperometric detection is selected at +1.0 V.

In addition to the detection potential, another potential is needed for the opposite electrode to obtain the current ratio for peak identification. As the peak currents of all five polyphenols are found to increase from +0.7 to +0.9 V, the current ratios of +1.0 V/+0.9 V, +1.0 V/+0.8 V, and +1.0 V/+0.7 V are determined, with results shown in Figure 8.3 [39]. Based on a combination of +1.0 V/+0.8 V and +1.0 V/+0.7 V, a large difference in the current ratios is observed among the five polyphenols. Despite the largest difference in the current ratio observed at +1.0 V/+0.7 V, the magnitude of the current at the working electrode held at +0.7 V is smaller than that at +0.8 V. Thus, +0.8 V is selected as the working potential to be held at the opposite electrode to achieve a higher detection sensitivity.

To illustrate the applicability of DOCME for assessing peak purity, a test solution was prepared by mixing five polyphenol standards in 20 mM sodium tetraborate buffer (pH 9.0). β-CD was not added into the running buffer so that (+)-catechin and (−)-epicatechin cannot be well separated and migrated out of the separation capillary as one peak. The electropherograms recorded by the two working electrodes controlled at +0.8 and +1.0 V are shown in Figure 8.4 [39]. The calculated E_1/E_2 current ratios for peaks 1, 2, 3, and 4 (2.5, 4.0, 1.9, and 3.2, respectively) are also shown in Figure 8.4. The current ratios for peaks 1, 3, and 4 are very close to those from corresponding standards. Therefore, peaks 4, 3, and 1 can be

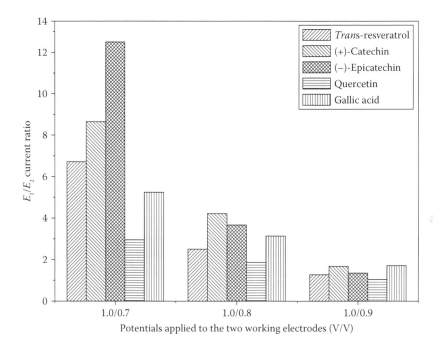

FIGURE 8.3
Effect for holding different potentials at DOCME on E_1/E_2 current ratio for the detection of gallic acid (20 mg/L), quercetin (10 mg/L), (–)-epicatechin (5.0 mg/L), (+)-catechin (10 mg/L), and *trans*-resveratrol (5.0 mg/L) at E_1 = +1.0 V and E_2 = +0.7, +0.8, and +0.9 V versus Ag/AgCl electrode. (From Du, F.Y. and Fung, Y.S. *Electrophoresis* 2010, 31, 2192–2199.)

identified as gallic acid, quercetin, and *trans*-resveratrol, respectively. However, a significant difference in current ratio is observed for peak 2 to those from (+)-catechin and (–)-epicatechin. Thus, the unresolved peak is suspected to be from comigrating interferents. The fact that both (+)-catechin and (–)-epicatechin have migrated as a single peak in the running buffer without β-CD provides supporting evidence. In this way, the applicability of DOCME for assessing peak purity is clearly shown by the use of a test solution containing mixed standards.

8.7.3 Real Sample Application of MC-CE/DOCME Devices for Determining Polyphenols in Red Wine

Commercial red wine was used as the real sample to test the application of DOCME for peak purity assessment. Red wine was purchased from a local supermarket. It was neutralized to pH 7.0 by adding NaOH and diluted with a running buffer before injection into the MC-CE device. The electropherograms obtained by the dual-electrode detector are shown in Figure 8.5 [39]; note the numerous peaks with close migration times.

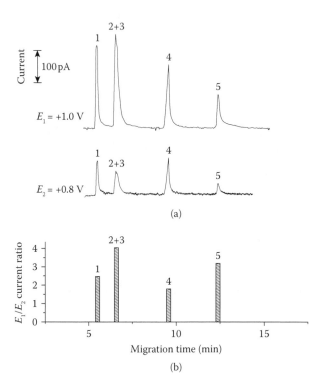

FIGURE 8.4
Current ratio for identification of comigrating peaks from a test solution containing
(–)-epicatechin and (+)-catechin with no β-CD added for their separation. (a) Electropherograms
at E_1 (+1.0 V) and E_2 (+0.8V) versus Ag/AgCl using mixed standard solution. (b) Current ratio
at E_1/E_2. Run voltage (12 kV); electrokinetic sample injection (12 kV for 8 s). Buffer (20 mM
sodium tetraborate at pH 9.0). Polyphenols: 5, gallic acid (12 mg/L); 4, quercetin (12 mg/L);
2+3 = (+)-catechin (4.0 mg/L) and (–)-epicatechin (4.0 mg/L); and 1, *trans*-resveratrol (4.0 mg/L).
(From Du, F.Y. and Fung, Y.S. *Electrophoresis* 2010, 31, 2192–2199.)

Thus, analytes of interest are likely to have a peak overlap with comigrated
impurities. Based only on matching migration time, peaks 5, II, 3, 2, and I
(Figure 8.5) may be identified as gallic acid, quercetin, (–)-epicatechin,
(+)-catechin, and *trans*-resveratrol, respectively.

As the E_1/E_2 current ratio for a compound provides a characteristic value
regardless of its concentration, it is used to identify eluting peaks. The
results obtained from the measurement of the peak current and the calculation of the current ratios for the two working electrodes controlled at
+0.8 V/+1.0 V versus Ag/AgCl for red wine samples are given in Table 8.1.
The current ratios of peaks 5, 3, and 2 are found closely matching those from
gallic acid, (–)-epicatechin, and (+)-catechin, respectively. Thus, peaks 5, 3,
and 2 are confirmed as gallic acid, (–)-epicatechin, and (+)-catechin, respectively. However, for peaks II and I for which the migration times match
well with quercetin and *trans*-resveratrol, respectively, their current ratios

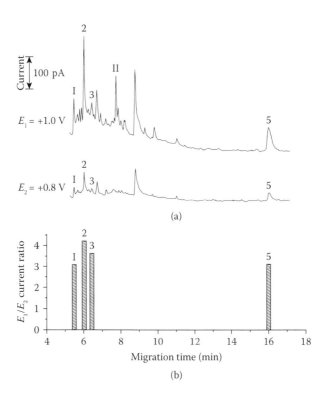

FIGURE 8.5
Application of MC-CE/DOCME device to determine polyphenols in red wine. (a) DOCME: E_1 = +1.0 V and E_2 = +0.8 V. (b) E_1/E_2 current ratio for designated peaks. Peaks: 5, gallic acid; II, unknown compound; 3, (–)-epicatechin; 2, (+)-catechin; and I, coeluted compounds. (From Du, F.Y. and Fung, Y.S. *Electrophoresis* 2010, 31, 2192–2199.)

TABLE 8.1

Identification of Polyphenols in Red Wine

Peak (See Figure 8.5)	Assignment	Current Ratio at E_1/E_2[a]	Migration Time (min)	Concn. Determined (mg/L)
I	Coeluents[b]	3.1 (1.5)	5.5 (1.6)	ND[c]
2	(+)-Catechin	4.2 (2.9)	6.0 (1.8)	26 (3.8)
3	(–)-Epicatechin	3.6 (2.1)	6.5 (1.9)	0.74 (2.2)
II	Unknown	ND	7.7 (2.2)	ND
5	Gallic acid	3.1 (3.2)	16 (2.9)	24 (3.0)

From Du, F.Y. and Fung, Y.S. *Electrophoresis* 2010, 31, 2192–2199.
[a] E_1 = +1.0 V and E_2 = +0.8 V versus Ag/AgCl. Numbers shown in parentheses are % relative standard deviation, n = 5.
[b] The coeluting peak contains *trans*-resveratrol and other electroactive impurities with matching migration times.
[c] Not detected.

TABLE 8.2

Recovery Test for Polyphenols Spiked to Wine Sample

Polyphenol	Original Concn.[a] (mg/L)	Added Concn. (mg/L)	Concn. Determined (mg/L)	Recovery %
(+)-Catechin	25.9	10.0	35.3	94
(−)-Epicatechin	0.738	1.00	1.79	105
Gallic acid	24.2	10.0	35.0	108

From Du, F.Y. and Fung, Y.S. *Electrophoresis* 2010, 31, 2192–2199.

[a] Values determined by the DOCME method.

are quite different from those of *trans*-resveratrol and quercetin, indicating the existence of coeluting impurities. Based on the change in current ratios observed at +0.8 and +1.0 V, peak I must contain an electrochemically active compound. Besides, this impurity has a greater current ratio at +1.0 V/+0.8 V than *trans*-resveratrol because of the increase in the current ratio of peak I. Thus, peak I is assigned as an unresolved peak with comigrating impurities. For peak II, a significant disappearance of peak was observed at +0.8 V. As quercetin is electrochemically active at +0.8 V, peak II is definitely not quercetin and should be assigned as an unknown compound that is only electrochemically active at +1.0 V. Therefore, the capability of DOCME to identify comigrating impurities and assess peak purity in a sample with a complex matrix is demonstrated.

The results for DOCME determination of gallic acid, (−)-epicatechin, and (+)-catechin in red wine are summarized in Table 8.1 [39]. All results agreed with normal concentrations expected in red wine [40]. To validate the results, different amounts of polyphenol standards were spiked into a red wine sample for a recovery test. Recoveries close to 100% are obtained, confirming the validity of the DOCME method for determination of polyphenols in red wine (Table 8.2) [39].

8.8 Summary

A newly designed MC-CE device fabricated with a DOCME detector was developed to determine polyphenols in red wine. The peak purity in a red wine sample was assessed by DOCME to further confirm the results obtained. The design of the dual-electrode detector makes it easy to align the electrode and capillary and to replace the plugged capillary and fouled working electrodes. The use of the current ratio to assess peak purity was investigated and measurement conditions were optimized. The dual currents obtained from the two working electrodes controlled at +1.0 and +0.8 V are found to provide the best detection sensitivity and selectivity. A test solution containing mixed polyphenol standards with no β-CD added to create comigrating peaks of (+)-catechin and (–)-epicatechin and a red wine sample from a local supermarket were used to demonstrate the capability of DOCME to pick up impurity peaks. Results show that the current ratio at +1.0 V/+0.8 V is capable of identifying pure peaks and overlapping peaks in both mixed standards and real red wine samples. The results obtained for determining polyphenol in red wine were found to be in good agreement with the expected values, and recoveries close to 100% for spiked polyphenols confirm the reliability of the MC-CE device developed. The use of both matching migration time and current ratio can reduce the possibility of giving erroneous results and enhance the accuracy of polyphenol quantitation in red wine containing a complex sample matrix.

References

1. Arts, I.C.W. and Hollman, P.C.H. Polyphenols and disease risk in epidemiologic studies. *Am J Clin Nutr* 2005, 81, 317S–325S.
2. Udenigwe, C.C., Ramprasath, V.R., Aluko, R.E. and Jones, P.J.H. Potential of resveratrol in anticancer and anti-inflammatory therapy. *Nutr Rev* 2008, 66, 445–454.
3. Fernandez-Panchon, M.S., Villano, D., Troncoso, A.M. and Garcia-Parrilla, M.C. Antioxidant activity of phenolic compounds: From in vitro results to in vivo evidence. *Crit Rev Food Sci Nutr* 2008, 48, 649–671.
4. Frankel, E.N., Kanner, J., German, J.B., Parks, E. and Kinsella, J.E. Inhibition of oxidation of human low-density lipoprotein by phenolic substances in red wine. *Lancet* 1993, 341, 454–457.
5. Preys, S., Mazerolles, G., Courcoux, P., Samson, A., Fischer, U., Hanafi, M., Bertrand, D. and Cheynier, V. Relationship between polyphenolic composition and some sensory properties in red wines using multiway analyses. *Anal Chim Acta* 2006, 563, 126–136.
6. Vinas, P., Lopez-Erroz, C., Marin-Hernandez, J.J. and Hernandez-Cordoba, M. Determination of phenols in wines by liquid chromatography with photodiode array and fluorescence detection. *J Chromatogr A* 2000, 871, 85–93.
7. Goncalves, J., Mendes, B., Silva, C.L. and Camara, J.S. Development of a novel microextraction by packed sorbent-based approach followed by ultrahigh pressure liquid chromatography as a powerful technique for quantification phenolic constituents of biological interest in wines. *J Chromatogr A* 2012, 1229, 13–23.
8. Molina-Garcia, L., Ruiz-Medina, A. and Fernandez-de Cordova, M.L. An automatic optosensing device for the simultaneous determination of resveratrol and piceid in wines. *Anal Chim Acta* 2011, 689, 226–233.
9. Careri, M., Corradini, C., Elviri, L., Nicoletti, I. and Zagnoni, I. Direct HPLC analysis of quercetin and trans-resveratrol in red wine, grape, and winemaking byproducts. *J Agric Food Chem* 2003, 51, 5226–5231.
10. Cavaliere, C., Foglia, P., Gubbiotti, R., Sacchetti, P., Samperi, R. and Lagana, A. Rapid-resolution liquid chromatography/mass spectrometry for determination and quantitation of polyphenols in grape berries. *Rapid Commun Mass Spectrom* 2008, 22, 3089–3099.

11. Novak, I., Janeiro, P., Seruga, M. and Oliveira-Brett, A.M. Ultrasound extracted flavonoids from four varieties of Portuguese red grape skins determined by reverse-phase high-performance liquid chromatography with electrochemical detection. *Anal Chim Acta* 2008, 630, 107–115.

12. Casella, I.G., Colonna, C. and Contursi, M. Electroanalytical determination of some phenolic acids by high-performance liquid chromatography at gold electrodes. *Electroanalysis* 2007, 19, 1503–1508.

13. Chen, J.J., He, S., Mao, H., Sun, C.R. and Pan, Y.J. Characterization of polyphenol compounds from the roots and stems of *Parthenocissus laetevirens* by high-performance liquid chromatography/tandem mass spectrometry. *Rapid Commun Mass Spectrom* 2009, 23, 737–744.

14. Cai, L.S., Koziel, J.A., Dharmadhikari, M. and van Leeuwen, J. Rapid determination of trans-resveratrol in red wine by solid-phase microextraction with on-fiber derivatization and multidimensional gas chromatography–mass spectrometry. *J Chromatogr A* 2009, 1216, 281–287.

15. Glavnik, V., Simonovska, B. and Vovk, I. Densitometric determination of (+)-catechin and (−)-epicatechin by 4-dimethylaminocinnamaldehyde reagent. *J Chromatogr A* 2009, 1216, 4485–4491.

16. Lambri, M., Jourdes, M., Glories, Y. and Saucier, C. High performance thin layer chromatography (HPTLC) analysis of red wine pigments. *JPC J Planar Chromatogr Mod TLC* 2003, 16, 88–94.

17. Peng, Y.Y., Chu, Q.C., Liu, F.H. and Ye, J.N. Determination of phenolic constituents of biological interest in red wine by capillary electrophoresis with electrochemical detection. *J Agric Food Chem* 2004, 52, 153–156.

18. Blasco, A.J., Barrigas, I., Gonzalez, M.C. and Escarpa, A. Fast and simultaneous detection of prominent natural antioxidants using analytical microsystems for capillary electrophoresis with a glassy carbon electrode: A new gateway to food environments. *Electrophoresis* 2005, 26, 4664–4673.

19. Cifuentes, A. Recent advances in the application of capillary electromigration methods for food analysis. *Electrophoresis* 2006, 27, 283–303.

20. Garcia-Canas, V. and Cifuentes, A. Recent advances in the application of capillary electromigration methods for food analysis. *Electrophoresis* 2008, 29, 294–309.

21. Jac, P., Polasek, M. and Pospisilova, M. Recent trends in the determination of polyphenols by electromigration methods. *J Pharm Biomed Anal* 2006, 40, 805–814.

22. Fung, Y.S. and Long, Y.H. Determination of phenols in soil by supercritical fluid extraction–capillary electrochromatography. *J Chromatogr A* 2001, 907, 301–311.

23. Crevillen, A.G., Avila, M., Pumera, M., Gonzalez, M.C. and Escarpa, A. Food analysis on microfluidic devices using ultrasensitive carbon nanotubes detectors. *Anal Chem* 2007, 79, 7408–7415.

24. Crevillen, A.G., Pumera, M., Gonzalez, M.C. and Escarpa, A. Towards lab-on-a-chip approaches in real analytical domains based on microfluidic chips/electrochemical multi-walled carbon nanotube platforms. *Lab Chip* 2009, 9, 346–353.

25. Wang, J., Pumera, M., Chatrathi, M.P., Escarpa, A., Konrad, R., Griebel, A., Dorner, W. and Lowe, H. Towards disposable lab-on-a-chip: Poly (methylmethacrylate) microchip electrophoresis device with electrochemical detection. *Electrophoresis* 2002, 23, 596–601.

26. Pumera, M. and Escarpa, A. Nanomaterials as electrochemical detectors in microfluidics and CE: Fundamentals, designs, and applications. *Electrophoresis* 2009, 30, 3315–3323.
27. Escarpa, A., Gonzalez, M.C., Crevillen, A.G. and Blasco, A.J. CE microchips: An opened gate to food analysis. *Electrophoresis* 2007, 28, 1002–1011.
28. Zhang, Q.L., Lian, H.Z., Wang, W.H. and Chen H.Y. Separation of caffeine and theophylline in poly(dimethylsiloxane) microchannel electrophoresis with electrochemical detection. *J Chromatogr A* 2005, 1098, 172–176.
29. Zhang, Q.L., Xu, J.J., Li, X.Y., Lian, H.Z. and Chen, H.Y. Determination of morphine and codeine in urine using poly (dimethylsiloxane) microchip electrophoresis with electrochemical detection. *J Pharmaceut Biomed Anal* 2007, 43, 237–242.
30. Ding, Y., Qi, Y. and Suo, X. Rapid determination of β2-agonists in urine samples by microchip micellar electrokinetic chromatography with pulsed electrochemical detection. *Anal Methods* 2013, 5, 2623–2629.
31. Zhao, X., You, T., Qiu, H., Yan, J., Yang, X. and Wang. E.K. Electrochemiluminescence detection with integrated indium tin oxide electrode on electrophoretic microchip for direct bioanalysis of lincomycin in the urine. *J Chromatogr B* 2004, 810, 137–142.
32. Nie, Z. and Fung, Y.S. Microchip capillary electrophoresis for frontal analysis of free bilirubin and study of its interaction with human serum albumin. *Electrophoresis* 2008, 29, 1924–1931.
33. Sun, H., Nie, Z. and Fung, Y.S. Determination of free bilirubin and its binding capacity by HSA using a microfluidic chip-capillary electrophoresis device with a multi-segment circular-ferrofluid-driven micromixing injection. *Electrophoresis* 2010, 31, 3061–3069.
34. Guo, W.P., Lau, K.M. and Fung, Y.S. Microfluidic chip-capillary electrophoresis for two orders extension of adjustable upper working range for profiling of inorganic and organic anions in urine. *Electrophoresis* 2010, 31, 3044–3052.
35. Wu, R.G., Yeung, W.S.B. and Fung, Y.S. 2D t-ITP/CZE determination of clinical urinary proteins using microfluidic-chip capillary electrophoresis device. *Electrophoresis* 2011, 32, 3406–3414.
36. Wu, R.G., Wang, Z.P., Zhao, W.F., Yeung, W.S. and Fung, Y.S. 2D t-ITP/CZE determination of clinical urinary proteins using a microfluidic-chip capillary electrophoresis device. *J Chromatogr A* 2013, 1304, 220–226.
37. Fermier, A.M., Gostkowski, M.L. and Colon, L.A. Rudimentary capillary-electrode alignment for capillary electrophoresis with electrochemical detection. *Anal Chem* 1996, 68, 1661–1664.
38. Zhong, M., Zhou, J.X., Lunte, S.M., Zhao, G., Giolando, D.M. and Kirchhoff, J.R. Dual-electrode detection for capillary electrophoresis/electrochemistry. *Anal Chem* 1996, 68, 203–207.
39. Du, F.Y. and Fung, Y.S. Development of CE–dual opposite carbon-fiber microdisk electrode detection for peak purity assessment of polyphenols in red wine. *Electrophoresis* 2010, 31, 2192–2199.
40. Hamoudova, R., Urbanek, M., Pospisilova, M. and Polasek, M. Assay of phenolic compounds in red wine by on-line combination of capillary isotachophoresis with capillary zone electrophoresis. *J Chromatogr A* 2004, 1032, 281–287.

9

Enhancing Detection Versatility: MC-CE Device with Serial Dual-Electrode Detection for Determining Glutathione Disulfide and Glutathione in Pharmaceutical Supplement

Fuying Du

Wuhan University
Wuhan, People's Republic of China

Ying Sing Fung

The University of Hong Kong
Hong Kong SAR, People's Republic of China

CONTENTS

9.1 Introduction

Among the various analytical detection methods, amperometric detection (AD) with a detection limit at the subnanomolar (~0.1 nM) level is one of the most sensitive detection methods suitable for coupling with capillary electrophoresis (CE) separation. Unlike optical detection methods, the miniaturization of the detection electrodes does not compromise the analytical performance of AD, an attractive feature for a CE detection system. However, AD is limited to the determination of electrochemically active analytes, which reduces its scope of application. Dual-electrode detection based on the reaction between analytes and electrogenerated bromine provides an attractive approach for indirect determination of electrochemically inactive analytes. In this chapter, fundamental issues on the design for arrangement of the two serial electrodes to the CE high-voltage path, working conditions for dual-electrode detection, and its integration with a microfluidic chip-capillary electrophoresis (MC-CE) device for indirect determination of glutathione (GSH) and glutathione disulfide (GSSG) are presented, with results discussed.

9.2 Needs and Requirements for GSSG and GSH Determination

GSH is a tripeptide thiol composed of glycine, glutamic acid, and cysteine. It can be converted to GSSG under specific conditions by cells and tissues under oxidative stress. In eukaryotic cells, the GSH–GSSG system is the most abundant redox system [1]. For signaling pathways associated with apoptosis for programmed cell death, GSH plays an important role [2–4]. It also plays a fundamental role in cell homeostasis [5,6]. The concentrations of both GSH and GSSG, as well as their relative molar ratios, are considered indicators for cell functionality and oxidative stress [7]. Thus, the determination of GSH and GSSG can provide useful information under various clinical states to assess redox homeostasis and to understand the role of GSH and GSSG in different physiological processes.

9.3 Detection of GSSG and GSH

Due to the very low levels of GSH and GSSG present in complex biofluids, methods such as CE and high-performance liquid chromatography are commonly used for their separation and quantitation [8–15].

CE is an attractive method for separation of GSH and GSSG because of the short analysis time, low sample consumption, high separation efficiency, and simple sample pretreatment. However, insufficient detection sensitivity remains the major obstacle limiting the scope of application of CE for GSH and GSSG determination in biofluids because both GSH and GSSG do not contain strong chromophores or fluorophores to enable direct detection by ultraviolet (UV) or fluorescence detectors. To make them amendable to UV or fluorescence detection, derivatization of GSH and GSSG by chromophores or fluorophores is necessary. The derivatization procedure is usually complicated, time-consuming, and labor-intensive [16–23]. Thus, the use of electrochemical detection has been explored, with details given in the next section.

9.3.1 Direct Electrochemical Detection

Electrochemical detection offers a highly sensitive, selective, and low-cost detection method for CE and other electrophoretic techniques [24–29]. Normally, one working electrode controlled at a desired potential is used for detection. Although the setup of the detection cell is simple, its performance is highly dependent on the electrochemical reversibility of the reaction and the electrokinetics of the analytes at the surface of the working electrode. For the determination of GSSG and GSH by direct electrochemical detection, the limiting factors are the slow electron-transfer kinetics at the solid electrode surface and the high overpotentials to initiate the oxidation reaction.

The use of chemically modified electrodes provides an efficient way to address these problems. Salimi and Hallaj [30] reported the fabrication method for a glassy carbon electrode (GCE) modified with multiwall carbon nanotubes and the use of the modified electrode for the determination of GSH, cysteine, and thiocytosine. Han and Tachikawa [31] had prepared a GCE coated with a single-wall carbon nanotube (SWNT) film for electrocatalytic oxidation of GSH, homocysteine, cysteine, *N*-acetyl cysteine, and other. Further modification of SWNT/GCE with pyrroloquinoline quinone showed an improvement in the catalytic behavior toward the oxidation of thiols. Salimi and Pourbeyram [32] reported the use of a three-dimensional sol-gel carbon ceramic electrode modified with $Ru[(tpy)(bpy)Cl]PF_6$ for electrochemical detection of GSH and cysteine. Long-term stability, short response time, and high detection sensitivity were shown using this modified electrode. However, the preparation of a specific and sensitive coating for this modified electrode highly depends on the coating technique and requires the use of buffer components that do not interfere with the coating material, hence introducing constraint for buffer selection. An indirect detection approach based on dual-electrode detection, as described in the next section, has been proposed to broaden the scope of application to detect analytes with similar functions by bare electrodes.

In summary, direct electrochemical detection, in general, shows high sensitivity and selectivity for analyte detection. However, the number of analytes to be monitored is limited by electroactive analytes and the detection potential imposed on the electrode, as a high potential leads to a high sensitivity and less selectivity and a low potential reduces interference but suffers lower detection sensitivity. Thus, the detection potential needs to be optimized. The use of indirect electrochemical detection via a redox mediator could expand the scope of application to nonelectroactive analytes and also enhance the detection selectivity via the control of electrode potential and chemical selectivity for analytes capable of sustaining a fast chemical reaction with the redox mediator. Details on indirect electrochemical detection are given in Section 9.5.2.

9.3.2 Indirect Electrochemical Detection

To solve the problem of slow electrokinetics for thiol detection, an indirect electrochemical detection scheme has been investigated. Dual electrodes placed in serial sequence are fabricated in the detector setup, with an upstream generation electrode to produce the redox mediators and a downstream detection electrode to detect the concentration of the redox mediators after reaction with the analytes that exit from the CE capillary. Holland and Lunte [33] demonstrated the use of electrogenerated bromine for indirect detection of glutathione, cysteine, and methionine using dual-electrode detection after CE separation.

The indirect detection route is shown in Figure 9.1. Dual electrodes in serial mode were employed to generate bromine by an upstream electrode before their detection by a downstream electrode. After the elution of analytes capable of reacting quickly with bromine, a decrease in the baseline reduction current at the detection electrode was observed, with a magnitude proportional to the concentration of the eluted analytes. In their work, a wire–wire dual-electrode configuration was used after CE separation, and a detection of 80 nM for cysteine was achieved. The high detection sensitivity makes the indirect electrochemical detection mode a promising method to determine low levels of thiols in biological samples.

The problem facing the use of the dual-electrode detection mode is broadening of the peak observed during CE detection, leading to poor separation efficiency. This broadening is attributed to the disturbance of the analytical stream by an uneven introduction of the bromide reagent by pressure, as well as the effect of the back pressure after the decoupling fracture of the capillary due to the reagent flow, reducing the analyte flow toward the detection electrode. Both factors lead to a serious peak broadening and poor separation efficiency, hampering the application of the indirect detection method for trace analyte determination in complex biological samples.

In summary, the dual-electrode indirect electrochemical detection approach offers a promising electrochemical method to detect thiols after CE separation.

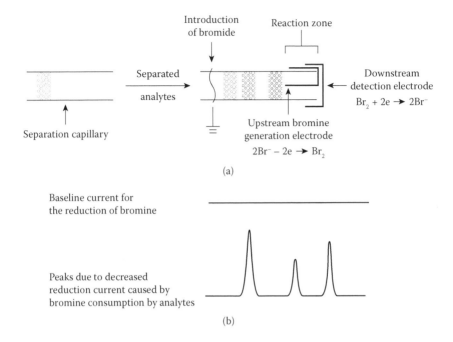

FIGURE 9.1
Schematic diagrams showing the flow direction and electrochemical reaction at the dual-serial electrodes using electrogenerated bromine for the detection of nonelectroactive analytes. (a) SDEs and associated flow direction and electrochemical reaction. (b) Detection peak and baseline at the downstream electrode.

Methods to solve the problems of peak broadening and its associated adverse effect on separation efficiency are addressed by the fabricated MC-CE device, with results discussed in Section 9.4.

9.4 MC-CE Device Integrated with Dual-Serial Electrode Detection for Indirect GSSG and GSH Determination

9.4.1 Problems for Generating Bromine during CE Separation

The serial dual-electrode (SDE) detection scheme based on electrogenerated halogen was first reported by King and Kissinger [34] for application in liquid chromatography (LC). After their pioneering work, several groups reported the use of electrogenerated bromine for the detection of proteins, amino acids, and thiols for flow injection analysis and LC [35–38]. Although the detection strategy is the same, the implementation of SDE detection based on electrogenerated bromine in CE is more difficult than that in LC. For application in LC, the bromide reagent used to electro-generate bromine at the upstream electrode can be introduced directly into the mobile phase without disturbing analyte separation. However, for application in CE, the bromide reagent cannot be introduced directly into the background electrolyte for two reasons. The first reason is that bromide can be oxidized at the high-voltage anode placed in the buffer vial, and the bromine generated can react with the analytes before their separation in the CE capillary, thereby interfering with the high-efficiency CE separation mode. The second reason is that the addition of bromide to the running buffer results in an increase in ionic strength in the separation capillary, leading to an undesirable increase in the background noise, affecting subsequent AD.

9.4.2 Design Considerations of MC-CE Device Integrated with SDE Detection

To prevent the effect of bromine generated before CE separation on analyte separation, it is most desirable to introduce the bromide reagent for generation of bromine at a position just after the exit of the CE separation capillary for reaction with the eluted analytes prior to detection of the residual bromine by the downstream detection electrode. Thus, a coaxial postcapillary reactor used in CE-laser-induced fluorescence was modified to introduce the bromide reagent to the SDE detector. The schematic diagram of the SDE detection system based on the coaxial postcapillary reactor is shown in Figure 9.2a [39], together with the enlarged view of the SDE configuration (Figure 9.2b). The coaxial postcapillary reactor also enables the construction of the upstream generation electrode by simple sputtering of a Pt film onto the outer surface of the separation capillary to oxidize bromide to bromine and to guide the oxidized bromine flow toward the downstream detection electrode to react with analytes eluted from the separation capillary. The reduction of bromine concentration in the analyte flow with respect to the background flow is then detected by the downstream Pt microdisk electrode. From this design, all bromine was generated and introduced to SDE at the capillary outlet after CE separation, and the background noise resulting from the introduction of bromide was reduced considerably. Furthermore, by optimizing the relative flowrates of the bromide reagent and the running buffer, satisfactory separation efficiency and high detection sensitivity can be achieved, as described in the next section.

The decoupling of bromine generation and detection electrodes from the high separation voltage path is an important novel design for the development of AD for eluted analytes from CE separation. Attributed to the use of a separation capillary with an inner diameter of 25 µm for CE run, the Pt microdisk detection electrode can be placed in a wall-jet configuration

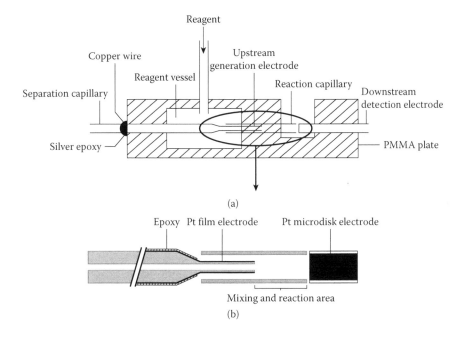

FIGURE 9.2
(a) Schematic diagram showing the integration of SDE detection with the coaxial postcapillary reactor. (b) Enlarged view of the SDE detector with the bromine/analyte mixing chamber. (From Du, F.Y., Cao, S.A. and Fung, Y.S. *Electrophoresis* 2014, 35, 3556–3563.)

without the need for a decoupler. Both the on-capillary Pt film working electrode and the reference electrode are not in the high-voltage path; thus, the potential imposed on the Pt film electrode will not be affected by the high separation voltage. This had been confirmed by the fact that the current of the Pt film electrode did not change after applying a high separation voltage between the buffer and the detection vials.

9.5 Feasibility Study of MC-CE/Dual-Serial Electrode Device for Indirect GSH Determination

The feasibility for determination of GSH using MC-CE device integrated with an SDE detector with electrogenerated bromine introduced after CE separation was studied. The electropherograms obtained after the injection of test samples containing 50 μM GSH to the MC-CE device with and without the operation of the generation electrode are given in Figure 9.3 [39]. When the generation electrode is turned off, no discernible signal for GSH

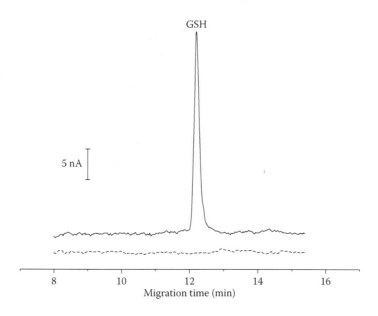

FIGURE 9.3
Electropherogram for detecting samples containing 50 μM GSH. The upper solid line was obtained with generation electrode on, whereas the lower dashed line was obtained with generation electrode off. Generation and detection electrodes controlled at +1.0 and +0.2 V, respectively, versus Ag/AgCl. KBr concentration, 5 mM. Height difference between bromide vial and buffer vial, 10 cm. CE conditions were sample injection, 6 kV for 12 s; separation voltage, 12 kV; and running buffer (pH 8.2), 20 mM sodium borate. (From Du, F.Y., Cao, S.A. and Fung, Y.S. *Electrophoresis* 2014, 35, 3556–3563.)

can be observed, indicating that GSH cannot be directly determined by the Pt disk electrode held at +0.2 V. When the generation electrode has turned on, a peak due to bromine depletion by reaction with GSH is observed as expected, demonstrating the feasibility for using the MC-CE/serial dual electrode (SDE) postcapillary detector for the determination of analytes that can react with bromine.

9.6 Optimization of Working Conditions for Bromine Generation

In the MC-CE integrated DSE detection scheme, the analyte of interest reacts with bromine produced at the generation electrode, and the concentration of analyte is determined indirectly by the loss of bromine detected at the downstream detection electrode. Therefore, high collection efficiency by the downstream platinum disk electrode and sufficient bromine generation by the upstream platinum film electrode are important to achieve desired detection sensitivity. Suitable generation and detection potentials should be used to generate a sufficient amount of bromine and to detect residual bromine with high detection sensitivity. Thus, the effect of potentials applied at both electrodes was investigated by monitoring the generation current and detection current under various generation and detection potentials.

As shown in Figure 9.4a, the oxidation of bromide starts at +0.9 V, and the generation current rises gradually with increasing generation potential [39]. Although the maximum generation current was achieved at +1.5 V versus Ag/AgCl reference electrode, the use of high generation potential can lead to the oxidation of other impurities and to solvent decomposition, which may interfere with bromine detection at the downstream electrode. Thus, +1.0 V was selected as the generation potential to generate a sufficient amount of bromine for postcapillary reaction detection and to avoid undesirable interference. With generation voltage controlled at +1.0 V, the variation of detection current on the detection potential is illustrated in Figure 9.4b. The detection current starts at +0.75 V and increases until a current plateau is reached with voltage controlled at +0.2 V versus Ag/AgCl reference electrode. Use of more negative potentials gives rise to the liberation of hydrogen. Thus, +0.2 V is selected as the detection potential in the present work.

In addition to the working potentials of the generation and detection electrodes, the flowrate of bromide reagent also greatly affects the analytical performance of the SDE detector as the residence time for mixing and reaction between the electrogenerated bromine and analytes exiting from the separation capillary is dependent on the flowrate, with faster flowrate for shorter residence times. The longer the residence time, the higher the yield of the postcapillary reaction between bromine and analytes. Although it is not necessary to achieve a reaction yield of 100%, it is desirable to have a high and constant yield to achieve a high depletion ratio with satisfactory repeatability and high detection sensitivity. The laminar coaxial flow consisting of an outer sheath reagent flow and an inner analyte flow is expected to be maintained inside the reaction chamber, enabling a fast mass transfer of reagents and analytes across the boundary between the two flows via their proximity and direct physical contact to achieve good repeatability and high detection sensitivity for the MC-CE device integrated with a DSE detector in the present design.

Undesirable peak broadening observed in the dual-electrode detector reported in the work of Holland and Lunte [33], with the detector setup shown in Figure 9.1, is the outcome of a poorly designed flow pattern. The analyte flow is obstructed by the dual-detection setup placed directly opposing the flow at the exit of the capillary. The introduction of the extra

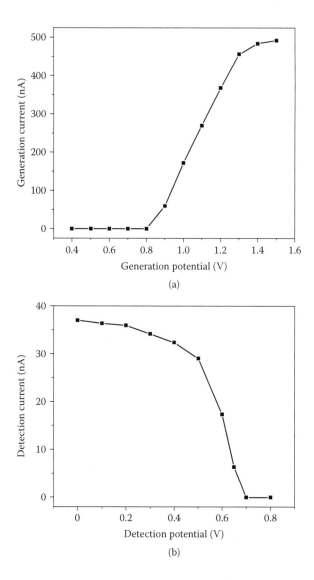

FIGURE 9.4
(a) Variation of generation current at different generation potentials. (b) Effect of detection potential on detection current with generation electrode controlled at a constant 1.0 V. (From Du, F.Y., Cao, S.A. and Fung, Y.S. *Electrophoresis* 2014, 35, 3556–3563.)

reagent flow through a crack in the separation capillary with irregular fracture to mix with the analyte flow creates a back pressure to push the analyte flow away from the detecting electrode, prolonging the residence time of the sample plug. As the results of a fixed inner diameter of the capillary and the lengthening of the total time for the enlarged and highly dispersed sample plug to pass through the downstream detection electrode, a broad peak is hence observed in conventional detector design, lowering the detection sensitivity.

In the present design (Figure 9.2), there are three special features adopted by the MC-CE/SDE device to minimize flow disturbance:

1. There is no obstruction to the analyte and reagent flow before the detection electrode.

2. The coaxial arrangement of the analyte flow and reagent flow enables a more efficient mass transfer of analyte and bromine by diffusion across the boundary between the two flows without flow disturbance.

3. The volume of the mixing chamber is considerably larger than the separation capillary as indicated by a significant difference of the inner diameters of the two capillaries. Thus, the change in flow-rate and flow pattern when the two flows encountered each other in the reaction chamber is minimized to allow a steady and repeatable flow pattern to impact onto the downstream detection electrode for bromine determination.

As the reagent flowrate affects the mixing and reaction efficiency between bromine and analytes, it is optimized to achieve the highest detection sensitivity. As the bromide reagent is introduced hydrodynamically, the height difference between the reagent and the buffer vials determines the hydrostatic pressure, and hence the bromide reagent flowrate. Therefore, to assess the bromide flowrate on the device's analytical performance, the effect of the height difference on the separation efficiency and detection sensitivity in terms of the number of theoretical plates (N) and peak height (nA) was investigated, with results shown in Figure 9.5 [39].

The increase in height difference was determined to lead to an increase in the number of theoretical plates, hence better separation efficiency and a decrease in the detection peak current for both GSH and GSSG. This is attributed to the increase of the reagent flowrate at increasing height difference. Slow reagent flowrate at a small height difference (2–3 cm) leads to insufficient reagent sheath flow to match the analyte flow, turbulent mixing in the reaction chamber, and a bigger and higher dispersed sample plug to reach the detection electrode for bromine determination, hence a broad peak with low peak height. In contrast, higher reagent flowrate at a larger height difference (14–15 cm) leads to less bromine generated and too high of a reagent sheath flow to disturb the analyte flow, leading to analyte dilution in the reaction chamber, and thus a smaller peak. With consideration of the various factors discussed above, 10 cm is chosen as the working height difference for the introduction of the bromide reagent.

As thiols are determined indirectly by the reduction of bromine concentration at the eluting peak consumed by stoichiometric reaction with thiols with respect to the background electrolyte, the detection limit and linear working range depend on the electrogenerated

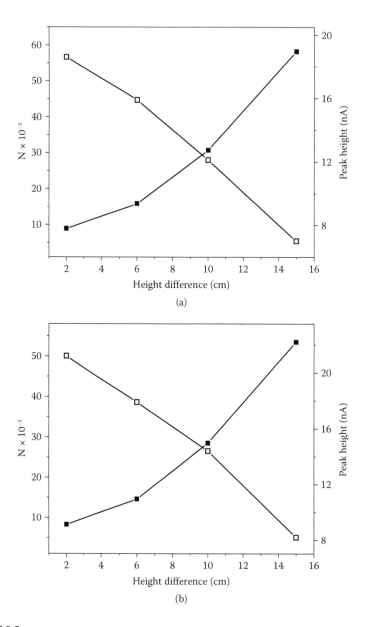

FIGURE 9.5
(a) Effect of the height difference between reagent vial and buffer vial on detection current peak height (□) and theoretical plates (N; ■) of 20 μM GSH. (b) Effect of the height difference between reagent vial and buffer vial on detection current peak height (□) and theoretical plates (■) of 20 μM GSSG. Generation and detection electrode controlled at +1.0 and +0.2 V, respectively, versus Ag/AgCl. KBr concentration, 2 mM. CE conditions were sample injection, 6 kV for 12 s; separation voltage, 12 kV; and running buffer (pH 8.2), 20 mM sodium borate. (From Du, F.Y., Cao, S.A. and Fung, Y.S. *Electrophoresis* 2014, 35, 3556–3563.)

TABLE 9.1

Effect of Bromide Concentration on Detection Limits and Linear Working Ranges of GSH and GSSG by MC-CE/SDE Detection

KBr Concn. (mM)	GSH		GSSG	
	Detection Limit[a] (µM)	Linear Working Range[b] (µM)	Detection Limit[a] (µM)	Linear Working Range[b] (µM)
1	0.096	0.3–30	0.085	0.3–25
2	0.16	0.5–50	0.14	0.4–40
5	0.54	2–100	0.46	1.5–80
10	2.2	5–200	1.9	4–150

From Du, F.Y., Cao, S.A. and Fung, Y.S. *Electrophoresis* 2014, 35, 3556–3563.

[a] Calculated based on S/N = 3.

[b] Correlation coefficient >0.995, peak height versus analyte concentration.

bromine concentration. The effect of different bromide concentrations on the detection limit and linear working range for GSH and GSSG is examined, with results summarized in Table 9.1 [39]. The lowest detection limits for GSH and GSSG were achieved at a bromide concentration of 1 mM, showing the lowest background noise and the highest signal-to-noise ratio (S/N). Both the lower and upper limits of the working range for GSH and GSSG are extended with the increase in the bromide concentration. However, the highest upper linear working ranges for GSH and GSSG were obtained at 2 mM bromide concentration. Due to the fact that GSSG is normally present at a level much lower than that of GSH in pharmaceutical samples, higher linear working ranges are preferred to achieve simultaneous determination of GSSG and GSH in one CE run. Thus, 2 mM bromide was used for subsequent study.

The experimental conditions were as follows. Generation and detection electrodes were controlled at +1.0 and +0.2 V, respectively, versus Ag/AgCl. The height difference between the bromide reagent vial and the detection vial was 10 cm. CE conditions were sample injection, 6 kV for 12 s; separation voltage, 12 kV; and running buffer (pH 8.2), 20 mM sodium borate. Under the working conditions described above, the detection limit, linearity, and repeatability were determined. The detection limit at S/N = 3 was 0.16 µM for GSH and 0.14 µM for GSSG. Based on a least square line with a slope of 0.64 nA/µM and a correlation coefficient of 0.998, the linear working range for GSH was 0.5–50 µM. Based on a least square line with a slope of 0.79 nA/µM and a correlation coefficient of 0.997, the corresponding linear working range for GSSG was 0.4–40 µM. The repeatability of the method was evaluated by analyzing a standard mixture of 20 µM GSH and 20 µM GSSG intraday repeatability in five replicates and interday repeatability in duplicates over five consecutive days. The intraday relative standard deviations (RSDs) for GSSG and GSH determination were 6.2 and 5.7%, respectively (*n* = 5). The interday RSDs for GSSG and GSH determination were 6.7 and 6.4%, respectively (*n* = 5).

9.7 Application of MC-CE/SDE Device for GSSG and GSH Determination in Pharmaceutical Supplement

9.7.1 Sample Description and Preparation

GSH is one of the most powerful antioxidants in biological systems, and it can help people stay healthy and prevent aging, cancer, heart disease, and other medical conditions. GSH in pharmaceutical supplement is mainly present in the reduced form. However, it may convert into the oxidized form (GSSG) during production; the GSSG form is considered an impurity and a marker for quality control for GSH preparation. Thus, simultaneous determination of GSH and its impurity GSSH in a pharmaceutical supplement is required for quality control and routine analysis. Although AD is sensitive, direct AD of these compounds, in particular for GSSG, is very difficult because of their high overpotentials for oxidation [40].

Pharmaceutical GSH supplement claiming to have 50 mg of GSH per tablet was purchased from a local drugstore. A tablet was weighed (0.6035 ± 0.0037 g, $n = 5$) and then ground thoroughly. A 120-mg portion of the powdered sample was dissolved in 100 mL of running buffer (pH 8.2, 20 mM sodium borate), diluted 10 times by the running buffer, and filtered to remove insoluble cellulose prior to injection to the MC-CE device.

9.7.2 GSSG and GSH Determination in Pharmaceutical Supplement

The application of the MC-CE/SDE device based on indirect thiol detection is studied for the determination of GSSG and GSH in a commercially available pharmaceutical supplement. Figure 9.6 shows a typical electropherogram of the supplement sample [39]. Five replicate determinations were made, with results shown in Table 9.2 [39]. The amount of the active ingredient GSH in the pharmaceutical supplement was found to be 50.8 ± 2.3 mg/tablet ($n = 5$), an amount in good agreement with the claimed value on the label of the supplement sample. The weight percentage of the impurity GSSG to GSH was 2.34 ± 0.07% ($n = 5$), indicating that a small portion of GSH had converted to GSSG during the production process. Recovery experiments were performed by spiking 10 μM GSH and 1 μM GSSG to the sample. Recoveries for GSH and GSSG are calculated to be 102 and 105%, respectively, indicating the reliability of the fabricated MC-CE/SDE device for thiol determination.

In brief, the MC-CE device fabricated for indirect determination of thiols based on the detection of the residual electrogenerated bromine after CE separation has been demonstrated to deliver a sensitive and reliable method for GSSG and GSH determination. Results matching the label value of a purchased pharmaceutical supplement and close to 100% recovery of GSH spiked to real samples were achieved to validate the procedure developed.

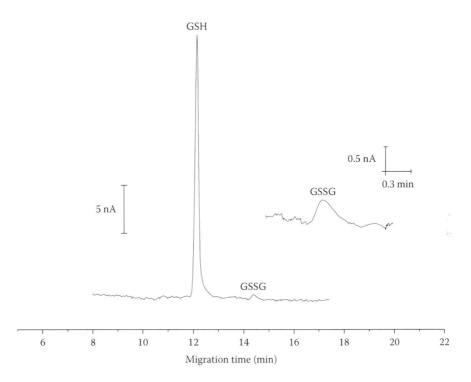

FIGURE 9.6
Typical electropherogram of the L-glutathione supplement sample. Inset is the enlarged electropherogram of the GSSG peak. Generation and detection electrodes were controlled at +1.0 and +0.2 V, respectively, versus Ag/AgCl. KBr concentration, 2 mM. Height difference between bromide vial and buffer vial, 10 cm. CE conditions were sample injection, 6 kV for 12 s; separation voltage, 12 kV; and running buffer (pH 8.2), 20 mM sodium borate. (From Du, F.Y., Cao, S.A. and Fung, Y.S. *Electrophoresis* 2014, 35, 3556–3563.)

TABLE 9.2

Amount of GSH and GSSG Present in GSH Supplement Tablet Determined by Fabricated MC-CE/SDE Device

Sample	GSSG$_{found}$ (mg/tablet)	GSH$_{found}$ (mg/tablet)	GSH$_{found}$/GSH$_{claimed}$ (%)	GSSG$_{found}$/GSH$_{found}$ (%)
1	1.13	50.1	100.2	2.26
2	1.12	48.3	96.6	2.31
3	1.16	49.4	98.8	2.35
4	1.31	53.5	107.0	2.45
5	1.22	52.9	105.8	2.31
Mean ± SD	1.19 ± 0.08	50.8 ± 2.3	101.7 ± 4.5	2.34 ± 0.07

From Du, F.Y., Cao, S.A. and Fung, Y.S. *Electrophoresis* 2014, 35, 3556–3563.

9.8 Summary

To solve the peak-broadening problem caused by the introduction of bromide reagent encountered in a previously developed dual-electrode detectors, an MC-CE device integrated with a novel coaxial postcapillary reactor design was developed with the separation capillary and reaction capillary sandwiched between two poly(methyl methacrylate) plates and a decoupling of the generation and detection electrodes from the high-voltage path. Operation conditions were optimized to achieve desired analytical performance, such as baseline separation of analyte peaks with no visible peak broadening, high separation efficiency, low detection limit, and wide linear working range for indirect determination of thiols.

The improvement of the performance of the MC-CE device integrated with SDE detection over previously developed dual-electrode detectors is attributed to the several novel design considerations for producing a repeatable flow pattern to deliver bromine to the reaction chamber at suitable concentrations in the form of laminar sheath flow to enable a fast and efficient reaction with the analyte exiting from the separation capillary prior to the detection of the residual bromine by the downstream detection electrode. First, no microstructure obstructs the analyte and reagent flow before the reaction chamber and the detection electrode. Second, the analyte and reagent flowrates are optimized to achieve a laminar coaxial flow to enable a quick mass transfer of bromine and thiols by diffusion across the liquid boundary between the outer sheath reagent flow and the inner analyte flow without flow disturbance. Third, the inner diameter of the mixing chamber is considerably larger than that of the separation capillary to enable nonturbulent dynamic mixing of the reagent and analyte flows when they have encountered each other to react in the reaction chamber before residue bromine determination by the detection electrode.

Under the optimized working conditions with potentials for upstream on-capillary Pt film electrode and downstream Pt disk electrode controlled at +1.0 and +0.2 V, respectively, to generate suitable bromine concentrations for reaction with thiols and leave behind sufficient residual bromine for sensitive detection by the detection electrode for the determination of GSSG and GSH, the detection limit, linearity, and reproducibility were determined and summarized as follows. The detection limit at S/N = 3 was 0.16 μM for GSH and 0.14 μM for GSSG. The linear working range for GSH with a slope of 0.64 nA/μM ranged from 0.5 to 50 μM, with a correlation coefficient of 0.998. The linear working range for GSSG with a slope of 0.79 nA/μM ranged from 0.4 to 40 μM, with a correlation coefficient of 0.997.

The MC-CE device fabricated for indirect determination of thiols based on the detection of residual electrogenerated bromine after CE separation has been demonstrated to deliver a sensitive and reliable method for GSSG and GSH determination. Successful application of the device developed has been shown for GSSG and GSH determination in a pharmaceutical supplement, with results validated by matching results with label values of a purchased sample and with close to 100% recovery of GSH spiked to a real sample. The integration of the simple and versatile MC-CE device with an SDE detector with novel reactor design to generate a stable and repeatable reagent flow to react with analytes eluted from CE separation capillary makes the fabricated MC-CE/SDE device attractive for clinical and biomedical applications for trace thiols determination in complex biological samples.

References

1. Meister, A. and Anderson, M.E. Glutathione. *Annu Rev Biochem* 1983, 52, 711–760.
2. Sies, H. Glutathione and its role in cellular functions. *Free Radic Biol Med* 1999, 27, 916–921.
3. Filomeni, G., Rotilio, G. and Ciriolo, M.R. Cell signalling and the glutathione redox system. *Biochem Pharmacol* 2002, 64, 1057–1064.
4. Circu, M.L. and Aw, T.Y. Glutathione and apoptosis. *Free Radic Res* 2008, 42, 689–706.
5. Valko, M., Leibfritz, D., Moncol, J., Cronin, M.T., Mazur, M. and Telser, J. Free radicals and antioxidants in normal physiological functions and human disease. *Int J Biochem Cell Biol* 2007, 39, 44–84.
6. Dröge, W. Free radicals in the physiological control of cell function. *Physiol Rev* 2002, 82, 47–95.
7. Forman, H.J. and Dickinson, D.A. Oxidative signaling and glutathione synthesis. *Biofactors* 2003, 17, 1–17.
8. Gawlik, M., Krzyzanowska, W., Gawlik, M.B. and Filip, M. Optimization of determination of reduced and oxidized glutathione in rat striatum by HPLC method with fluorescence detection and pre-column derivatization. *Acta Chromatogr* 2014, 26, 335–345.
9. Bai, S.L., Chen, Q.S., Lu, C. and Lin, J.M. Automated high performance liquid chromatography with on-line reduction of disulfides and chemiluminescence detection for determination of thiols and disulfides in biological fluids. *Anal Chim Acta* 2013, 768, 96–101.
10. Khan, A., Khan, M.I., Iqbal, Z., Shah, Y., Ahmad, L., Nazir, S., Watson, D.G., Khan, J.A., Nasir, F. and Khan, A. A new HPLC method for the simultaneous determination of ascorbic acid and aminothiols in human plasma and erythrocytes using electrochemical detection. *Talanta* 2011, 84, 789–801.
11. Moore, T., Le, A., Niemi, A.K., Kwan, T., Cusmano-Ozog, K., Enns, G.M. and Cowan, T.M. A new LC-MS/MS method for the clinical determination of reduced and oxidized glutathione from whole blood. *J Chromatogr B* 2014, 929, 51–55.
12. Squellerio, I., Caruso, D., Porro, B., Veglia, F., Tremoli, E. and Cavalca, V. Direct glutathione quantification in human blood by LC–MS/MS: Comparison with HPLC with electrochemical detection. *J Pharmaceut Biomed* 2012, 71, 111–118.

13. Bronowicka-Adamska, P., Zagajewski, J., Czuak, J. and Wrobel, M. RP-HPLC method for quantitative determination of cystathionine, cysteine and glutathione: An application for the study of the metabolism of cysteine in human brain. *J Chromatogr B* 2011, 879, 2005–2009.

14. Conlan, X.A., Stupka, N., McDermott, G.P., Francis, P.S. and Barnett, N.W. Determination of intracellular glutathione and cysteine using HPLC with a monolithic column after derivatization with monobromobimane. *Biomed Chromatogr* 2010, 24, 455–457.

15. McDermott, G.P., Terry, J.M., Conlan, X.A., Barnett, N.W. and Francis, P.S. Direct detection of biologically significant thiols and disulfides with manganese(IV) chemiluminescence. *Anal Chem* 2011, 83, 6034–6039.

16. Wu, J.F., Ferrance, J.P., Landers, J.P. and Weber, S.G. Integration of a precolumn fluorogenic reaction, separation, and detection of reduced glutathione. *Anal Chem* 2010, 82, 7267–7273.

17. Bouligand, J., Deroussent, A., Paci, A., Morizet, J. and Vassal, G. Liquid chromatography-tandem mass spectrometry assay of reduced and oxidized glutathione and main precursors in mice liver. *J Chromatogr B* 2006, 832, 67–74.

18. Camera, E. and Picardo, M. Analytical methods to investigate glutathione and related compounds in biological and pathological processes. *J Chromatogr B* 2002, 781, 181–206.

19. Sakhi, A.K., Blomhoff, R. and Gundersen, T.E. Simultaneous and trace determination of reduced and oxidized glutathione in minute plasma samples using dual mode fluorescence detection and column switching high performance liquid chromatography. *J Chromatogr A* 2007, 1142, 178–184.

20. Huang, K.J., Xie, W.Z., Wang, H. and Zhang, H.S. Sensitive determination of S-nitrosothiols in human blood by spectrofluorimetry using a fluorescent probe: 1,3,5,7-tetramethyl-8-(3',4'-diaminophenyl)-difluoroboradiaza-s-indacene. *Talanta* 2007, 73, 62–67.

21. Tang, B., Xing, Y., Li, P., Zhang, N., Yu, F. and Yang, G. A rhodamine-based fluorescent probe containing a Se–N bond for detecting thiols and its application in living cells. *J Am Chem Soc* 2007, 129, 11666–11667.

22. Pires, M.M. and Chmielewski, J. Fluorescence imaging of cellular glutathione using a latent rhodamine. *Org Lett* 2008, 10, 837–840.

23. Shibata, A., Furukawa, K., Abe, H., Tsuneda, S. and Ito, Y. Rhodamine-based fluorogenic probe for imaging biological thiol. *Bioorg Med Chem Lett* 2008, 18, 2246–2249.

24. Martin, R.S., Ratzlaff, K.L., Huynh, B.H. and Lunte, S.M. In-channel electrochemical detection for microchip capillary electrophoresis using an electrically isolated potentiostat. *Anal Chem* 2002, 74, 1136–1143.

25. Lacher, N.A., Garrison, K.E., Martin, R.S. and Lunte, S.M. Microchip capillary electrophoresis/electrochemistry. *Electrophoresis* 2001, 22, 2526–2536.

26. Wang, J., Chen, G., Chatrathi, M.P. and Musameh, M. Capillary electrophoresis microchip with a carbon nanotube-modified electrochemical detector. *Anal Chem* 2004, 76, 298–302.

27. Wang, J. Electrochemical detection for capillary electrophoresis microchips: A review. *Electroanalysis* 2005, 17, 1133–1140.

28. Chen, B., Zhang, L. and Chen, G. Determination of salidroside and tyrosol in *Rhodiola* by capillary electrophoresis with graphene/poly(urea-formaldehyde) composite modified electrode. *Electrophoresis* 2011, 32, 870–876.

29. Trojanowicz, M. Recent developments in electrochemical flow detections—A review: Part I. Flow analysis and capillary electrophoresis. *Anal Chim Acta* 2009, 653, 36–58.

30. Salimi, A. and Hallaj, R. Catalytic oxidation of thiols at preheated glassy carbon electrode modified with abrasive immobilization of multiwall carbon nanotubes: Applications to amperometric detection of thiocytosine, l-cysteine and glutathione. *Talanta* 2005, 66, 967–975.

31. Han, H. and Tachikawa, H. Electrochemical determination of thiols at single-wall carbon nanotubes and PQQ modified electrodes. *Front Biosci* 2005, 10, 931–939.

32. Salimi, A. and Pourbeyram, S. Renewable sol–gel carbon ceramic electrodes modified with a Ru-complex for the amperometric detection of l-cysteine and glutathione. *Talanta* 2003, 60, 205–214.

33. Holland, L.A. and Lunte, S.M. Postcolumn reaction detection with dual-electrode capillary electrophoresis-electrochemistry and electrogenerated bromine. *Anal Chem* 1999, 71, 407–412.

34. King, W.P. and Kissinger, P.T. Liquid chromatography with amperometric reaction detection involving electrogenerated reagents: Applications with in-situ generated bromine. *Clin Chem* 1980, 26, 1484–1491.

35. Fung, Y.S. and Mo, S.Y. Determination of amino acids and proteins by dual-electrode detection in a flow system. *Anal Chem* 1995, 67, 1121–1124.

36. Sato, K., Takekoshi, Y., Kanno, S., Kawase, S., Jin, J.Y., Takeuchi, T. and Miwa, T. Plastic film ring-disk carbon electrodes for the indirect amperometric detection of underivatized amino acids in liquid chromatography. *Anal Sci* 1999, 15, 957–961.

37. Sato, K., Jin, J.Y., Takeuchi, T., Miwa, T., Takekoshi, Y., Kanno, S. and Kawase, S. Indirect amperometric detection of underivatized amino acids in microcolumn liquid chromatography with carbon film based ring–disk electrodes. *Analyst* 2000, 125, 1041–1043.

38. Zhou, H., Holland, L.A. and Liu, P. An integrated electrochemical capillary liquid chromatography-dual microelectrode system for bromine based reaction detection. *Analytst* 2001, 126, 1252–1256.

39. Du, F.Y., Cao, S.A. and Fung, Y.S. A Serial dual-electrode detector based on electrogenerated bromine for capillary electrophoresis. *Electrophoresis* 2014, 35, 3556–3563.

40. Safavi, A., Maleki, N., Farjami, E. and Mahyari, F.A. Simultaneous electrochemical determination of glutathione and glutathione disulfide at a nanoscale copper hydroxide composite carbon ionic liquid electrode. *Anal Chem* 2009, 81, 7538–7543.

10

Detecting Optically Inactive Analyte: MC-CE Device Integrated with QDs/LIF Detection for Determination of Acrylamide in Food

Qidan Chen

Jilin University, Zhuhai College
Zhuhai, People's Republic of China

Ying Sing Fung

The University of Hong Kong
Hong Kong SAR, People's Republic of China

CONTENTS

10.1 Introduction

Determination of contaminants at trace levels in food has become a hot topic for analytical method development due to the increasing demand to cover various analytes at trace levels which are often beyond the detection limits of conventional methods.

Acrylamide (AA) is present in many food products attributed to high temperature treatment. The needs, requirements and current methods for its determination in food is given and discussed, with insufficient detection sensitivity identified for real sample application. To enhance detection sensitivity, quantum dots (QDs) medicated LIF detection is integrated with MC-CE device for indirect determination of AA in food. The design and layout of the device, the optimized operation conditions, and the application of the MC-CE/QDs-LIF device for indirect determination of AA in Potato Chips are given and discussed in light of the results presented.

10.1.1 Needs and Requirements

2-Propenamide, also know as acrylamide (AA), is a compound classified as "probably carcinogenic in humans" [1]. AA has recently raised health concerns due to its appearance in food samples cooked at high temperature, such as fried and roasted food products. Thus, potato chips, French fries, bread biscuits, and even coffee and common snacks may contain AA. Moreover, exposure to harmful dosages of AA have found to cause cancer in animals, and long-term exposure to AA can damage the nervous system.

Monomeric AA is commonly used in the laboratory for preparation of gel electrophoresis for protein separation [2,3]. It is also used to produce dyes, grout, contact lenses, and as additives for construction of dams, tunnels, and sewers. AA in the form of a monomer is needed for the preparation of polyacrylamide materials. Polyacrylamide has been used for the treatment of waste water and purification of drinking water. It is an important additive for the production of papers, glues, cosmetics, soil conditioning agents, textiles, and numerous organic chemicals in crude oil processing. Although polyacrylamide is not toxic, it may contain a very small amount of residual AA monomer [4].

After its classification as a potential hazardous compound, AA has been found in a large variety of cooked foods, such as fried and oven-baked food products, and it has caused worldwide health concern due to its carcinogenic effect in laboratory animals, suspected carcinogenic effect in humans, and probable role as a reproductive nerve toxin [5,6]. Attributed to its highly toxic effects, including reproductive neurotoxicity, genotoxicity, and carcinogenicity for both somatic and germ cells [7], the European Union Scientific Committee on Toxicity, Ecotoxicity and the Environment has estimated the risk for exposure to AA and recommended that exposure to AA, which might be absorbed quickly through skin contact, should be limited to a level as low as possible.

Recently, active research on AA has contributed to the development of rapid analytical methods, the discovery of new biological formation pathways, and mitigation methods via additives. Simultaneously, new mechanisms have been established indicating that the Maillard reaction may be the most probable mechanism for AA formation [8–10]. To meet the demand for a better analytical method for the determination of AA in complex food products, current methods for assay of AA in food are discussed in Section 10.1.2.

10.1.2 Current Analytical Methods for Determination of AA in Food

Since a report in 2002 showing the presence of AA in food products after thermal food processing, numerous methods have been investigated for AA analysis in different food products, including mushroom samples [11], field crops [12], and sugar [13]. Analytical methods, such as gas chromatography (GC), liquid chromatography (LC), micellar electrokinetic capillary chromatography (MEKC), capillary zone electrophoresis (CZE) with electrochemical detection [14], near-infrared spectrometry [15], time-of-flight mass spectrometry [16,17], and computer image analysis [18,19], have been developed for AA determination in food. However, due to the low detection sensitivity of these conventional methods, most are not able to detect AA in thermally processed food samples at trace levels. Thus, more sensitive methods, such as mass spectrometry (MS)-based chromatographic methods, are being developed for determination of AA at trace levels [20–22].

To tackle complex food samples, various MS-based procedures developed for AA determination in foods have to be coupled with a separation method, such as high-performance liquid chromatography [8,23–30] and GC [31–35], for sample cleanup before MS determination. Comparing the LC and GC separation procedures, LC-MS is better than GC-MS for AA determination because the use of a cumbersome bromination process to increase the volatility of AA intermediates is not required in LC-MS. Although MS detection is preferred for AA determination, the fragmentation pattern for AA is not unique because of the presence of other coextractives that can yield intermediates with the same charge-to-mass (m/z) ratio as AA [36].

In 2005, the World Health Organization (WHO) and the Food and Agriculture Organization of the United Nations (FAO) established that certain food samples cooked at high temperature, especially western-style snacks, may contain AA at levels harmful to human health [37].

Subsequent research on determination of AA in other food products has indicated a much bigger problem of AA contamination in foods, and a general monitoring method to detect the presence of AA at trace levels in a large variety of foods is needed. The high capital and operational costs of MS instrumentation have prevented its use for monitoring purposes in the food industry. Thus, non–MS-based methods that are easy to use, widely applicable, and have affordable capital and running costs are urgently needed for quantitative determination of AA in food.

Most alternative methods for routine analysis of AA are based on direct ultraviolet (UV) [38,39] and diode array detection (DAD) [40,41] or indirect UV detection after separation, such as by LC-UV [42–45]. However, AA is a highly polar molecule [46], and it is hard to select a suitable mobile phase for achieving a reasonable retention time by conventional LC reverse-phase columns. Despite the use of tandem MS, the LC method presents limitations because of coelution of interfering compounds. Capillary electrophoresis (CE) methods [29,47,48], such as MEKC, offer a possibility for the separation of acidic, basic, and neutral compounds, and hence a promising method for determination of AA in food samples.

Despite the advantages of the CE methods, such as high separation efficiency, relatively short analysis time, and low cost of analysis compared to other separation methods, their application for the determination of AA was only first realized in 2004, due to the problems and issues related to the separation and detection of AA. Details are discussed in Section 10.2.

10.2 CE Methodologies for Detecting Optically Inactive AA

10.2.1 Choice of Separation Mode

The most common CE methodology, CZE, is based on the differential migration of analytes governed by their m/z ratios. Thus, most CZE procedures are carried out in aqueous background electrolytes (BGEs). However, the aqueous CZE procedure cannot be used to separate AA due to the polar and uncharged state of the molecule. To solve this problem, Jezussek and Schieberle [29] proposed a procedure using precolumn derivatization of AA with 2-mercaptobenzoic acid [49] prior to CZE separation. However, the derivatization procedure requires working for 3 h in darkness under an N_2 atmosphere; this procedure is highly inconvenient, tedious, and time-consuming, and thus introduces the possibility for analyte loss during the long and complicated procedure.

Another method is making a significant change in the acid–base characteristics of AA by using BGE containing a suitable nonaqueous solvent to impact a charge to AA to assist its separation. A CZE method using nonaqueous BGE has been proposed [47] as an alternative method to traditional aqueous-organic BGE. However, there are other issues and problems for using organic solvents in CZE to handle aqueous samples containing AA, as critically discussed in the literature [50], making the nonaqueous procedure unpopular for AA determination.

Another CE methodology, MEKC, had been developed for AA determination without derivatization by coupling with DAD [48]. However, the MEKC-DAD method does not give a good detection limit. An improved MEKC-UV method [49] was then developed for the determination of AA. However, the method is difficult to operate and the detection sensitivity is still poor. Therefore, research is currently focused on developing methods that could be easily used in most laboratories, while delivering high-sensitivity, low-cost, rapid, and convenient operation for AA determination.

10.2.2 Operations and Procedures

Cleanup of the separation capillary is necessary for a new as well as a used capillary to achieve satisfactory repeatability for CE separation. A new capillary has to be conditioned by overnight washing with 1 M NaOH prior to use. A conditioned capillary is washed with deionized H_2O for 1 h and then BGE for 5 min before sample injection. For used and conditioned capillaries, the following washings were used at the start of the working day by flushing the capillary in the following sequence: 1 M NaOH for 10 min, deionized H_2O for 2 min, and BGE for 3 min before use. Between runs, the separation capillaries were washed first with deionized H_2O for 1 min and then rinsed by BGE for 1 min. Between two consecutive runs, BGE was used to wash the capillary for 1 min.

For sample introduction, three modes are commonly used: hydrodynamic injection, electrokinetic injection, and pressure-time injection. For hydrodynamic injection, the sample vial was typically held at an 8-cm height difference for 15 s. There are no biases for introducing different analytes using hydrodynamic injection. However, the process is too slow for automation of CE injection. For electrokinetic injection, small and charged analytes are introduced more than neutral analytes, leading to biases for sample introduction. Pressure-time injection, typically in pressure-time (psi × seconds) units, is commonly used for sample introduction by commercial CE instruments as it can be easily automated for sample injection.

For the microfluidic chip-capillary electrophoresis (MC-CE) device developed for the present study, both hydrodynamic injection and pressure-time injection introduce too large of a sample size for the microchannel segment, although no biases have been shown for introducing different analytes. For electrokinetic injection, biases in sample injection are serious, in particular for introducing nanosized quantum dots (QDs) particles, together with small analytes, such as AA, in the sample. The double-T injection mode with details given in Chapter 3 is thus adopted in the present study.

10.3　Integrating QDs with LIF for Indirect Determination of AA

Known for its high detection sensitivity, laser-induced fluorescence (LIF) detection can be integrated with CE separation for developing CE-LIF procedures to deliver zeptomole (10^{-21}) detection limits [51]. Extremely high detection sensitivities have been shown using CE-LIF in various application areas [52–55]. As AA does not possess any chromophoric group in its structure, it has to be determined indirectly via a photoluminescent agent that has exhibited a strong interaction with AA.

Photoluminescent QDs are commonly used as nanoparticle fluorescent labels for biomedical research and diagnostic application. QDs made up by group II/VI and III/V nanoparticles are found to exhibit diameters between 1 and 6 nm. The development of QDs materials is governed by the capability to prepare particles with desired optical properties. Thus far, the most common QDs materials used for detection applications are CdSe and CdTe. Although group III/V QDs, such as InP and InGaP, are probable alternatives because of their lower toxicity due to the absence of cytotoxic cadmium ions, QDs made up of InP and InGaP with a high monodispersity and luminescent power have not yet been successfully synthesized. Compared to the traditional organic dyes, QDs made up of CdSe or CdTe, which exhibit strong quantum confinement, have exhibited excellent optical properties, such as photobleaching resistance, long fluorescence half-life (5–40 ns), and a large excitation spectral range [56]. These advantages make QDs a good agent as a mediator for indirect determination of analytes that have shown strong interaction with QDs.

In this chapter, the development of an MC-CE device integrated with an innovative, low-cost, simple, and fast method based on MEKC separation coupled with indirect LIF detection mediated by QDs is presented, with results discussed for determination of AA in potato chips [57]. To optimize the analytical procedure for AA determination, a systematic investigation of the applied voltage, pH, and buffer components, including buffer additives, has been conducted, with results discussed in Section 10.4.

10.3.1　Preparation of CdTe QDs and QDs-AA Complexes

Water-soluble CdTe QDs coated with mercaptopropyl acid (MPA) at different sizes were synthesized as reported previously [58,59]. A brief description of the procedure is given below. First, NaHTe was prepared by reacting $NaBH_4$ with tellurium powder by mixing in an aqueous solution up to a molar ratio of 2:1, respectively. Second, under vigorous stirring, freshly prepared oxygen-free NaHTe solution was added to the 1.25×10^{-3} mol/L $CdCl_2$ aqueous solution saturated by nitrogen. After adjusting to pH 11.4, MPA was added as a stabilizing agent until the molar ratio of Cd^{2+}:MPA:NaHTe reached 1:2.4:0.5, respectively. The resulting mixture was heated to 96–100°C and then refluxed for a suitable duration to control the size of CdTe QDs. To get rid of excess MPA and cadmium ions, the CdTe QDs were precipitated using the same procedure as described previously [60]. QDs-AA complexes were prepared by mixing AA with prepared CdTe QDs (3.2 nm, 6.25×10^{-5} mol/L, with reference to Te at pH 8.0). The resulting solutions with no visible aggregation were found to contain stable QDs-AA complex which was used for subsequent studies.

10.3.2 Characterization of Prepared QDs and QDs-AA Complex

Different sizes of CdTe QDs were synthesized in solutions showing different colors (Figure 10.1). The spectra of absorption and emission were recorded from pure QDs in phosphate-buffered saline (PBS) (pH 8.0) (Figure 10.2). The particle sizes were estimated by the absorption spectrum of the QDs samples. The wavelength (λ) and the absorbance (A) (optical density value) were obtained at the first excitonic absorption peak. The particle sizes (D) of the QDs sample were obtained from the sizing curves or by using the following fitting equation to calculate their sizes [61] (for CdTe), $D = (9.8127 \times 10^{-7}) \lambda_3 - (1.7147 \times 10^{-3}) \lambda_2 + 1.0064 \lambda - 194.84$, where D (in nanometers) is QDs particle size, λ (in nanometers) is wavelength of the first QDs excitonic absorption peak, and λ_n (in nanometers) is wavelength of the nth QDs excitonic absorption peak.

Thus, QDs with 568 nm emission wavelength were selected in this work because their fluorescence spectra had shown a higher symmetric shape

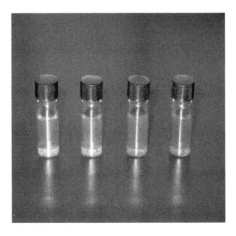

FIGURE 10.1
Solutions containing different sizes of QDs: four differential emission colors of CdTe QDs under irradiation by a near-UV lamp. From left to right: the maximum of fluorescence emission wavelengths are changing from 526, 568, 610, and 650 nm.

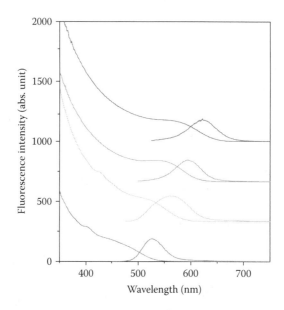

FIGURE 10.2
Fluorescence spectra of solutions containing CdTe QDs at different sizes marked by the corresponding colors. The excitonic position in the absorption peak has shifted from 441 to 580 nm, with corresponding shift in the maximum emission peak from 526 to 650 nm, while the particle size is increasing from ~2.8 to 4.0 nm.

suitable for LIF detection. The particle size was calculated to be ~3.2 nm according to the equation relating particle diameter to fluorescence wavelength as reported previously [61]. Comparing the fluorescence with Rhodamine 6G, ~25% luminescence quantum yield was estimated for the CdTe QDs at room temperature [62]. The concentration of the aqueous QDs solution based on Te content was estimated to be 6.25×10^{-4} mol/L [63]. The maximum fluorescence emission wavelength of the selected CdTe QDs was centered at 568 nm upon 473 nm excitation.

10.3.3 Study of QDs-AA Interaction

To assist the discussion on QDs-AA interaction, the molecular formula for AA is shown in Figure 10.3.

AA is a highly polar and water-soluble molecule. Some characteristic properties, such as easy protonation of its amide group by the solvent medium as well as strong acid and rapid reaction with Br_2 and R-SH, respectively [64], indicate strong interaction with specific functional groups. Among these interactions, hydrogen bonding between –CO–NH– from AA to –COO– group from MPA capped CdTe QDs was the most likely mode of interaction [46] (Figure 10.4).

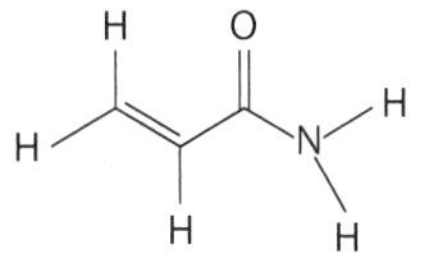

FIGURE 10.3
Chemical structure of AA.

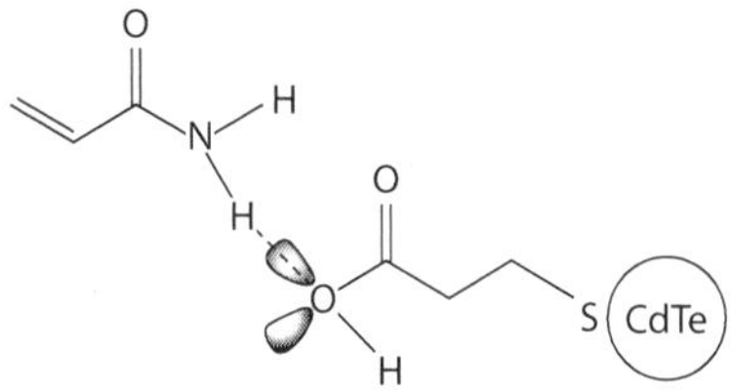

FIGURE 10.4
Formation of the AA-QDs complex through hydrogen bonding.

An investigatory study using spectrophotometric and spectrofluoromet-ric techniques on the QDs-AA interaction was conducted, with results given below. The absorption spectra were measured by a UV-visible-near-infrared spectrophotometer (model 3100, Shimadzu, Kyoto, Japan). The QDs-AA interaction was studied using a fluorescence spectrophotometer (F-2500, Hitachi, Tokyo). All optical experiments were conducted under ambient conditions in an air-conditioned room (20 ± 1°C). The solution pH was mea-sured by a pH meter (model 59002-02, Cole-Parmer Instrument Co., Vernon Hills, IL).

The AA absorption spectra before and after addition of QDs are shown in Figure 10.5 [57]. With AA alone, a weak peak was found at 204 nm. After the addition of QDs, a rather complex spectrum was produced with a notice-able shoulder (~300 nm) and two strong peaks (209 and 240 nm). The peak at 209 nm is the original AA peak, showing almost four times enhancement after the addition of QDs. The 240-nm peak and the 300-nm shoulder are from the QDs-AA complex produced by a strong interaction between AA and QDs.

To investigate the fluorescence spectra of the QDs-AA complex, fluores-cence experiments were conducted using buffers with a pH range from 7.0 to 10.0 at a constant QDs concentration (0.1 mM), while AA was increased from 0.5 to 100 mg/kg. Figure 10.6 shows that lowering of the fluorescence was observed at higher AA concentrations within the pH range under investigation [57].

A linear relationship was found for lowering of fluorescence at increas-ing AA concentration, indicating the possibility of using QDs for AA deter-mination. However, Figure 10.5 shows that the fluorescence of the QDs-AA

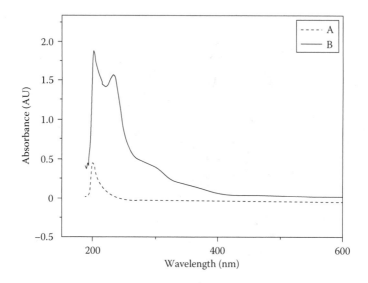

FIGURE 10.5
Addition of QDs to PBS on UV spectrum of AA at pH 8.0. (A) 0.1 mg/kg AA alone.
(B) 0.1 mg/kg AA + 0.1 mM QDs. (From Chen, Q.D., Zhao, W.F. and Fung, Y.S. *Electrophoresis* 2011, 32, 1252–1257.)

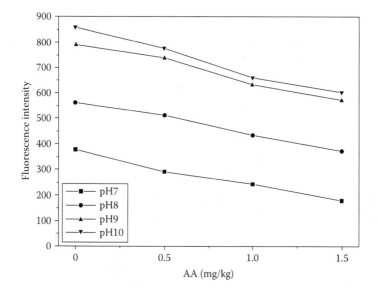

FIGURE 10.6
Effect of buffer pH (pH 7.0–10.0) on the fluorescence of the QDs-AA complex. [QDs] = 0.1 mmol/L; [AA] = 0, 5, 20, and 100 mg/kg; Ex = 473 nm; Em = 568 nm; spectral band width (SBW) = 2.5 nm. (From Chen, Q.D., Zhao, W.F. and Fung, Y.S. *Electrophoresis* 2011, 32, 1252–1257.)

complex was strongly affected by pH, especially at high pH. This may due to the electrostatic interaction of QDs with the negatively charged hydroxide anions, leading to the increase of QDs fluorescence as the result of higher QDs fluorescent efficiency in alkaline buffers.

The enhanced QDs fluorescence at high pH may be attributed to a replacement of hydroxide anions to the QDs-AA complex. As shown by the UV results, the QDs-AA complex is capable of absorbing emissions at multiple wavelengths. However, QDs are unstable in solution with pH <6.5. Therefore, the pH range from 8.0 to 10.0 was selected for optimization of the analytical procedure for AA determination using MC-CE device integrated with QDs-mediated LIF detection.

10.4 MC-CE Device Integrated with QDs-LIF Detection for Indirect AA Determination

10.4.1 Design and Layout of the MC-CE Device Integrated with QDs-LIF Detection

The layout of the MC-CE device integrated with QDs-LIF detection is shown in Figure 10.7. A fused silica capillary (physical and effective length at 515 and 500 mm, respectively, and inside diameter [i.d.] at 50 μm) was bonded between two poly(methyl methacrylate) plates with a hole drilled at the detection zone and the exit end of the embedded capillary dipping into an external buffer waste vial. The other capillary end was inserted into a buffer reservoir fabricated at the MC device. Usually, 18 kV (CZE1000R high-voltage power supply, Spellman, Hauppauge, NY, USA) was used for the CE run.

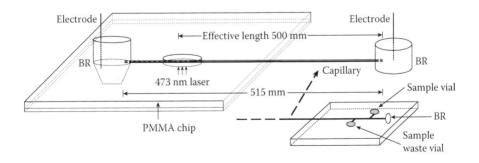

FIGURE 10.7
Schematic diagram showing the fabricated MC-CE device. BR, buffer reservoir; DZ, detection zone for LIF with 568 nm emission under 473 nm excitation. (a) On-capillary sample intake by gravity feed, (b) On-capillary sample intake by double T Injector.

Samples were introduced via gravity feed or a double-T injector fabricated on-chip, and the analytes separated were detected at the detection zone by LIF. To carry out different functions, the MC-CE/LIF device consisted of a capillary/microfluidic device and a laser/detection system. The laser/detection system is equipped with a digital camera for selection of position for laser excitation and a diode laser to generate 473 nm laser for irradiation at the selected position. The detection system was positioned underneath a horizontal detection plane where the capillary/microfluidic device was placed. The two parts can be adjusted independently in three directions to optimize the position for laser irradiation and the collection of fluorescence light. An interference filter was inserted to remove scattered light that can give rise to an undesirable high-fluorescence background.

The design of the MC-CE device with separate sample and detector chip enables the position of the separation capillary in horizontal at the detection zone for laser irradiation at 90° to the separation capillary. The arrangement has the following advantages. First, the scatter light from the capillary wall is minimized by the confocal optical arrangement. This is made possible by precise arrangement of the excitation spot of laser light onto the solution part at the middle of the capillary. Second, a double-T injector was used to introduce the sample containing extract with AA with QDs and other buffer constituents, while the main buffer flow was stopped by sealing off buffer reservoir and buffer waste vial. Thus, a repeatable and constant volume of sample segment can be introduced to the MC-CE device. Third, the off-chip buffer waste vial arrangement enables the introduction of a low but adjustable buffer flow during electrophoretic separation by lowering the buffer waste vial to a suitable height with respect to the MC during separation. Thus, sufficient time for interaction of AA from sample with QDs from buffer could be used to achieve the maximum detection sensitivity.

10.4.2 Optimizing Operation Conditions of MC-CE/QDs-LIF Device for Indirect Determination of AA

To enhance the AA signal and shorten the analytical time, a simple buffer containing QDs, PBS, and sodium dodecyl sulfate (SDS) was optimized for MEKC separation. The interactions among QDs, AA, and other constituents, such as SDS, inside the capillary are given in Figure 10.8 [57]. As no interaction between QDs and other constituents was found [65], the interaction is attributed to a competition of AA by SDS and QDs. The effects of buffer pH and buffer concentration on AA migration time are investigated. The results show a slight increase of AA migration time with increasing pH. This effect may be due to anion adsorption on QDs surface under alkaline pH. Therefore, buffer pH at 8.0 is selected for subsequent study.

As the concentration of BGE strongly affects current heating, detection sensitivity and column resolution, PBS with concentrations ranging from 20 to 60 mM was used to investigate the effect of BGE on AA separation.

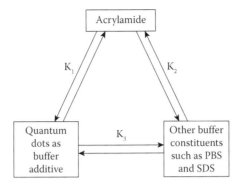

FIGURE 10.8
Competitive interaction among QDs, AA, and other buffer constituents in the MC-CE/QDs-LIF separation system. (From Chen, Q.D., Zhao, W.F. and Fung, Y.S. *Electrophoresis* 2011, 32, 1252–1257.)

A large current and a noticeable increase of the retention time were observed using 40 nm BGE or higher concentrations. This is attributed to the increase in the ionic strength of the buffer, which affects the micelles electrophoretic velocities and solutes concentration [66,67]. The same trend was observed in previous studies for polar contaminants separated by MEKC [65,68], indicating that the addition of QDs to BGE did not increase the current or affected AA separation. Satisfactory separation and low background current were obtained for AA determination using 40 mM buffer.

K_1, K_2 and K_3 refer to the equilibrium constants established amongst the various interactions inside the capillary among QDs, AA, SDS, and other buffer constituents. The main interaction is a competition of AA by QDs and SDS, both of which are under strongly influenced by buffer pH and buffer concentration. It is expected that the sequence is as follows: $K_1 > K_2 \gg K_3$.

The effect of SDS on AA separation was investigated using SDS concentrations ranging from 10 to 30 mmol/L. The results showed that the AA peak was increased significantly with increase in SDS concentration, giving rise to an obvious peak broadening, as reported in previous work [65,68]. As SDS exhibited a strong effect on peak profile and analyte detection, SDS at 30 mM was adopted for separation and determination of AA.

In summary, indirect LIF mediated by QDs-AA interaction has successfully integrated with MC-CE device for the determination of neutral AA in food after MEKC separation. The optimized MEKC separation conditions are given below: BGE consisting of 0.1 mM QDs, 30 mM SDS, 40 mM PBS with pH adjusted to 8.0; separation voltage at 18 kV; physical and effective length for separation capillary at 515 and 500 mm, respectively, i.d. at 50 µm; and LIF detection at 568 nm emission under 473 nm excitation. The detection mechanism may be attributed [67] to the quenching of the fluorophore by the analytes, giving rise to a reduction of the quantum efficiency of QDs by hydrogen bond interaction in the sample zone. The effect of quenching by AA on the sensitivity of detection is discussed in Section 10.5.1.

10.5 Application of the MC-CE/QDs-LIF Device for Indirect Determination of AA in Potato Chips

For real sample application of MC-CE device, removal of potential interferents from the sample matrix is important, in particular, for determination of trace analytes in a complex food sample matrix. Thus, procedures for sample preparation and cleanup are investigated, with results given in Section 10.5.3.1. For real sample application, the content of AA in potato chips, a food product commonly sold in supermarkets, is chosen for study because trace levels of AA are suspected to be present due to the use of high temperature for preparation of potato chips. The important analytical issue for real sample application is validation of the results. Recovery test is used to confirm the validity of the method by spiking a given amount of AA to potato chip samples purchased from supermarkets for determination of the percentage of recovery of spiked AA.

10.5.1 Sample Preparation and Cleanup Procedures

The sample matrix for potato chips is very complex. In addition to common ingredients such as protein, starch, and lipids, it contains numerous food additives. Most of them can be coextracted with AA during various sample pretreatment operations that are needed before the separation and detection of AA from potato chip samples.

The following procedures were used for food sample preparation. AA was extracted by a previously published method [69] from potato chip samples purchased from supermarket. Detailed operations are given in Chen et al. [57] and essential steps are summarized below. The potato chips sample was ground and weighed before the extraction and purification procedures. For recovery study, standard AA solution was spiked to samples and was then shaken for 1 h; ultrasonicated for 10 min; and centrifuged at 4000 rpm for 30 min. The supernatant was filtered and defatted prior to use.

For the cleanup process, a solid phase extraction cartridge was conditioned by passing 3 mL each of deionized H_2O and MeOH prior to loading with 3 mL of extract and then eluted with 1 mL of a mixed solution (MeOH:H_2O, 3:2 [v/v]). The eluent was collected and filtered through a 0.2-μm microfilter before CE separation. All solutions were kept at 4°C.

10.5.2 Optimized Analytical Performance

A baseline separation of the target analytes is shown in Figure 10.8. A noticeable quenching of QDs fluorescence was shown due to the interaction of QDs with AA extracted from the potato chips sample (Figure 10.9a–c) [57]. The AA peak is obviously separated from other peaks from the sample matrices.

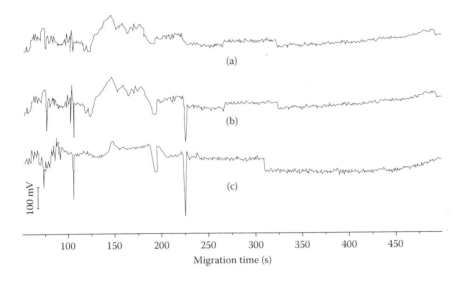

FIGURE 10.9
Electropherograms of AA extract obtained from potato chips by MC-CE/QDs-LIF device. Capillary = 50/51.5 cm × 50 μm i.d. Injection = 8 cm × 15 s. Separation voltage = 18 kV. Running buffer (pH 8.0) = QDs (0.1 mM), PBS (40 mM), SDS (30 mM). LIF = excitation (473 nm) and emission (568 nm). (a) Nonspiked sample. (b) Extract spiked with AA (10 mg/kg). (c) Extract spiked with AA (20 mg/kg) (From Chen, Q.D., Zhao, W.F. and Fung, Y.S. *Electrophoresis* 2011, 32, 1252–1257.)

In addition, a linear relationship was found between the quenching fluorescence of QDs and AA concentration (Figure 10.9b and c). Under optimized conditions, no interference peaks were found near the AA peak.

The analytical performance of the MC-CE device under optimized working conditions is summarized below. The detection limit (signal-to-noise ratio [S/N] = 2) was 0.1 mg of AA per kilogram of food sample, and working linear range (peak area) was 0.5–100 mg/kg based on 3% deviation from linearity (0.999 correlation coefficient) of calibration curve. Satisfactory relative standard deviation (%RSD, $n = 3$) for retention time (0.18%) and peak height measurement (6.4%) for AA determination.

10.5.3 Applicability for AA Determination in Real Samples

The recovery for AA spiked to real sample was studied using the device developed. Different amounts of AA were spiked to potato chips at three different levels for recovery test. The blank potato chips sample was spiked by AA standard solution to the following concentrations: 100, 20, and 10 mg/kg food sample. The recoveries achieved based on peak area measurement were ranging from 90 to 95%, with RSD from 3.4 to 5.7% ($n = 3$). According to the maximum level (4.080 mg/kg) required by the Joint FAO/WHO Expert

Committee [70], the method developed with 0.1 mg/kg detection limit and 1–100 mg/kg working range is adequate for AA determination. The recovery of close to 100% of spiked AA confirmed the applicability of the method for determination of AA in potato chip samples.

10.6 Summary

AA is a water-soluble and toxic contaminant [64] causing worldwide health concern due to its extensive presence in various food samples cooked by high temperature. As AA does note possess any electrochemically active functional group and exhibits poor UV absorption, an assay using MC-CE device integrated with QDs-LIF indirect detection is shown in the present work to provide a fast and simple method for AA determination in food.

The quenching of the background fluorescence from QDs by AA gives rise to a linear response suitable for analytical use upon excitation by a 473 nm laser to induce fluorescence at 568 nm. Under the optimized conditions, the method has been shown to be capable of determining AA at sub-ppm concentration in potato chips with complex sample matrices. With the use of sample cleanup procedure typically employed in analytical laboratories, the electropherograms have shown no interfering peaks that can affect the sharp AA peak. Close to 100% recoveries were obtained for AA spiked to potato chips, indicating the reliability of the measurement by MC-CE/QDs-LIF device. The limit of detection and working range of the developed method are suitable for AA determination in accordance with the WHO specification. The indirect detection mode adopted by the MC-CE/QDs-LIF device is in general applicable to the determination of other analytes present in complex samples at trace levels if a suitable interaction with QDs existed. Thus, it offers an alternative and sensitive detection method to analytes that are electrochemically and optically inactive for detection.

References

1. International Agency for Research on Cancer (IARC), 1994. Monographs on the Evaluation of Carcinogenic Risks to Humans. Some Industrial Chemicals. IARC, Lyon, France.
2. Giese, J. Acrylamide in foods. *Food Technol* 2002, 56, 71–72.
3. Simonne, H. and Archer, L. Acrylamide in foods: A review and update. *Univ Florida Extension* 2002, 10, 1–3.

4. Lopachin, R.M. and Lehning, E.J. Acrylamide-induced distal axon degeneration: A proposed mechanism of action. *Neurotoxicology* 1994, 15, 247–259.
5. Costa, L.G., Deng, H., Greggotti, C., Manzo, L., Faustman, E.M., Bergmark, E. and Calleman, C.J. Comparative studies on the neuro and reproductive toxicity of acrylamide and its expoxide metabolite glycidamide in the rat. *Neurotoxicology* 1992, 13, 219–224.
6. Dearfield, K.L., Abernathy, C.O., Ottley, M.S., Brantner, J.H. and Hayes, P.F. Acrylamide: Its metabolism, developmental and reproductive effects, genotoxicity, and carcinogenicity. *Mutat Res* 1988, 195, 45–77.
7. Scientific Committee on Toxicity Ecotoxicity and the Environment (CSTEE). *Opinion on the Results of the Risk Assessment of Acrylamide*. Report version: October 2000 carried out in the framework of Council Regulation (EEC) 793/93 on the evaluation and control of the risks of existing substances. Opinion expressed at the 22nd CSTEE Plenary Meeting, Brussels, March 6/7, 2001.
8. Becalski, A., Lau, B.P.Y., Lewis, D. and Seaman, S.W. Acrylamide in foods: Occurrence, sources, and modeling. *J Agric Food Chem* 2003, 51, 802–808.
9. Amrein, T.M., Bachmann, S., Noti, A., Biedermann, M., Ferraz Barbosa, M., Biedermann-Brem, S., Grob, K., et al. Potential of acrylamide formation, sugars, and free asparagine in potatoes: A comparison of cultivars and farming systems. *J Agric Food Chem* 2003, 51, 5556–5560.
10. Yaylayan, V.A. and Stadler, R.H. Acrylamide formation in food: A mechanistic perspective. *J AOAC Int* 2005, 88, 262–267.
11. Castle, L.J. Determination of acrylamide monomer in mushrooms grown on polyacrylamide gel. *Agric Food Chem* 1993, 41, 1261–1263.
12. Bologna, L.S., Andrawes, F.F., Barvenik, F.W., Lentz, R.D. and Sojka, R.E.J. Analysis of residual acrylamide in field crops. *J Chromatogr Sci* 1999, 37, 240–244.
13. Tekel, J., Farkas, P. and Kova, C.M. Determination of acrylamide in sugar by capillary GLC with alkaliflame-ionization detection. *Food Addit Contam* 1989, 6, 377–381.
14. Stobiecka, A. and Radecka, H. Novel voltammetric biosensor for determining acrylamide in food samples. *J Biosens Bioelectr* 2007, 22, 2165–2170.
15. Segtnan, V.H., Kita, A., Mielnik, M., Jørgensen, K. and Knutsen, S.H. Screening of acrylamide contents in potato crisps using process variable settings and near-infrared spectroscopy. *Mol Nutr Food Res* 2006, 50, 811–817.
16. Dunovska, L., Cajka, T., Hajslova, J. and Holadova, K. Direct determination of acrylamide in food by gas chromatography-high-resolution time-of-flight mass spectrometry. *Anal Chim Acta* 2006, 578, 234–240.
17. Brathen, E. and Knutsen, S.H. Effect of temperature and time on the formation of acrylamide in starch-based and cereal model systems, flat breads and bread. *Food Chem* 2005, 92, 693–700.
18. Gokmen, V., Senyuva, H.Z., Dulek, B. and Cetin, E. Computer vision based analysis of potato chips—A tool for rapid detection of acrylamide level. *Mol Nutr Food Res* 2006, 50, 805–810.
19. Gokmen, V., Senyuva, H.Z., Dulek, B. and Cetin, E. Computer vision-based image analysis for the estimation of acrylamide concentrations of potato chips and french fries. *Food Chem* 2007, 101, 791–798.
20. Govaert, Y., Arisseto, A., van Loco, J., Scheers, E., Fraselle, S., Weverbergh, E., Degroodt, J.M. and Goeyens, L. Optimisation of a method for the determination of acrylamide in foods. *Anal Chim Acta* 2006, 556, 275–280.

21. Wenzl, T., Lachenmeier, D.W. and Gokmen, V. Analysis of heat-induced contaminants (acrylamide, chloropropanols and furan) in carbohydrate-rich food. *Anal Bioanal Chem* 2007, 389, 119–137.
22. Zhang, Y., Zhang, G.Y. and Zhang, Y. Occurrence and analytical methods of acrylamide in heat-treated foods: Review and recent developments. *J Chromatogr A* 2005, 1075, 1–21.
23. Rosen, J. and Hellenas, K.E. Analysis of acrylamide in cooked foods by liquid chromatography tandem mass spectrometry. *Analyst* 2002, 127, 880–882.
24. Tareke, E., Rydberg, P., Karlsson, P., Eriksson, S. and Tornqvist, M. Analysis of acrylamide, a carcinogen formed in heated foodstuffs. *J Agric Food Chem* 2002, 50, 4998–5006.
25. Ahn, J.S., Castle, L., Clarke, D.B., Lloyd, A.S., Philo, M.R. and Speck, D.R. Verification of the findings of acrylamide in heated foods. *Food Addit Contam* 2002, 19, 1116–1124.
26. Zyzak, D.V., Sanders, R.A., Stojanovich, M., Tallmadge, D.H., Eberhart, B.L., Ewald, D.K., Gruber, D.C., et al. Acrylamide formation mechanism in heated foods. *J Agric Food Chem* 2003, 51, 4782–4787.
27. Murkovic, M. Acrylamide in Austrian foods. *J Biochem Bioph Meth* 2004, 61, 161–167.
28. Roach, J.A.G., Andrzejewski, D., Gay, M.L., Nortrup, D. and Musser, S.M. Rugged LC-MS/MS survey analysis for acrylamide in foods. *J Agric Food Chem* 2003, 51, 7547–7554.
29. Jezussek, M. and Schieberle, P. A new LC/MS-method for the quantitation of acrylamide based on a stable isotope dilution assay and derivatization with 2-mercaptobenzoic acid. Comparison with two GC/MS methods. *J Agric Food Chem* 2003, 51, 7866–7871.
30. Riediker, S. and Stadler, R.H. Analysis of acrylamide in food by isotope-dilution liquid chromatography coupled with electrospray ionization tandem mass spectrometry. *J Chromatogr A* 2003, 1020, 121–130.
31. Castle, L., Campos, M.J. and Gilbert, J.J. Determination of acrylamide monomer in hydroponically grown tomato fruits by capillary gas chromatography-mass spectrometry. *Sci Food Agric* 1991, 54, 549–555.
32. Biedermann, M., Biedermann-Brem, S., Noti, A. and Grob, K. Two GC-MS methods for the analysis of acrylamide in foodstuffs. *Mitt Lebensm Hyg* 2002, 93, 638–652.
33. Tateo, F. and Bononi, M. A GC/MS method for the routine determination of acrylamide in food. *Ital J Food Sci* 2003, 15, 149–151.
34. Ono, H., Chuda, Y., Ohnishi-Kameyama, M., Yada, H., Ishizaka, M., Kobayashi, H. and Yoshida, M. Analysis of acrylamide by LC-MS/MS and GC-MS in processed Japanese foods. *Food Addit Contam* 2003, 20, 215–220.
35. Nemoto, S., Takatsuki, S., Sasaki, K. and Maitani, T. Determination of acrylamide in foods by GC/MS using 13C-labeled acrylamide as an internal standard. *J Food Hyg Soc Jpn* 2002, 43, 371–376.
36. Bermudo, E., Nunez, O., Moyano, E., Puignou, L. and Galceran, M.T. Field amplified sample injection–capillary electrophoresis–tandem mass spectrometry for the analysis of acrylamide in foodstuffs. *J Chromatogr A* 2007, 1159, 225–232.
37. International Food Safety Authorities Network. *Acrylamide in Food is a Potential Health Hazard.* Information Note No. 2/2005. International Food Safety Authorities Network, Geneva, Switzerland, 2005.

38. Paleologos, E.K. and Kontominas, M.G. Determination of acrylamide and methacrylamide by normal phase high performance liquid chromatography and UV detection. *J Chromatogr A* 2005, 1077, 128–135.

39. Wang, H.Y., Lee, A.W.M., Shuang, S.M. and Choi, M.M.F. SPE/HPLC/UV studies on acrylamide in deep-fried flour-based indigenous Chinese foods. *Microchem J* 2008, 89, 90–97.

40. Geng, Z.M., Jiang, R. and Chen, M.J. Determination of acrylamide in starch-based foods by ion-exclusion liquid chromatography. *Food Compos Anal* 2008, 21, 178–182.

41. Gokmen, V., Senyuva, H.Z., Acar, J. and Sarioglu, K. Determination of acrylamide in potato chips and crisps by high-performance liquid chromatography. *J Chromatogr A* 2005, 1088, 193–199.

42. Wenzl, T., de la Calle, M.B. and Anklam, E. Analytical methods for the determination of acrylamide in food products: A review. *Food Addit Contam* 2003, 20, 885–902.

43. JIFSAN, Acrylamide Infonet. 2002. http://www.acrylamide-food.org/

44. Cavalli, S., Maurer, R. and Hoefler, F. Fast determination of acrylamide in food samples using accelerated solvent extraction followed by ion chromatography with UV or MS detection. *J LC-GC Eur* 2003, 2, 9–11.

45. Terada, H. and Tamura, Y. Determination of acrylamide in processed foods by column-switching HPLC with UV detection. *J Food Hyg Soc Jpn* 2003, 44, 303–309.

46. Friedman, M. Chemistry, biochemistry, and safety of acrylamide. A review. *J Agric Food Chem* 2003, 51, 4504–4526.

47. Baskam, S. and Erimbedla, F. NACE for the analysis of acrylamide in food. *Electrophoresis* 2007, 28, 4108–4113.

48. Bermudo, E., Ruiz-Calero, V., Puignou, L. and Galceran, M. Microemulsion electrokinetic chromatography for the analysis of acrylamide in food. *Electrophoresis* 2004, 25, 3257–3262.

49. Bermudo, E., Nuñez, O., Puignou, L. and Galceran, M.T. Analysis of acrylamide in food samples by capillary zone electrophoresis. *J Chromatogr A* 2006, 1120, 199–204.

50. Porras, S.P. and Kenndler, E. Are the asserted advantages of organic solvents in capillary electrophoresis real? A critical discussion. *Electrophoresis* 2005, 26, 3203–3220.

51. Diebold, G.J. and Zare, R.N. Laser fluorimetry: Subpicogram detection of aflatoxins using high-pressure liquid chromatography. *Science* 1977, 196, 1439–1441.

52. Banks, P.R. Fluorescent derivatization for low concentration protein analysis by capillary electrophoresis. *Trends Anal Chem* 1998, 17, 612–622.

53. Bardelmeijer, H.A., Lingeman, H., de Ruiter, C. and Underberg, W.J.M. Derivatization in capillary electrophoresis. *J Chromatogr A* 1998, 807, 3–26.

54. Zhu, R. and Kok, W.T. Post-column derivatization for fluorescence and chemiluminescence detection in capillary electrophoresis. *J Pharm Biomed Anal* 1998, 17, 985–999.

55. Huang, T. and Pawliszyn, J. Axially illuminated fluorescence imaging detection for capillary isoelectric focusing on Teflon capillary. *J Pawliszyn Analyst* 2000, 125, 1231–1233.

56. Bruchez, M., Jr., Moronne, M. and Gin, P. Semiconductor nanocrystals as fluorescent biological labels. *Science* 1998, 281, 2013–2015.

57. Chen, Q.D., Zhao, W.F. and Fung, Y.S. Determination of acrylamide in potato crisps by capillary electrophoresis with quantum dot-mediated LIF detection. *Electrophoresis* 2011, 32, 1252–1257.

58. Gao, M.Y., Kirstein, S., Möhwald, H., Rogach, A.L., Kornowski, A., Eychmüller, A. and Weller, H. Strongly photoluminescent CdTe nanocrystals by proper surface modification. *J Phys Chem B* 1998, 102, 8360–8363.

59. Zhang, H., Zhou, Z., Yang, B. and Gao, M.Y. The influence of carboxyl groups on the photoluminescence of mercaptocarboxylic acid-stabilized CdTe nanoparticles. *J Phys Chem B* 2003, 107, 8–13.

60. Tang, Z.Y., Kotov, N.A. and Giersig, M. Spontaneous organization of single CdTe nanoparticles into luminescent nanowires. *Science* 2002, 297, 237–240.

61. Yu, W., Qu, L., Guo, W. and Peng, X. Experimental determination of the extinction coefficient of CdTe, CdSe, and CdS nanocrystals. *Chem Mater* 2003, 15, 2854–2860.

62. Georges, J., Arnaud, N. and Parise, L. Limitations arising from optical saturation in fluorescence and thermal lens spectrometries using plused laser excitation: Application to the determination of the fluorescence quantum yield of rhodamine 6G. *Appl Spectrosc* 1996, 50, 1505–1511.

63. Gaponik, N., Talapin, D., Rogach, A., Eychmuller, A. and Weller, H. Efficient phase transfer of luminescent thiol-capped nanocrystals: From water to nonpolar organic solvents. *Nano Lett* 2002, 2, 803–806.

64. Weisshaar, R. Acrylamide in heated potato products–analytics and formation routes. *Eur J Lipid Sci Technol* 2004, 106, 786–792.

65. Chen, Q.D. and Fung, Y.S. Capillary electrophoresis with immobilized quantum dot fluorescence detection for rapid determination of organophosphorus pesticides in vegetables. *Electrophoresis* 2010, 31, 3107–3114.

66. Cai, J. and Rassi, Z.E. Micellar electrokinetic capillary chromatography of neutral solutes with micelles of adjustable surface charge density. *J Chromatogr* 1992, 608, 31–45.

67. Garrison, A.W., Schmitt, P. and Kettrup, A. Separation of phenoxy acid herbicides and their enantiomers by high-performance capillary electrophoresis. *J Chromatogr A* 1994, 688, 317–327.

68. Zhou, X., Fan, L.Y., Zhang, W. and Cao, C.X. Separation and determination of acrylamide in potato chips by micellar electrokinetic capillary chromatography. *Talanta* 2007, 71, 1541–1545.

69. Bermudo, E., Moyano, E., Puignou, L. and Galceran, M.T. Determination of acrylamide in foodstuffs by liquid chromatography ion-trap tandem mass-spectrometry using an improved clean-up procedure. *Anal Chim Acta* 2006, 559, 207–214.

70. *Evaluation of Certain Food Contaminants (Sixty-Fourth Report of the Joint FAO/WHO Expert Committee on Food Additives).* WHO Technical Report Series, 2005. http://www.who.int/ipcs/food/jecfa/summaries/summary_report_64_final.pdf

11

Quantum Dots–Enhanced Fluorescence Detection: Immobilized Quantum Dots for Laser-Induced Fluorescence Detection of Organophosphorus Pesticides in Vegetables

Qidan Chen

Jilin University, Zhuhai College
Zhuhai, People's Republic of China

Ying Sing Fung

The University of Hong Kong
Hong Kong SAR, People's Republic of China

CONTENTS

11.1 Introduction

The worldwide concern on pesticide contamination of vegetables leads to tightening control and monitoring, in particular for the toxic organophosphorus pesticides quantum dots (QDs) which are found to be present in recently reported pesticide contaminated cases.

The needs and requirements for the control of QDs in food are discussed in light of the limitation of current methods for their determination. The integration of sensitive quantum-dot enhanced laser-induced fluorescence detection with MC-CE devices has been studied with results showing detection limits of OCs at trace levels in complex food products. The applicability of the MC-CE device for QDs detection in tomato has been investigated, showing detection limits capable to meet the recently tightened requirements for detection of QDs in vegetables.

11.1.1 Needs and Requirements

Although the use of pesticides for pest control improves food production, their adverse environmental and health impacts have led to worldwide concern and the need to monitor their levels in all food products. Of the two main types of pesticides, organochlorines (OCs) and organophosphorus pesticides (OPs), the OCs are persistent in the environment, leading to their accumulation in the food chain; hence, most OCs are banned in most countries. In contrast, OPs are less persistent in the environment and can be degraded over a few days after spraying onto vegetables for pest and weed control. Thus, most OPs are allowed to be used under strict control to allow sufficient time for them to decay after application before crop harvest or marketing. However, under poor climatic conditions that destroy much of a vegetable crop and give rise to high vegetable prices, growers have a high incentive to harvest and sell their products before allowing sufficient time for the decay of OPs after spraying. This leads to an excess amount of OP residues in vegetables and the associated problem that the public will not buy vegetables where even one or two samples were found to contain excess OP residue. As vegetables are perishable products and cannot be kept more than a few days after harvest, growers who have followed the rules and produced noncontaminated products often experience huge economic losses.

To avoid excessive intake of OPs-contaminated vegetables, the Codex Alimentarius Commission has issued guidelines for maximum residue level (MRL) of specified OPs in food samples [1]. In recent years, the MRL values have been lowered to sub-ppm concentrations to meet the rising requirements in food safety regulation. Thus, the need for a highly sensitive method that can produce quick results has arisen, preferably for methods with the capability for on-site monitoring of OPs, with fast turnaround of results, to stop contaminated vegetables from entering the market after their sampling and inspection at a control point.

11.1.2 Current Analytical Methods for Determination of OPs

Most analytical methods for OPs determination are developed for plant, water, soil, and food samples [2–10]. Due to the fact that the presence of highly toxic OPs in vegetables causes intense public health concern, the levels of OPs in vegetables need to be determined rapidly, with results released to the public as soon as possible before release of the perishable vegetables kept by the authorities. Thus, a rapid analysis of the widely used OPs in a food sample, especially vegetables, is needed. The World Health Organization (WHO) requires the use of the whole plant as the basis for calculation of the control levels for OPs in vegetables. Such whole plant samples create a complex test sample containing all of ingredients extracted from these vegetables, thereby contributing to problems with sample matrix interference in the determination of OPs at trace levels. Thus, a separation method is needed prior to the use of sensitive detection mode for OPs quantitation.

Of two common separation methods, gas chromatography (GC) and liquid chromatography (LC), GC can integrate with sensitive detectors, such as flame ionization detectors and mass spectrometry (MS). However, GC methods require volatile analytes for separation and most OPs are nonvolatile. Although the use of derivatization can increase analyte volatility, this extra transformation step creates the associated problems of analyte loss and lengthy sample preparation. Although high-performance liquid chromatography can handle nonvolatile analysts, its separation efficiency is insufficient to separate OPs at trace levels from a complex sample matrix, and most detection modes, except expensive MS, do not have sufficient detection sensitivity to meet control requirements. The recently developed microscale electrophoresis and related separation techniques, such as capillary electrophoresis (CE) and microfluidic chip-capillary electrophoresis (MC-CE), offer excellent separation efficiency and high-sensitivity detection modes to meet the application requirements for determination of OPs in vegetables. Details are given in Section 11.2.

11.2 CE and Related Techniques for Separation and Determination of OPs

To meet the need for determination and quantification of pesticides in food, there are currently two major electrophoresis techniques applied: CE and MC-CE. CE is used for laboratory analyses and MC-CE is used for field applications. Of the various CE modes for analysis of pesticides [11–13], the micellar electrokinetic chromatography (MEKC) mode has been shown to be suitable for the analysis of pesticides extracted from lettuce and grape samples [14]. MCs have been developed to expand the scope of CE for portable applications, such as DNA and protein analyses and clinical and food analyses [15–17].

However, the number of studies published on the determination of pesticides by CE in fruit or vegetable samples is relatively small, and most of these studies are limited to the determination of pesticides in a standard mixture, probably due to the problems of insufficient detection sensitivity and selectivity for determination of OPs at trace levels in complex sample matrixes that are required for real sample application. To extend the scope, sensitivity, and selectivity of analytical methods to real sample application for OPs determination, various sample preparation techniques for sample cleanup and analyte enrichment have been investigated. Two of these techniques are solid phase extraction (SPE) [8,18,19] and liquid-liquid extraction [9,20]. However, traditional sample pretreatment procedures are often laborious, involve substantial organic solvent consumption, operate over a long period, and can give rise to analyte loss due to incorrect procedures being carried out during preparation of the samples.

In this chapter, the focus is on the development of a rapid analysis method for the determination of OPs in vegetables. A novel detection method is discussed that can achieve high detection sensitivity and adequate analyte selectivity to satisfy the required limits of detection (LODs) and to meet the demanding requirements for OPs residue determination imposed by the Codex Alimentarius Commission and WHO.

11.3 Quantum Dots–Enhanced LIF for Detection of OPs

11.3.1 LIF for Detection of OPs

As most OPs are electrochemically inactive with poor ultraviolet-visible absorbance, the choice for their sensitive detection is limited to MS and laser-induced fluorescence (LIF), both of which are known for their high instrument costs. In addition, the operational cost for MS is high due to the use of a high-vacuum operation environment and the need for skilled workers to operate the machine. LIF is much easier to work with under ambient atmospheric pressure, and the recent development of a solid-state diode laser to replace gas lasers makes LIF detection affordable as a stand-alone detector for OP determination after MEKC separation. Thus, LIF offers a good choice for high-sensitivity detection of OPs after MEKC separation.

The operational lifetime of the solid-state diode laser in LIF is much longer than that of the gas laser, thus its operational cost is greatly reduced.

The consumption of laser-induced fluorescent organic dye during operation under high-intensity laser irradiation becomes the main operation cost for LIF methods. In addition, many highly fluorescent organic dyes can easily undergo photobleaching under continuous irradiation during operation, and the associated LIF method can be subjected to interference by coextracts from the sample matrix.

QDs, with photochemical properties such as high luminescence and tunable excitation and emission wavelengths [21–25], are an excellent optically stable fluorescence agent. The good photochemical stability of QDs [23] provides a promising analytical fluorescence agent for OP detection in real samples, as discussed in the next section.

11.3.2 Fluorescence Investigation for Interaction of QDs with OPs

Due to the special properties of QDs, with enhanced fluorescence intensity upon interaction with specific compounds rather than quenching normally expected from fluorescence study [26], the interaction of QDs with OPs is investigated as a unique, selective, and sensitive detection mode. Preliminary results from fluorescence study indicated insufficient detection sensitivity; therefore, the coated QDs-LIF detection technique was investigated in the present study to enhance the detection sensitivity and to establish a simple, fast, economic, and selective method by coupling with CE for detecting OP residues in vegetable samples. The water-soluble QDs were synthesized as reported previously [27]. Thiourea was added directly to the original mercaptopropionic acid (MPA)–capped CdTe solution as the sulfur source, and the CdTe seed acted as the core for the CdS shell to grow around it, with the concentration ratio thiourea:MPA:Cd:Te of 4:2:1:0.2, respectively.

A fluorescence spectrophotometer (F-2500, Hitachi, Tokyo, Japan) was used to investigate the interaction of CdTe/CdS core-shell QDs capped with MPA with the commonly found OPs diazinon, phosalone, methidathion, and mevinphos. The fluorescence spectra of QDs alone and QDs–OPs complexes formed upon interaction were measured and recorded under excitation at 473 nm (spectral band width = 2.5 nm). The fluorescence peak of QDs was found to increase significantly when OPs were added. The fluorescence enhancement from QDs upon interaction with OPs may be attributed to the rearrangement of the surface charge at the QDs-OPs conjugate formed by the hydrogen bond or other interactions between OPs and QDs [28]. According to the fluorescence theory of semiconductor QDs, the enhancement was attributed to the nonadiabatic effects exhibited by the exciton–phonon system [29]. Another enhancement factor may be attributed to the attendant effects from the inorganic core which improve the exciton recombination efficiency [30].

Different fluorescence enhancement factors were exhibited by the four QDs as shown by the difference in slopes of their linear curves with OPs from 10 to 0.5 ppm in buffer (pH 9.6) containing 6.7×10^{-5} mol/L QDs and 10 mmol/L borate. The linear relationship between the fluorescence intensity and OPs concentration at a constant QDs concentration indicated the applicability of QDs as a fluorescent agent for determination of OPs in vegetable samples.

Attributed to the protonation of the sulfur atom (S→SH) and the loss of coordination to the QDs surface, QDs capped with thiol ligands can aggregate at low pH [31]. Thus, the pH effects on the QDs-OPs conjugates were investigated. pHs 8.0–10.0 were studied, and the fluorescence intensity showed no significant variation. At pH 9.6, maximum fluorescences were obtained for mevinphos, methidathion, and phosalone; at pH 8.8, the fluorescence of diazinon was only slightly improved. Therefore, in the present work, pH 9.6 was selected.

Under pH 9.6, the effect of the ionic strength of the buffer was investigated using borate buffer at 10 mM, with fluorescence intensity recorded at NaCl concentrations between 0.2 and 0.01 mol/L. The results indicated a negligible fluorescence change while the ionic concentration was varied from 10 to 200 mM. Therefore, borate buffer of 10 mmol/L was selected in the present study.

To further enhance the detection sensitivity to eliminate a tedious SPE step, a new procedure has been developed for immobilizing QDs on the inside wall surface of the capillary by coupling with 3-mercaptopropyltrimethoxysilane (MPTS) to produce a three-dimensional (3D) network coating. Details on the immobilization procedure and its effect on the analytical performance of the MC-CE device fabricated are given in the following section.

11.4 MC-CE Device Integrated with Immobilized QDs for LIF Detection of OPs

11.4.1 Design and Fabrication of MC-CE Device

A schematic diagram showing the integration of MC-CE with immobilized QDs for LIF detection of OPs is given in Figure 11.1 [32]. The device consists of an MC-CE unit, with QDs immobilized onto the inner wall surface of the capillary within the detection zone, and an LIF detection system. The device is fabricated by bonding a fused silica capillary (physical length 45 cm, effective length 43.5 cm, and inside diameter [i.d.] 50 μm) between two poly(methyl methacrylate) (PMMA) plates (width 27 mm, length 85 mm, and thickness 0.15 mm), each with a hole drilled at a suitable region to expose the detection zone of the capillary with QDs immobilized onto the inner wall surface.

Details of the experimental procedures are given in Chen and Fung [32], with the major steps summarized below. The MC-CE device (part I) was placed horizontally at the laser irradiation plane with fluorescence light recorded by the LIF detector (part II). The two parts could be adjusted in

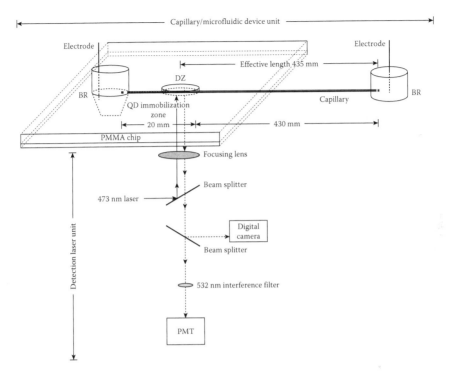

FIGURE 11.1

Layout of MC-CE device showing immobilized QDs detection zone and LIF detection system. DZ, QDs-immobilized detection zone; BR, buffer reservoir; LIF detection at 532 nm under 473 nm excitation. (From Chen, Q.D. and Fung, Y.S. *Electrophoresis* 2010, 31, 3107–3114.)

all three directions. A digital camera was used to locate the laser irradiation spot within the separation capillary. A solid-state diode laser was used for irradiation at 473 nm and an interference filter (532 ± 2 nm) was used for light collection. Portable CE equipment (model CE-P2, CE Resources Pte Ltd., Singapore) was used for sample introduction and CE runs.

QDs were introduced to the MC-CE device via the capillary inlet for immobilization at the inner wall surface at the detection zone, after conditioning of the inner capillary wall. After a quality assurance check to ensure the status of coating of the MPTS network, sample extracts were introduced to the MC-CE device via the capillary inlet for determination of OPs in the vegetable extract. Details on the immobilization procedure and quality assurance check are given in the next section.

11.4.2 Immobilization of QDs at Capillary Inner Wall Surface

Recently, a range of capillary immobilization methods have been proposed and developed to enhance the selectivity of CE separation [33–35]. The immobilized coatings can enhance, eliminate, or reverse the electro-osmotic flow (EOF) and improve the detection sensitivity. In addition, the band-broadening effect that may occur due to solute–wall interactions can be greatly reduced using suitable coatings.

Various methods [21,36] had been used for the immobilization of QDs onto different substrates: covalent binding by functional reagents [24,25], electrostatic interactions with reversibly charged polymers [22,23], bioconjugated binding [26,37–39], and cross-linking by functionalized sol-gel 3D network [40,41]. However, during preliminary studies, the coatings by electrostatic interactions, covalent binding, and bioconjugated binding were found not to be sufficiently stable upon continuous laser irradiation. Therefore, the main efforts were focused on cross-linking by 3D network to produce a stable and repeatable QDs coating. The use of an MPTS-functionalized sol-gel method for immobilizing QDs onto the capillary inner wall surface was investigated. A schematic diagram showing the MPTS procedure for immobilization of QDs is shown in Figure 11.2.

The procedures for immobilization of QDs can be summarized in the following steps. (1) Conditioning of the surface of capillary inner wall. A new capillary was rinsed with 0.1 M NaOH solution and immersed in

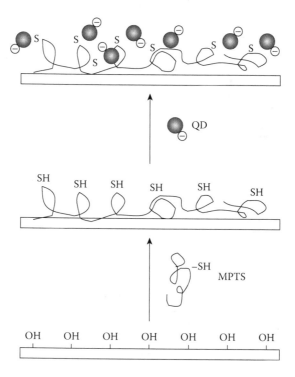

FIGURE 11.2
Schematic diagrams showing the procedures for immobilization of QDs onto the inner wall surface of the capillary via the MPTS network.

the solution overnight. (2) MPTS immobilization. A programed syringe pump (SP120, WPI, Stevenage, UK) was then used to introduce the 5×10^{-3} mol/L MPTS toluene solution into the capillary until a length of 2 cm has reached before standing for 6 h. (3) Stabilizing the 3D network of MPTS. To get rid of the physically absorbed molecules, the capillary was washed by chloroform, acetone, and deionized water before drying by N_2 for 1 h at 120°C. (4) QDs immobilization. A solution with 6.7×10^{-4} mol/L QDs was used to fill the capillary, and then allowed to stand for 2 h before being washed with deionized water and dried with N_2.

The QDs-MPTS–coated capillary exhibited a stable fluorescence when filled with the buffer, whereas the uncoated capillary showed no fluorescence. The progress for immobilizing QDs into MPTS was monitored by fluorescence intensity until a stable fluorescence was obtained using the LIF detector. Under the optimized conditions, the fluorescence intensity was changed markedly after coating of QDs by MPTS. To optimize the QDs immobilization time, the saturation-time (coating time to reach the maximum relative fluorescence intensity) was recorded while the coating capillary was filled with the following buffer: 50 mM sodium dodecyl sulfate (SDS), 3% methanol, and 30 mM borate (pH 9.6). As shown in Figure 11.3, a linear and steep increase in fluorescence intensity was observed by QDs immobilization from 10 to 90 min [32]. Afterward, the increase in fluorescence became more gradual and was flat until 180 min. Thus, 2 h is selected as the QDs immobilization saturation-time for the present work.

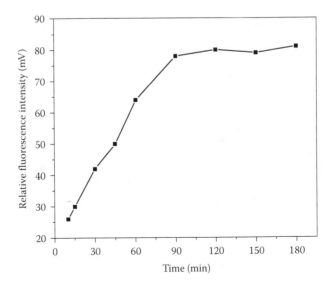

FIGURE 11.3
Saturation-time for immobilization of QDs onto the inner wall surface of the separation capillary. Buffer (pH 9.6) composition: 3% (v/v) methanol, 50 mM SDS, and 30 mM borate. (From Chen, Q.D. and Fung, Y.S. *Electrophoresis* 2010, 31, 3107–3114.)

Successful immobilization is confirmed by recording the fluorescence intensity from the immobilized inner wall surface of the capillary three times, with relative standard deviation (RSD) <3% before it is used for CE separation. The morphology of the inner capillary wall surface, revealed by scanning electron microscopy (SEM) before treatment and after successful QDs-MPTS immobilization, is shown in Figure 11.4 [32]. The bare

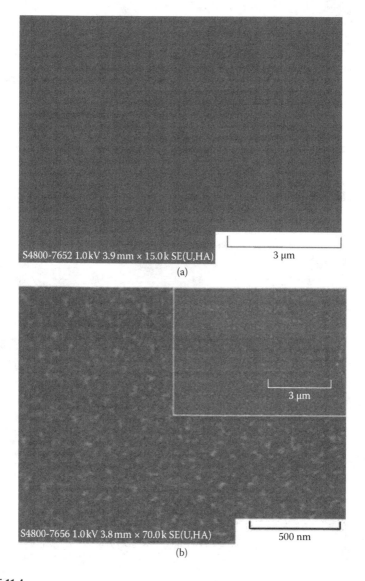

FIGURE 11.4

Morphology of the capillary inner wall surface revealed by SEM before (a) and after (b) QDs-MPTS coating. (From Chen, Q.D. and Fung, Y.S. *Electrophoresis* 2010, 31, 3107–3114.)

capillary wall surface showed a smooth surface (Figure 11.4a), whereas the QDs-MPTS coating surface (Figure 11.4b) showed a coating with evenly and compactly distributed nanoparticles on its surface. The SEM results indicated a repeatable coating upon satisfactory QDs immobilizing by the QDs-MPTS network.

In summary, the use of thermal sol-gel transition is shown to be able to immobilize QDs into a stable coating by coupling MPTS with the silane groups at the inner capillary wall surface to produce 3D organic silica-gel surface networks. At the same time, the terminal mercapto groups from MPTS can capture QDs by the formation of a stable fluorescent label beacon. The QDs-MPTS coating network produced was found to sustain a stable LIF fluorescence not affected by constituents of the CE separation buffer, with pH ranging from 8.0 to 10.0, methanol <5% (v/v), and borate buffer from 10 to 50 mM.

11.5 Application of MC-CE Device Integrated with QDs-LIF Detection for Determination of OPs in Vegetables

11.5.1 Sample Preparation and Conditioning of MC-CE Device

Details on the sample preparation procedure are given in the paper published by Chen and Fung [32], with the major steps summarized below. Fresh vegetables purchased from local markets were weighed and homogenized before ultrasonication and filtration. The extract was dried by anhydrous Na_2SO_4 and centrifuged by a high-speed centrifuge at 2,800 rpm for 30 min. The supernatant was evaporated to dryness and redissolved in a 0.5-mL solution (background electrolyte [BGE]: MeOH = 9:1 by volume ratio) for MEKC separation. For MC-CE conditioning, the QDs-immobilized capillaries were washed with deionized water (1 min) and BGE (1 min) before CE runs under 18 kV. The samples were introduced into the capillary by pressure (0.3 psi × 15 s). Recovery test was conducted by spiking OPs standard mixture to vegetable samples prior to analysis.

11.5.2 Optimizing MEKC for Separation of OPs

To optimize the working conditions for MEKC separation of the four OPs under study, QDs-LIF detection was investigated by LIF with 532 nm fluorescence and 473 nm excitation. For OPs separation, the sodium tetraborate buffer system was selected. As the use of a higher pH gives a better resolution for OPs, the optimized pH value of 9.6 was selected to achieve baseline separation of four OPs within a reasonable migration time.

The retention times of OPs increased at a higher SDS concentration; this increase was attributed to the associated increase in the viscosity of the electrolyte [40,41]. In addition, higher SDS gave to lower mobility for both micelles and OPs due to the adsorption of SDS onto the inner wall surface of the capillary, efficient solubilization of OPs, and enhanced solute–micelle interaction upon increasing micellar phase volume, leading to the lengthening of the retention times [42]. However, the use of a higher SDS concentration can increase the retention factors of the analytes, and consequently an improved stacking efficiency [43]. To maintain a compromise between resolution and sensitivity, SDS at 50 mM was selected to obtain a satisfactory sensitivity with an acceptable resolution.

However, at higher borate concentrations, a rapid increase in the OPs migration times was observed. This increase may be attributed to the effect of a higher ionic strength of the buffer solution at higher borate concentration [40,41,44]. In addition, when the borate concentration was increased to >30 mM, peak height was reduced. Therefore, borate at 30 mM was selected.

The results showed an obvious increase in retention times for all solutes by increasing methanol concentration. However, the EOF and mobility of micelles were found to reduce at the same time. This effect may due to the increase in viscosity of the electrolyte [45] as the relative viscosity of the aqueous organic solvent is increased at a higher concentration of the organic modifier [46]. As the concentration range of methanol was small and within 3%, it would not have any significant effect on the retention time.

Thus, the composition of the optimized MEKC separation buffer at pH 9.6 for OPs separation is as follows: 50 mM SDS, 3% methanol, and 30 mM borate. The performance and application of the MC-CE device developed using the optimized buffer are given in the next section.

11.5.3 Analytical Performance and Applicability Study

Figure 11.5 shows typical MEKC electropherograms, showing baseline separation of the four standard OPs (10 ppm each) for QDs-LIF detection [32]. A significant increase of the fluorescence signals was found due to the interaction between the immobilized QDs with the separated OPs. All four OPs—phosalone, methidathion, diazinon, and mevinphos—were baseline separated in <12 min. The results were shown to deliver LODs from 0.05 to 0.18 ppm (signal-to-noise ratio [S/N] = 2, n = 3), linear ranges from 30 to 0.1 ppm, and repeatability (%RSD, n = 3) for retention time 0.75 to 0.42% and for peak height 5.7 to 2.9% for the four OPs investigated. The linear ranges for all the four OPs were shown to be sufficient to meet the MRL values given in the guidelines for tomato and other vegetable samples issued from the Codex Alimentarius Commission.

According to the recent requirements of the U.S. Food and Drug Administration [47], the whole vegetable should be used for sample preparation in pesticide determination. This sample type could give rise to a more complex sample and hence a more rigorous and time-consuming sample preparation procedure to remove sample matrix interferents. As a selective

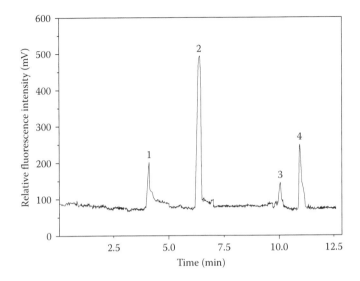

FIGURE 11.5
MEKC electropherograms for separation of a standard mixture of OPs under optimized conditions. Capillary: 43.5/45 cm × 50 μm i.d., with detection zone of QDs immobilized at 2 cm to capillary end. Run voltage: 18 kV. Injection: 0.3 psi × 15 s. Separation buffer (pH 9.6, 3% methanol [v/v]) contains 30 mM borate and 50 mM SDS. QD/LIF detection = 532 nm fluorescence/473 nm excitation. OPs (10 ppm each): phosalone (4), diazinon (3), methidathion (2), and mevinphos (1).

interaction between OPs and QDs was utilized in tandem with sensitive LIF detection, a simple sample pretreatment procedure is sufficient for OPs determination by the MC-CE/QD-LIF device. Only centrifugation and organic solvent extraction were used, with no interfering substances detected. Therefore, the time-consuming SPE procedure is not required. With the use of a rapid and simple sample preparation procedure, the device is able to detect OPs in vegetables on-site, for example, at the border when they are entering Hong Kong, with results known within a couple of hours and thus in time to stop the distribution of contaminated vegetables to the public.

Figure 11.6 shows MEKC electropherograms of the control (nonspiked sample) and tomato samples spiked with a standard mixture containing four OPs at 0.5 mg/kg each for recovery test [32]. The recoveries ranged from 88.7 to 96.1% (%RSD, $n = 3$), with repeatability from 2.96~4.5% for the four OPs investigated, indicating that the loss of analyte was within variation of the MEKC-QDs/LIF method for whole vegetable analysis. To meet the Codex Alimentarius Commission requirements for OPs detection in tomato samples (1, 0.5, 0.1, and 1.0 ppm for mevinphos, methidathion, phosalone, and diazinon, respectively) [1], the detection limits of OPs using the MC-CE/QDs-LIF method are >10 times lower than the MRLs, hence satisfying the Codex Alimentarius Commission requirements.

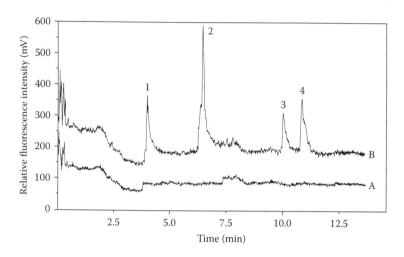

FIGURE 11.6

MEKC electropherograms for tomato samples without spiked (A) and with spiked (B) by a standard mixture of OPs. Capillary: 43.5/45 cm × 50 μm i.d., with detection zone (2 cm to exposed end) coated with QDs. Run voltage, 18 kV. Injection: 0.3 psi × 15 s. BGE (pH 9.6): 50 mM SDS, 3% methanol, and 30 mM borate. QDs/LIF = 532 nm fluorescence/473 nm excitation. Spiked OPs (0.5 ppm each): phosalone (4), diazinon (3), methidathion (2), and mevinphos (1). (From Chen, Q.D. and Fung, Y.S. *Electrophoresis* 2010, 31, 3107–3114.)

11.6 Summary

A MC-CE device integrated with immobilized QDs for LIF determination of residual pesticides in vegetable was fabricated and tested for application to determine OPs in tomato samples. A new two-step method by sol-gel 3D network was developed to immobilize the water-soluble core-shell QDs by QDs-MPTS coating onto the capillary inner wall surface at the detection zone, with results showing a great enhancement in the LIF detection sensitivity for OPs using diode laser QDs fluorescence at 532 nm and excitation at 473 nm.

The sample preparation process prior to sample injection to the MC-CE device is convenient and simple, using procedures such as solvent extraction and centrifugation commonly available in the laboratory. Based on the specific fluorescence intensity enhancement interaction between OPs and QDs and the significant enhancement in detection sensitivity by immobilized QDs technology at the detection zone, a time-consuming pretreatment process such as SPE is not necessary using the developed MC-CE/QDs-OPs device, making it to offer a simple and quick method with capability for on-site determination of OPs extracted from vegetables to meet the demanding WHO requirements.

For applicability study, four commonly used OPs were determined in tomato samples, as the fluorescence intensity of CdTe/CdS core-shell QDs capped with MPA was found to increase with increasing OPs concentration. The optimum conditions are as follow: run voltage at 18 kV; sample injection by pressure at 0.3 psi × 15 s; QDs-LIF detection at room temperature by 532 nm emission under excitation at 473 nm; and the optimized BGE containing 30 mM SDS, 3% MeOH, and 50 mM borate (pH 9.6). All four OPs showed baseline separation within 12 min in the MEKC electropherograms. The recoveries of spiked OPs to tomato samples (%RSD, $n = 3$) ranged from 88.7 to 96.1% and repeatability from 2.96 to 4.5% for the four OPs investigated, verifying the applicability of the device fabricated for real vegetable application.

In summary, the fabricated MC-CE/QDs-LIF device satisfies the requirement of a quick, selective methodology for determining trace OPs in complex a food matrix to meet the demanding MRLs specified by the Codex Alimentarius Commission for control of OPs in vegetables.

References

1. FAO/WHO Food Standards Programme and Codex Alimentarius Commission. http://www.codexalimentarius.net/
2. Kaur, I., Mathur, R.P., Tandon, S.N. and Dureja, P. Identification of metabolites of malathion in plant, water and soil by GC-MS. *Biomed Chromatogr* 1997, 11, 352–355.
3. Tolosa, I., Douy, B. and Carvalho, F.P. Comparison of the performance of graphitized carbon black and poly(styrene–divinylbenzene) cartridges for the determination of pesticides and industrial phosphates in environmental waters. *J Chromatogr A* 1999, 864, 121–136.
4. Psathaki, M., Manoussaridou, E. and Stephanou, E.G. Determination of organophosphorus and triazine pesticides in ground- and drinking water by solid-phase extraction and gas chromatography with nitrogen-phosphorus or mass spectrometric detection. *J Chromatogr A* 1994, 667, 241–248.
5. Aguilar, C., Borrull, F. and Marcé, R.M. Identification of pesticides by liquid chromatography–particle beam mass spectrometry using electron ionization and chemical ionization. *J Chromatogr A* 1998, 805, 127–135.
6. Schenck, F.J., Wagner, R., Hennessy, M.K. and Okrasinski, J.L., Jr. Screening procedure for organochlorine and organophosphorous pesticide residues in eggs using a solid-phase extraction cleanup and gas chromatographic detection. *J AOAC Int* 1994, 77, 1036–1040.
7. Bennett, D.A., Chung, A.C. and Lee, S.M. Multiresidue method for analysis of pesticides in liquid whole milk. *J AOAC Int* 1997, 80, 1065–1077.
8. Baynes, R.E. and Bowen, J.M. Rapid determination of methyl parathion and methyl paraoxon in milk by gas chromatography with solid-phase extraction and flame photometric detection. *J AOAC Int* 1995, 78, 812–815.
9. Muccio, A.D., Pelosi, P., Camoni, I., Barbini, D.A., Dommarco, R., Generali, T. and Ausili, A. Selective, solid-matrix dispersion extraction of organophosphate pesticide residues from milk. *J Chromatogr A* 1996, 754, 497–506.

10. Gillespie, A.M., Daly, S.L., Gilvydis, D.M., Schneider, F. and Walters, S.M. Multicolumn solid-phase extraction cleanup of organophosphorus and organo-chlorine pesticide residues in vegetable oils and butterfat. *J AOAC Int* 1995, 78, 431–436.

11. Pico, Y., Rodriguez, R. and Manes, J. Capillary electrophoresis for the determination of pesticide residues. *Trends Anal Chem* 2003, 22, 133–151.

12. Ravelo-Perez, L.M., Hernandez-Borges, J. and Rodriguez-Delgado, M.A. Pesticides analysis by liquid chromatography and capillary electrophoresis. *J Sep Sci* 2006, 29, 2557–2577.

13. Kodama, S., Saito, Y., Chinaka, S., Yamamoto, A. and Hayakawa, K. Chiral capillary electrophoresis of agrochemicals in real samples. *J Health Sci* 2006, 52, 489–494.

14. Juan-Garcia, A., Font, G. and Pico, Y. On-line preconcentration strategies for analyzing pesticides in fruits and vegetables by micellar electrokinetic chromatography. *J Chromatogr A* 2007, 1153, 104–113.

15. Waters, L.C., Jacobson, S.C., Kroutchinina, N., Khandurina, J., Foote, R.S. and Ramsey, J.M. Multiple sample PCR amplification and electrophoretic analysis on a microchip. *Anal Chem* 1998, 70, 5172–5176.

16. Colyer, C.L., Tang, T., Chiem, N. and Harrison, D.J. Clinical potential of microchip capillary electrophoresis systems. *Electrophoresis* 1997, 18, 1733–1741.

17. Qin, J.H., Leung, F.C., Fung, Y.S., Zhu, D.R. and Lin, B.C. Rapid authentication of ginseng species using microchip electrophoresis with laser-induced fluorescence detection. *Anal Bioanal Chem* 2005, 381, 812–819.

18. Liu, J., Suzuki, O., Kumazawa, T. and Seno, H. Rapid isolation with Sep-Pak C18 cartridges and wide-bore capillary gas chromatography of organophosphate pesticides. *Forensic Sci Int* 1989, 41, 67–72.

19. Futagami, K., Narazaki, C., Kataoka, Y., Shuto, H. and Oishi, R. Application of high-performance thin-layer chromatography for the detection of organophosphorus insecticides in human serum after acute poisoning. *J Chromatogr B* 1997, 704, 369–373.

20. Sharma, V.K., Jadhav, R.K., Rao, G.J., Saraf, A.K. and Chandra, H. High performance liquid chromatographic method for the analysis of organophosphorus and carbamate pesticides. *Forensic Sci Int* 1990, 48, 21–25.

21. Wang, L., Wang, X., Xu, M., Chen, D. and Sun, J. Layer-by-layer assembled microgel films with high loading capacity: Reversible loading and release of dyes and nanoparticles. *Langmuir* 2008, 24, 1902–1909.

22. Smith, J.C., Lee, K.B., Wang, Q., Finn, M.G., Johnson, J.E., Mrksich, M. and Mirkin, C.A. Nanopatterning the chemospecific immobilization of cowpea mosaic virus capsid. *Nano Lett* 2003, 3, 883–886.

23. Shavel, A., Gaponik, N. and Eychmüller, A. Covalent linking of CdTe nanocrystals to amino-functionalized surfaces. *Chem Phys Chem* 2005, 6, 449–451.

24. Mamedov, A., Belov, A., Giersig, M., Mamedova, N. and Kotov, N. Nanorainbows: Graded semiconductor films from quantum dots. *J Am Chem Soc* 2001, 123, 7738–7739.

25. Sinani, V., Koktysh, D., Yun, B., Matts, R., Pappas, T., Motamedi, M., Thomas, S. and Kotov, N. Collagen coating promotes biocompatibility of semiconductor nanoparticles in stratified LBL films. *Nano Lett* 2003, 3, 1177–1182.

26. Sapsford, K.E., Pons, T., Medintz, I.L. and Mattoussi, H. Biosensing with luminescent semiconductor quantum dots. *Sensors* 2006, 6, 925–953.

27. Gu, Z.Y., Zou, L., Zhu, W.H. and Zhong, X.H. One-pot synthesis of highly luminescent CdTe/CdS core/shell nanocrystals in aqueous phase. *Nanotechnology* 2008, 19, 135604–135610.

28. Gong, Y.J., Gao, M.Y., Wang, D.Y. and Mohwald, H. Incorporating fluorescent CdTe nanocrystals into a hydrogel via hydrogen bonding: Toward fluorescent microspheres with temperature-responsive properties. *Chem Mater* 2005, 17, 2648–2653.

29. Devreese, J.T., Fomin, V.M., Gladilin, V.N. and Klimin, S.N. Photoluminescence spectra of quantum dots: Enhanced efficiency of the electron-phonon interaction. *Phys Stat Sol* 2001, 224, 609–612.

30. Wan, Y., Wang, L., Lin, Z., Chen, Q., Zhang, H., Yang, B., Su, X. and Jin, Q. Studies on quantum dots synthesized in aqueous solution for biological labeling. *Can J Anal Sci Spect* 2004, 49, 99–104.

31. Aldana, J., Lavelle, N., Wang, Y. and Peng, X. Size-dependent dissociation pH of thiolate ligands from cadmium chalcogenide nanocrystals. *J Am Chem Soc* 2005, 127, 2496–2504.

32. Chen, Q.D. and Fung, Y.S. Capillary electrophoresis with immobilized quantum dot fluorescence detection for rapid determination of organophosphorus pesticides in vegetables. *Electrophoresis* 2010, 31, 3107–3114.

33. Xu, L. and Lee, H.K. Preparation, characterization and analytical application of a hybrid organic–inorganic silica-based monolith. *J Chromatogr A* 2008, 1195, 78–84.

34. Zheng, F. and Hu, B. MPTS-silica coated capillary microextraction on line hyphenated with inductively coupled plasma atomic emission spectrometry for the determination of Cu, Hg and Pb in biological samples. *Talanta* 2007, 73, 372–379.

35. Liu, Y., Shu, C. and Lamb, J.D. Characterization of bitline stress effects on flash cell after program/erase cycle. *J Capillary Electrophoresis* 1997, 4, 97–103.

36. Hezinger, A.F.E., Teßmar, J. and Göpferich, A. Polymer coating of quantum dots—A powerful tool toward diagnostics and sensorics. *Eur J Pharm Biopharm* 2008, 68, 138–152.

37. Klostranec, J.M. and Chan, W.C.W. Quantum dots in biological and biomedical research: Recent progress and present challenges. *Adv Mater* 2006, 18, 1953–1964.

38. Zhang, Q., Zhang, L., Liu, B., Lu, X. and Li, J. Assembly of quantum dots-mesoporous silicate hybrid material for protein immobilization and direct electrochemistry. *Biosens Bioelectron* 2007, 23, 695–700.

39. Zajac, A., Song, D.S., Qian, W. and Zhukov, T. Protein microarrays and quantum dot probes for early cancer detection. *Colloid Surface B* 2007, 58, 309–314.

40. Yang, P., Li, C. and Murase, N. Highly photoluminescent multilayer QD-glass films prepared by LbL self-assembly. *Langmuir* 2005, 21, 8913–8917.

41. Cai, J. and Rassi, Z.E. Micellar electrokinetic capillary chromatography of neutral solutes with micelles of adjustable surface charge density. *J Chromatogr* 1992, 608, 31–45.

42. Bechet, I., Fillet, M., Hubert, P. and Crommen, J. Determination of benzodiazepines by micellar electrokinetic chromatography. *Electrophoresis* 1994, 15, 1316–1321.

43. Wang, S.F., Wu, Y.Q., Ju, Y., Chen, X.G., Zheng, W.J. and Hu, Z.D. On-line concentration by field-enhanced sample injection with reverse migrating micelles in micellar electrokinetic capillary chromatography for the analysis of flavonoids in *Epimedium* brevicornum Maxim. *J Chromatogr A* 2003, 1017, 27–34.

44. Garrison, A.W., Schmitt, P. and Kettrup, A. Separation of phenoxy acid herbicides and their enantiomers by high-performance capillary electrophoresis. *J Chromatogr A* 1994, 688, 317–327.
45. Chen, N., Terabe, S. and Nakagawa, T. Effect of organic modifier concentrations on electrokinetic migrations in micellar electrokinetic chromatography. *Electrophoresis* 1995, 16, 1457–1462.
46. Pyell, U. and Bütehorn, U. Optimization of resolution in micellar electrokinetic chromatography via computer-aided simultaneous variation of concentrations of sodium dodecyl sulfate and urea as modifier. *J Chromatogr A* 1995, 716, 81–95.
47. Food and Drugs Administration, U.S. Government, U.S.A., Investigations Operation Manual, Sample, Schedule Chart 3, 2010, 188–189. http://www.fda.gov/downloads/ICECI/Inspections/IOM/UCM123523.pdf

Section IV

Integration to Achieve Intended Application

12

On-Site Assay by Portable UV Detector:
MC-CE Device for Authenification
of Chinese Medicine

Zhou Nie

Guangzhou Amway (China) Co., Ltd.
Guangzhou, People's Republic of China

Ying Sing Fung

The University of Hong Kong
Hong Kong SAR, People's Republic of China

CONTENTS

12.1 Introduction

The increasing use of herbs for medicinal purpose in recent years leads to the requirement for assessing their quality to assist international trading of herbal products. Herbs are complex products with variable contents, depending on sources and season for their harvesting. The analytical issue for assessing the quality of herbal medicine is addressed in this chapter and current practices using fingerprinting are reviewed in Section 12.1.1.

The role of MC-CE device for quality assessment of herbal medicine is discussed in light of requirements and practice carried out in the trade. The operation of MC-CE devices for fingerprinting of herbal medicine in the laboratory as well as in the field are presented with results given to assess its merits and limitation for authentication of herbal medicine in Hong Kong to meet legal requirements and for in-house quality assessment.

12.1.1 Assessing Quality of Herbal Medicine by Fingerprinting

Herbs have a long history of medicinal use in China and elsewhere, and they have offered effective treatment for many illness and medical conditions for which traditional, current medical treatments do not deliver a satisfactory solution. Although the export of herbal preparations in the form of drugs outside China is limited, such preparations are sold in other forms, such as dietary supplements, or directly as medicinal plants for international trade for use in herbal preparations. The reason that most herbal preparations or drugs cannot be accepted for trade is the lack of quality control (QC) and quality assurance (QA) programs to demonstrate the consistency of their content according to a specific formulation during preparation.

Most herbal drugs are complex mixtures, typically 6 to 8, but sometimes >10, herbal ingredients. An herbal drug has often undergone vigorous thermal procedures during the manufacturing process which may involve roasting, high-temperature extraction, blending, and other procedures that can change the composition of the final preparation compared to the original constituent herbs. In addition, there may be variability in the raw materials, an important factor to consider for maintaining the consistency of a given herbal preparation. The composition of a given herb is highly dependent on conditions such as exposure to rainfall, sun, and soil during growth and before it is harvested for use as herbal medicine [1–3].

To satisfy the increasing demand and regulations for the consistency of pharmaceutical products produced as herbal medicine, considerable analytical work has to be conducted on the characterization of individual herbs. First, variability due to the botanical source or time of harvest of individual herbs has to be assessed to establish the specifications for medicinal use. Second, the makeup of the effective ingredients of a given herbal preparation in an established formulation must be ascertained by QC and QA measures. This chapter focuses on use of a portable microfluidic chip-capillary electrophoreses (MC-CE) device to assess the quality of a commercially available herbal preparation to ensure that its makeup is in accordance with specific formulations consisting of several constituent herbs.

The formulation of a typical herbal drug often contains a mixture of six or more herbs. Thus, it is a more complicated process to assess the consistency of the final preparation than the quality of a single individual herbal constituent. The pharmaceutical action of herbal drugs is typically a combined action of all constituent herbs, with some herbs contributing to the medical effect and the others compensating for their side effects, such as toxicity of potent herbs. Thus, the determination of the composition of individual herbs for a given herbal preparation is essential for QA of the products prepared by the manufacturer and for regulatory control of herbal medicine by government agencies.

The fingerprinting approach, based on various analytical methods, has been advocated to provide a tool for assessing the quality of herbal drugs to ensure product consistency. The Hong Kong Special Administrative Region (SAR) Government implemented a 10-year modernization program, started 1997, that required all herbal manufacturers in Hong Kong to set up QA measures to ensure product consistency [4]. By the end of the 10-year plan starting in 2007, all herbal medicinal manufacturers in Hong Kong were required to submit results and methodology to the government as proof for their QC/QA efforts. Fingerprinting has been recognized as the best way to provide the assurance for product quality and consumer protection [5].

12.1.2 Analytical Methods for Fingerprinting of Herbal Medicine

Various analytical techniques have been studied for characterization of herbal drugs, including direct methods such as infrared [6–10], nuclear magnetic resonance [11], X-ray fluorescence [12], and various DNA phenotyping techniques [13–15], as well as indirect separation methods. Although direct methods provide quick results, they are restricted to specific applications. The various indirect separation methods provide a more general methodology for chromatographic fingerprinting of herbal medicines. The Chinese Food and Drug Administration and the Hong Kong SAR Government have accepted the use of chromatographic fingerprinting techniques as suitable methods for assessing the variability of traditional Chinese medicine (TCM).

Chromatographic fingerprinting techniques include gas chromatography (GC) [16–19], thin-layer chromatography (TLC) [20–23], high-performance liquid chromatography (HPLC) [2,24–28], and CE [29–37]. Although the resolution of capillary GC is high and there are many sensitive GC detectors available for qualitative and quantitative analyses, it can only separate volatile constituents extracted from herbal drugs. This provides a severe limitation for the analysis of herbal drugs, because >80% of their constituents are nonvolatile [16]. Although TLC and high-performance TLC have the advantage for keeping the plates after development as a future record, their resolution is highly insufficient. Due to the demand for high-purity organic solvent, the maintenance costs associated with high-pressure pumps, and the limited lifetime of expensive separation columns, the cost of HPLC for routine monitoring is high despite it giving better resolution compared to TLC alone. Moreover, its resolution is less than that of GC.

Traditional CE and the recently developed MC-CE device provide promising applicability due to the high separation efficiency, short analysis time, low capital costs for using high-voltage source as the driving force, and a large reduction in the running cost due to the extremely low reagent consumption for using microliter amounts of electrophoretic buffer during a CE run. The lab-on-a-chip design adopted by an MC-CE device provides an analytical tool with the capability to work on-site in the field for chemical characterization of a complex herbal mixture, making it particularly suitable for QC/QA application. Details on the application of MC-CE device for chromatographic fingerprinting of a selected herbal preparation are given in the following sections.

Current work on chromatographic fingerprinting for herbal analysis concentrates on the identification of a specific botanical source of a given herb by determination of known active principles for variability assessment. The approach works well for unprocessed herbs or for a single-herb product in its original form. However, for multiherbal preparations, the situation becomes more complicated as the number of herbs in the formulation increases and most of the constituents have unknown chemical composition and specific actions. Moreover, the herbal mixture has often undergone thermal or other mechanical processing during preparation that can alter chemical characteristics, or convert the chemicals into other forms prior to medicinal use. A suitable method is lacking at present for assessment of the variability of a multiherbal preparation.

12.1.3 Herbal Preparation Selected for the Present Investigation

As nearly all over-the-counter prepackaged TCM preparations sold in Hong Kong pharmaceutical stores are multiherbal in composition, we focused our attention on developing a high-resolution, low-cost CE and MC-CE device for fingerprinting of multiherbal preparations in the lab and factory, respectively, to quantify their original ingredients compared to a given formulation in an attempt to develop a methodology to satisfy the requirements of the Hong Kong SAR Government for QA of herbal products.

To demonstrate the applicability of CE and MC-CE devices developed for chromatographic fingerprinting of a complex multiherbal preparation, a well-established herbal formulation, *Liu Wei Di Huang Wan*, is selected. It is an herbal preparation made up of six ingredients in different amounts [38]. It is taken to improve general body conditions such as vertigo and dizziness; to relieve weakness and soreness of waist and knees; and to reduce tiredness, fatigue, and feverish sensations in palms and soles. Commercial products based on *Liu Wei Di Huang Wan* have been marketed by pharmaceutical companies in China, Hong Kong, Japan, and other Asian countries. Recently, it has been used to treat patients with asthma to relieve the direct effect on cytokine gene expression from activated human peripheral blood mononuclear cells [39].

12.1.4 Methodology for Fingerprinting of *Liu Wei Di Huang Wan* by CE and Portable MC-CE Device

The methodology developed for fingerprinting of *Liu Wei Di Huang Wan* by MC-CE device includes the following steps: (1) a suitable extraction and sample preparation procedure; (2) optimization of the CE separation procedure for its applicability in fingerprinting in the laboratory; (3) design and fabrication of MC-CE device with portable ultraviolet (UV) detector; and (4) optimization of MC-CE operations for its applicability in finger-printing in the field.

12.2 Procedures and Operations

12.2.1 Reagent and Buffer Preparation

Sodium tetraborate, sodium dodecyl sulfate (SDS), sodium phosphate, and all organic solvents were obtained from Aldrich Chemical Company (Milwaukee, WI, USA). Methanol, ethanol, propanol, and acetonitrile were used directly as received without further purification. The running buffer was adjusted to pH 9.3 by the addition of 0.1 M NaOH. All buffer solutions were prepared daily, filtered by 0.45-μm membranes, and degassed before use. All chemicals were purchased as analytical grade.

12.2.2 Preparation of *Liu Wei Di Huang Wan* Herbal Medicine

Two batches of the individual dried herbs of *Liu Wei Di Huang Wan* were purchased from Tong Ren Tang Pharmaceutical shop in Hong Kong in

2005 and 2007 for study using CE and MC-CE. The names of the individual herbs are as follows, listed as pharmaceutical Latin name (Chinese herbal name): *Cortex Moutan* (Mu dan pi), *Radix Rehmanniae* (Di huang), *Fructus Corni* (Shan zhu yu), *Poria* (Fu ling), *Rhizoma Alismatis* (Ze xie), and *Rhizoma Dioscoreae* (Shan yao). The six dried herbs were ground to a powder and mixed thoroughly according to the well-established formula of *Liu Wei Di Huang Wan* in the following ratios by weight: 3:8:4:3:3:4 for *C. Moutan:R. Rehmanniae:F. Corni:Poria:R. Alismatis:R. Dioscoreae*, respectively [39].

According to the above-mentioned formula, a test sample of *Liu Wei Di Huang Wan* was made up to 5 g using the following weight in grams for the same order of the six herbs named above: 0.6:1.6:0.8:0.6:0.6:0.8. To test the extraction efficiency of different solvent systems, 25 mL of each solvent system was added to the test sample in a beaker, and then the sample was sonicated for 50 min at room temperature and centrifuged at 10,000 rpm for 15 min prior to separation of the supernatant liquid from the solid residue. The supernatant liquid was then mixed with the running buffer at 1:4 (liquid:buffer) volume ratios to make up the laboratory samples. Benzoic acid was used as the internal standard and was added to the laboratory sample to a concentration of 1.0×10^{-4} M. A flowchart showing the preparation procedure for *Liu Wei Di Huang Wan* herbal mixture is given in Figure 12.1.

12.2.3 Extraction Procedures

For effective fingerprinting, the percentage of extraction for given chemical components from an herbal preparation should be as high as possible. As herbal medicines are normally prepared by aqueous extraction, the aqueous extracts of the six ingredients of *Liu Wei Di Huang Wan* were studied. Strong UV absorption was shown by the aqueous extract of *C. Moutan, R. Rehmanniae,* and *F. Corni,* whereas *Poria, R. Alismatis,* and *R. Dioscoreae* showed poor UV absorption. Most of the UV spectra of the extracts showed more or less a flat absorption plateau from 200 to 300 nm, with no other feature except for *Poria, R. Alismatis,* and *R. Dioscoreae* spectra that showed a broad and strong absorption peak at 210 nm. Thus, the detection wavelength was selected at 210 nm to study the extraction efficiency using different solvent systems.

To improve the extraction efficiency, ultrasonication was used during extraction. Organic solvent systems commonly used for herbal extraction were investigated for their extraction efficiency based on UV absorption at 210 nm. A methanol:water solution (1:1, v/v) was found to give the best extraction efficiency compared to other organic solvent systems using ethanol, propanol, and acetonitrile, as it showed a consistently high absorption for all six ingredients with maximum absorbance. Thus, a methanol:water solution was selected as the solvent system for the extraction of *Liu Wei Di Huang Wan* with its six ingredients mixed in weight ratios according to the traditional recipe of *C. Moutan* given in Section 12.2.2.

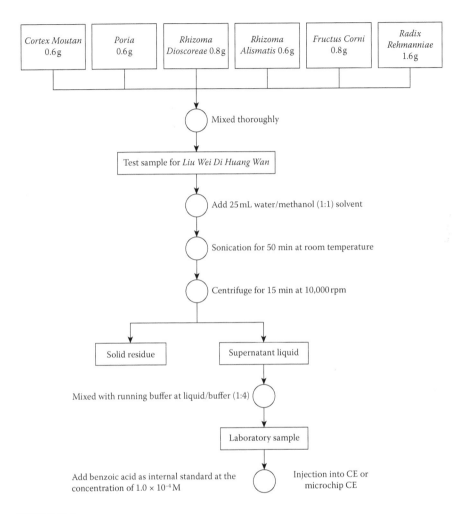

FIGURE 12.1
Flowchart for preparation of *Liu Wei Di Huang Wan*.

To test the homogeneity of the powders for the determination of the minimum amount of ingredients to be sampled, the ingredient with the weakest UV absorption (*R. Dioscoreae*) was selected. It was weighted to 0.6 g and was extracted by 25 mL methanol:water (1:1, v/v) under ultrasonication. The sample preparation procedure as given in Figure 12.1 was then followed. The difference was that the total dissolved matter was determined instead of injection into CE for separating individual constituents for quantitation. The relative standard deviation (RSD) of the dry weight after extraction is 0.74% (*n* = 3). The variation was considered to be small compared to the analytical error. Thus, the minimum sample size of 0.6 g was adopted from each of the six ingredients of *Liu Wei Di Huang Wan* for subsequent studies.

12.2.4 Operation Procedures for CE and MC-CE Device

The CE-P2 capillary electrophoresis system (CE Resources Pte Ltd., Singapore) was used to conduct a CE run. A capillary column obtained from the Yongnian Optical Fiber Factory (Yongnian, Hebei, People's Republic of China) was used directly without coating. A typical capillary is 60 cm in length, with an inner diameter of 50 μm. An on-column detection window was created at 45 cm from the injection end. The new capillary was treated by 1 M NaOH for 1 h, and then washed by distilled water for another hour prior to use. During routine analysis, a washing cycle consisting of a 2-min wash by 0.1 M NaOH, 2-min wash by distilled water, and 5-min wash by running buffer was conducted between each CE run. The injection of sample was performed by pressure mode at 0.3 psi for 15 s. The CE run was performed at room temperature ($25 \pm 1°C$) under a separation voltage of +15 kV for detection at 210 nm.

12.3 Optimization of CE Separation Procedures for Fingerprinting of *Liu Wei Di Huang Wan*

For optimization of the CE separation procedure for fingerprinting of *Liu Wei Di Huang Wan*, the electropherogram should be repeatable and contain as many peaks as possible, with a stable baseline under operational conditions. Moreover, for a small change in the concentration of each of the six ingredients of *Liu Wei Di Huang Wan*, detectable features should be shown in the electropherogram from the final product to demonstrate that the CE technique could be used for QC and QA of each ingredient.

12.3.1 Effect of Methanol Content

Methanol was added to the buffer to enhance sample resolution in CE [40] by increasing the viscosity of the buffer, thereby modifying the charge and hydrophobicity of the capillary inner wall, and leading to a reduction in electro-osmotic flow. In addition, it could increase the solubility of nonpolar analytes extracted from herbal medicines and be compatible with the extraction solution containing 50% methanol by volume. The effect of methanol on the separation of *Liu Wei Di Huang Wan* is given in Figure 12.2; the separation between the 24 diagnostic peaks was increased at a higher methanol concentration, in particular for methanol concentration >15% (v/v). As a compromise, 15% (v/v) methanol was used to achieve satisfactory separation efficiency within a shorter separation time.

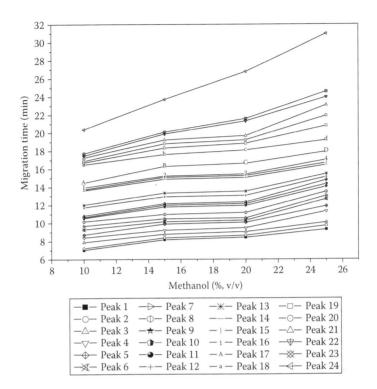

FIGURE 12.2
Effect of methanol on the separation of *Liu Wei Di Huang Wan*. CE conditions were buffer composed of 0.01 M borate and 0.05 M SDS, with pH adjusted to 9.3; run, +15 kV; detection, 210 nm; injection, 0.3 psi for 15 s; and capillary, 0.05 mm i.d × 60 cm (window 45 cm).

12.3.2 Effect of SDS Concentration

Different concentrations of SDS were found to affect the resolution for the separation of *Liu Wei Di Huang Wan*, with results shown in Figure 12.3. The results indicated that the same trend was observed as for methanol, with a longer migration time at a higher SDS concentration. When the SDS concentration was >50 mM, the last peak 24 disappeared. Thus, 50 mM SDS was selected for subsequent studies.

12.3.3 Effect of Borate Concentration

The effect of borate on the electropherogram for the separation of *Liu Wei Di Huang Wan* is shown in Figure 12.4. Borate exhibited a strong effect on the separation of ingredients of *Liu Wei Di Huang Wan*. The separation between the 24 diagnostic peaks increased at high borate concentrations. However, high concentration of borate can cause high current and peak broadening. Moreover, with an increase to 40 mM borate, the last peak 24 disappeared. As a compromise, 10 mM borate was selected for subsequent studies.

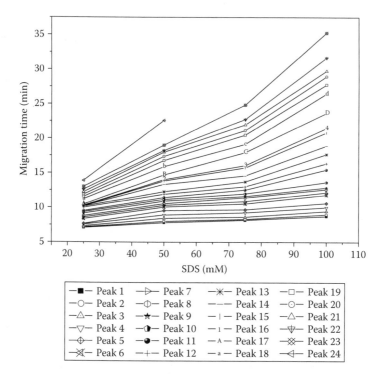

FIGURE 12.3
Effect of SDS on the separation of *Liu Wei Di Huang Wan*. CE conditions were buffer composed of 0.01 M borate and methanol:water, 15% (v/v), with pH adjusted to 9.3. Other conditions are the same as in Figure 12.2.

12.3.4 Effect of pH

The effect of pH on the migration time is shown in Figure 12.5. Migration time was shown to increase at a lower pH. As the original pH of the aqueous buffer containing 10 mM borate and 50 mM SDS was 9.3, this buffer was selected as the background electrolyte to reduce the frequency for pH adjustment. In addition, it gave a total analysis time <30 min, and the migration times for 24 diagnostic peaks were fairly constant at pH >8.0.

12.3.5 Effect of CE Operation Conditions

The effect of the running high voltage on the migration time is shown in Figure 12.6. The migration time was found to decrease at a higher voltage. However, the separation between different diagnosed peaks was reduced at a higher voltage. In consideration of the above-mentioned factors and the heating effect at higher voltages, 15 kV was selected to achieve satisfactory separation efficiency within a reasonable separation time. Although an increase in the injection time led to a higher sensitivity, peak broadening

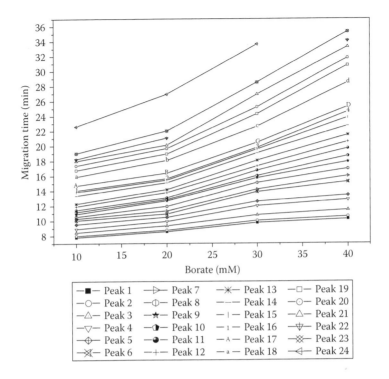

FIGURE 12.4

Effect of borate on the separation of *Liu Wei Di Huang Wan*. CE conditions were buffer composed of 0.05 M SDS and methanol/water, 15% (v/v), with pH adjusted to 9.3. Other conditions are the same as in Figure 12.2.

occurred at a longer injection time, leading to poor separation. Thus, the injection time was selected at 0.3 psi for 15 s.

A summary of the optimized conditions for CE separation of *Liu Wei Di Huang Wan* is as follows: pressure injection at 0.3 psi for 15 s; running voltage at +15 kV; and the running buffer containing 10 mM borate, 50 mM SDS, and 15% methanol (v/v), with pH adjusted to 9.3.

12.4 Results for Fingerprinting of *Liu Wei Di Huang Wan* by CE Methods

12.4.1 Qualitative Fingerprinting by Electropherogram

Under the optimized CE conditions, 24 diagnostic peaks were identified from the electropherogram of *Liu Wei Di Huang Wan* (Figure 12.7). Before application of CE for fingerprinting, the variability of these 24 peaks from

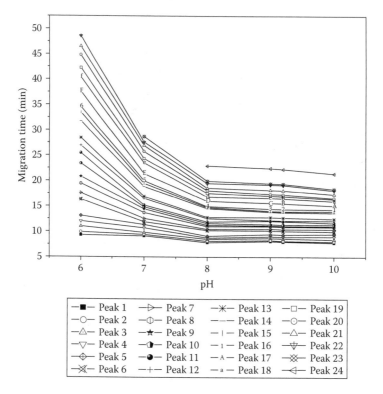

FIGURE 12.5

Effect of pH on the separation of *Liu Wei Di Huang Wan*. CE conditions were buffer composed of 0.01 M borate, 0.05 M SDS, and methanol/water, 15% (v/v). Other conditions are the same as in Figure 12.2.

the same batch of sample had to be assessed. Using the sample preparation procedure given in Figure 12.1, the repeatability of the migration time and peak height was determined (Tables 12.1 and 12.2, respectively). To reduce the variability and improve the repeatability of the results, the migration time and peak height were normalized to those from the internal standard benzoic acid, an acid that showed a peak with suitable intensity and migration time clearly apart from the diagnostic peaks.

Migration time was normalized using the internal standard peak recorded in the electropherogram (peak * in Figure 12.7). The relative migration time of all other peaks were calculated relative to the internal standard peak by the following formula:

$$t_n' = |t_n - t_*| \tag{12.1}$$

where t_n' is the relative migration time of peak n, t_n is the original migration time of peak n, and t_* is the original migration time of the internal standard peak.

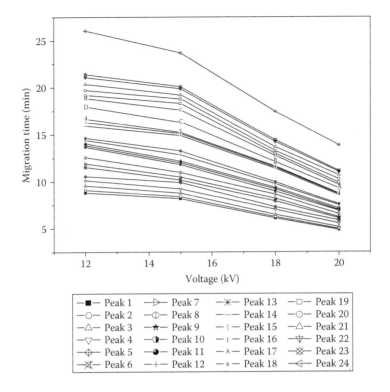

FIGURE 12.6
Effect of applied voltage on the separation of *Liu Wei Di Huang Wan*. CE conditions were buffer composed of 0.01 M borate, 0.05 M SDS, and methanol/water, 15% (v/v), with pH adjusted to 9.3. Other conditions are the same as in Figure 12.2.

The results on the adjusted standard deviation (SD′) and adjusted relative standard deviation (RSD′) are given in Table 12.1. Variation in the RSD′ for the peaks observed was within 1.43–4.42%, as normally expected. Based on the deviation of ±1.96 SD (95% confidence interval) as the criterion to detect a significant change in the ingredients of *Liu Wei Di Huang Wan*, the variation of the relative migration time provides a good parameter for peaks with short migration times up to that of the internal standard peak.

To normalize the peak height, the peaks in the electropherogram were corrected relative to the internal standard peak according to the following formula:

$$H_n' = H_n/H_*$$ (12.2)

where H_n' is the corrected height of peak n, H_n is the original height of peak n, and H_* is the original height of internal standard peak.

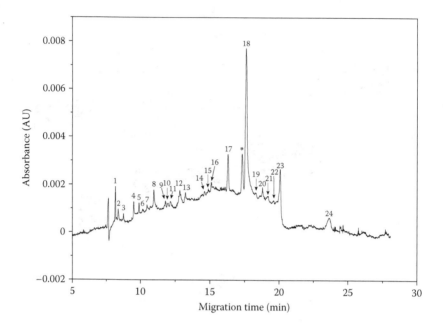

FIGURE 12.7

(A) CE electropherogram of *Liu Wei Di Huang Wan* without normalization. CE conditions were buffer composed of 0.01 M borate, 0.05 M SDS, and methanol/water, 15% (v/v), with pH adjusted to 9.3; run, +15 kV; detection, 210 nm; injection, 0.3 psi for 15 s; and capillary, 0.05 mm i.d. × 60 cm (window 45 cm). (B) CE electropherogram of *Liu Wei Di Huang Wan* with normalization. See Figure 12.11. * indicates internal standard peak (1.0×10^{-4} M benzoic acid).

After normalization with the internal standard peak, most of the peaks showed an appreciable reduction in RSD, except peaks 17 and 22; these two peaks gave a small increase in RSD. As the use of a smaller RSD could lower the noise and increase the chance to detect a small but significant difference in RSD due to small changes in the ingredients of *Liu Wei Di Huang Wan*, the corrected peak height was used as one of the diagnostic parameters for fingerprinting. Due to its much smaller RSD, it provided a more sensitive parameter compared to the change in the migration time.

Each electropherogram of the six ingredients of *Liu Wei Di Huang Wan* is shown in Figure 12.8a–f, respectively. Compared to Figure 12.7 for *Liu Wei Di Huang Wan* containing all six ingredients, a marked difference in the electropherograms was observed in both the number of peaks and their positions. The six ingredients could be divided into two groups. The first group included *C. Moutan* (Figure 12.8a), *R. Rehmanniae* (Figure 12.8b), and *F. Corni* (Figure 12.8c). These three ingredients were found to contribute to most of the peaks found in *Liu Wei Di Huang Wan*. However, the migration time of peaks obtained from the electropherograms of these three herbs alone were too far away from the peaks found in Figure 12.7 for *Liu Wei Di Huang Wan*, creating difficulty in assigning the source of the peaks. This may be

TABLE 12.1

Repeatability for Migration Times of 24 Diagnostic Peaks
from Capillary Electropherograms of *Liu Wei Di Huang Wan*

Peak	t_n (min)	t_n' (min)	SD' (min)	RSD' (%)
1	8.14	8.96	0.1827	2.04
2	8.34	8.76	0.1869	2.13
3	8.74	8.36	0.1732	2.07
4	9.47	7.63	0.1713	2.25
5	9.85	7.25	0.1683	2.32
6	10.25	6.85	0.2844	4.15
7	10.56	6.54	0.2888	4.42
8	10.91	6.19	0.1545	2.50
9	11.77	5.33	0.2126	3.99
10	11.95	5.15	0.1899	3.69
11	12.11	5.00	0.2019	4.04
12	12.73	4.37	0.1196	2.74
13	13.14	3.96	0.1183	2.99
14	14.35	2.75	0.0519	1.89
15	14.62	2.48	0.0686	2.76
16	14.89	2.21	0.0330	1.49
17	16.07	1.03	0.0320	3.11
18	17.39	0.29	0.0058	2.02
19	18.06	0.96	0.0169	1.77
20	18.58	1.48	0.0212	1.43
21	18.96	1.86	0.0364	1.96
22	19.44	2.34	0.0524	2.24
23	19.82	2.72	0.0522	1.92
24	23.42	6.32	0.0930	1.47

Note: t_n, migration time for peak n; t_n', relative migration time for peak
$n = t_n$ (migration time for peak n) − t. (migration time of internal
standard peak); SD', adjusted standard deviation; RSD', adjusted
relative standard deviation of t_n' (relative migration time), $n = 3$.

attributed to the occurrence of a strong interaction when each of these three
herbs was added to the *Liu Wei Di Huang Wan*, leading to a significant shift in
the migration time and the associated difficulty in assigning the peaks. The
only confirmed peaks were the strongest peak 18 and the last peak 24, which
were definitely coming from *C. Moutan* (Figure 12.8a).

The second group of herbs included *Poria* (Figure 12.8d), *R. Alismatis*
(Figure 12.8e), and *R. Dioscorea* (Figure 12.8f). All three herbs showed no
detectable peaks in the electropherograms when each herb was presented
alone. Although they did not contribute to peaks for identification, they could
be detected indirectly via the interaction of their constituents on detectable
peaks from the other three herbs, leading to observable shift in the relative
migration times and the corrected peak heights.

TABLE 12.2

Effects of Using Internal Peak to Normalize Peak Heights from Capillary Electropherograms of *Liu Wei Di Huang Wan*

	Peak Height[a]			Corrected Peak Height[b]			Diff.[c]
Peak	H_n (cm)	SD	RSD(%)	H_n' (cm)	SD'	RSD'(%)	(%)
1	14.43	0.3957	2.74	0.86	0.0215	2.51	0.23
2	3.81	0.0781	2.05	0.23	0.0041	1.80	0.25
3	2.83	0.0608	2.15	0.17	0.0033	1.95	0.20
4	7.68	0.1457	1.90	0.46	0.0075	1.65	0.25
5	4.96	0.1823	3.67	0.30	0.0101	3.41	0.26
6	1.41	0.0525	3.71	0.08	0.0029	3.49	0.22
7	2.84	0.1014	3.57	0.17	0.0056	3.32	0.25
8	7.52	0.1965	2.61	0.45	0.0116	2.60	0.01
9	3.46	0.1106	3.19	0.21	0.0060	2.93	0.26
10	1.70	0.0404	2.37	0.10	0.0021	2.11	0.26
11	2.43	0.0585	2.41	0.14	0.0031	2.15	0.26
12	7.66	0.1814	2.37	0.46	0.0103	2.26	0.11
13	4.48	0.0802	1.79	0.27	0.0041	1.52	0.27
14	1.82	0.1053	5.79	0.11	0.0060	5.53	0.26
15	2.58	0.1266	4.90	0.15	0.0074	4.79	0.11
16	3.70	0.1357	3.67	0.22	0.0076	3.43	0.24
17	15.69	0.2577	1.64	0.93	0.0178	1.91	−0.27
18	57.99	1.6254	2.80	3.45	0.0890	2.58	0.22
19	0.75	0.0251	3.34	0.04	0.0014	3.07	0.27
20	1.64	0.0651	3.96	0.10	0.0036	3.69	0.27
21	1.62	0.0416	2.56	0.10	0.0022	2.30	0.26
22	1.34	0.0360	2.69	0.08	0.0024	2.95	−0.26
23	13.29	0.4540	3.42	0.79	0.0250	3.15	0.27
24	4.75	0.1735	3.65	0.28	0.0096	3.38	0.27

Note: SD (standard deviation) and RSD (relative standard deviation) of original peak height, $n = 3$. H_n, original height of peak n; SD' (adjusted standard deviation) and RSD' (adjusted relative standard deviation) of corrected peak height, $n = 3$; H_n' (corrected height of peak n) = H_n/H. (height of internal standard peak); Diff., RSD − RSD'.

12.4.2 Quantitative Fingerprinting of *Liu Wei Di Huang Wan*

To provide a quantitative measure for fingerprinting of *Liu Wei Di Huang Wan*, six preparations were made using only five full ingredients and one preparation with 85% of the ingredients compared to the original recipe. The samples made for each of the six ingredients were extracted using the same procedure as shown in Figure 12.1 prior to CE analysis. The diagnostic pattern obtained for fingerprinting of *Liu Wei Di Huang Wan* is given in Table 12.3. Three test criteria are used in Table 12.3 based on the change in the migration time and peak height. The relative migration time, t_n'', and corrected

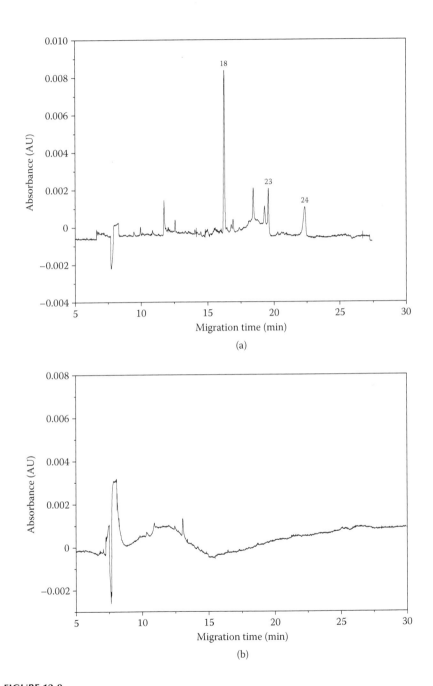

FIGURE 12.8
CE electropherograms for each of the six herbs. *C. Moutan* (a), *R. Rehmanniae* (b), *F. Corni* (c), *Poriae* (d), *R. Alismatis* (e), and *R. Dioscoreae* (f). CE conditions same as in Figure 12.7. (B) *R. Rehmanniae* alone, (C) *F. Corni* alone, (D) *Poriae* alone, (E) *R. Alismatis* alone, (F) *R. Dioscoreae* alone. *(Continued)*

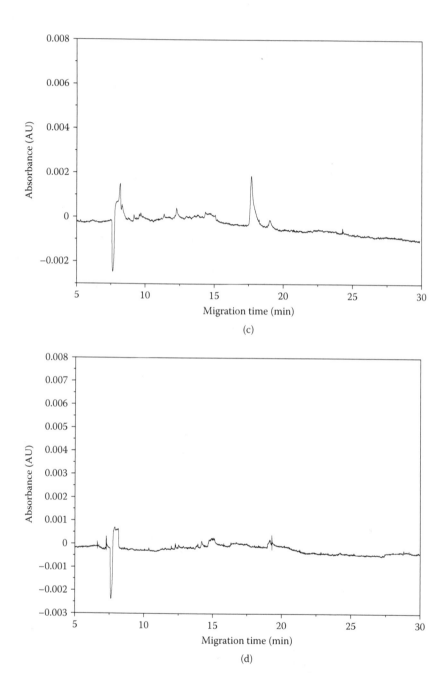

FIGURE 12.8 (Continued)
CE electropherograms for each of the six herbs. *C. Moutan* (a), *R. Rehmanniae* (b), *F. Corni* (c), *Poriae* (d), *R. Alismatis* (e), and *R. Dioscoreae* (f). CE conditions same as in Figure 12.7. (B) *R. Rehmanniae* alone, (C) *F. Corni* alone, (D) *Poriae* alone, (E) *R. Alismatis* alone, (F) *R. Dioscoreae* alone.   *(Continued)*

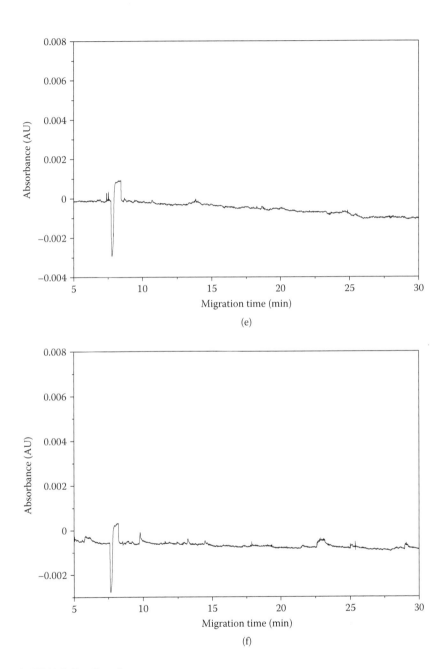

FIGURE 12.8 (Continued)
CE electropherograms for each of the six herbs. *C. Moutan* (a), *R. Rehmanniae* (b), *F. Corni* (c), *Poriae* (d), *R. Alismatis* (e), and *R. Dioscoreae* (f). CE conditions same as in Figure 12.7. (B) *R. Rehmanniae* alone, (C) *F. Corni* alone, (D) *Poriae* alone, (E) *R. Alismatis* alone, (F) *R. Dioscoreae* alone.

TABLE 12.3

Diagnostics Pattern for Fingerprinting of *Liu Wei Di Huang Wan* by CE

Peak	A t_n''	A H_n''	B t_n''	B H_n''	C t_n''	C H_n''	D t_n''	D H_n''	E t_n''	E H_n''	F t_n''	F H_n''
1	−	−	−	−	−	−		−	+	−	+	−
2	−	+	−	+	−	−		+	+	+	+	−
3	−	−	−	−	−	−		+	−		+	−
4	−	−	−	−	−	−		−	+	−	+	−
5	−	−	−	+	−				−		+	−
6	−	+	−	+	−	+		+	+			+
7	−		−	+	−	−		−				
8	−	−			−	−		+			−	
9	−	+	−	+	−	+		+	+			
10	−	+	−	−	−	−		−		−	−	
11	−	−	−		−				−		−	
12	−	−	−	−	−	−		−	−		−	
13	−	+	−	−	−	−	+	+		−	−	
14	−	+	−	+	−	+	+	+	−	+	+	
15	−	+	−	−	−	+	+		−	+	−	
16	−	−	−	−	−	−	+		−	−	−	
17	−	−	−	−	−	−		−	−	−	−	−
18	−	−	+	+	−	−		+	+	+	+	−
19	−	−	−	+	−	+		+	+	+	+	+
20	−	+	−	+	−	+		+	−	+	+	+
21	+	−	+	+	+			−	+	+	+	+
22	+	−	+	+	+	+		+	+	+	+	+
23	+	−	+	+	+			−	+	+	+	
24	+	−	+	+	+			+	+	+	+	+

Note: Samples: Five ingredients of *Liu Wei Di Huang Wan* plus (A) 85% *C. Moutan*, (B) 85% *R. Rehmanniae*, (C) 85% *F. Corni*, (D) 85% *Poriae*, (E) 85% *R. Alismatis*, and (F) 85% *R. Dioscoreae*. t_n'' (relative migration time of peak n from sample A–F) = t_n^{85} (original migration time of peak n from sample A–F) − t_* (migration time of internal standard peak), where t_n^{85} refers to migration time for peak n at 85%. H_n'' (relative peak height of peak n from sample A–F) = H_n^{85} (original peak height of peak n from sample A–F)/H_* (peak height of internal standard peak), where H_n^{85} refers to peak height for peak n at 85%. The symbol + or − refers to deviation of t_n'' or H_n'' beyond the range of ±1.96 SD of t_n' or H_n' (100% *Liu Wei Di Huang Wan*). The space is left as blank when t_n'' or H_n'' was found to be within the range of ±1.96 SD of t_n' or H_n', respectively.

peak height, H_n'', for peak n from sample A to F with five ingredients of *Liu Wei Di Huang Wan* plus 85% of the targeted ingredient is calculated by the following formula:

$$t_n'' = \left| t_n^{85} - t_* \right| \tag{12.3}$$

$$H_n'' = H_n^{85}/H_* \tag{12.4}$$

where $t_n{}^{85}$ and $H_n{}^{85}$ refer to the migration time and peak height, respectively, for peak n at 85% of one of the ingredients. The range of ±1.96 SD for both the relative migration time and corrected peak height of 100% *Liu Wei Di Huang Wan* is used as the standard for testing. The symbol + or − refers to deviation of $t_n{}''$ or $H_n{}''$ beyond the range of ±1.96 SD of $t_n{}'$ or $H_n{}'$ (100% *Liu Wei Di Huang Wan*), respectively. The space is left as blank when $t_n{}''$ or $H_n{}''$ was found to be the within the range of ±1.96 SD of $t_n{}'$ or $H_n{}'$, respectively.

Most of the peaks for *R. Dioscoreae* (Figure 12.8f) were shifted positively or do not shift when their concentrations were lowered to 85% in *Liu Wei Di Huang Wan*, expect for peak 17, which was shifted negatively. To the peak height of the strongest peak 18 and the last peak 24, peak 18 was shifted negatively and peak 24 was shifted positively.

The pattern shown in Table 12.3 indicated that the migration time for most peaks from *C. Moutan* (Figure 12.8a) were shifted negatively, except peaks 21–24 were shifted positively. The peak height of the strongest peak 18 and the last peak 24 were both shifted negatively.

For *R. Rehmanniae* (Figure 12.8b), the change in the migration time for all the peaks was the same as for *C. Moutan* (Figure 12.8a), except the strongest peak 18, which was shifted positively. The peak height of the strongest peak 18 and the last peak 24 were both shifted positively in contrast to that of *C. Moutan*.

The migration time of most peaks from *F. Corni* (Figure 12.8c) were shifted negatively, the same as for *C. Moutan* (Figure 12.8a), except the strongest peak 18 and the last peak 24. Those two peaks were considered as unalterable based on the ±1.96 SD testing criterion. To the peak height of the strongest peak 18 and the last peak 24, peak 18 was shifted negatively and peak 24 was shifted positively.

For *Poria* (Figure 12.8d), the migration time of most peaks did not shift, except peaks 13–16 were shifted positively, whereas peaks 19–21 were shifted negatively. The peak height of the strongest peak 18 and the last peak 24 were both shifted positively.

For *R. Alismatis* (Figure 12.8e), the migration time of half of all the 24 peaks did not shift. Some peaks were shifted positively, such as peaks 1, 2, 21, and 22–24, whereas some peaks were shifted negatively, including peaks 14–17, 19, and 20. The peak height of the strongest peak 18 and the last peak 24 were both shifted positively.

The change to 85% of one of the ingredients of *Liu Wei Di Huang Wan* was shown to produce a change in the fingerprinting pattern that can be identified. For some changes, such as a selective shift in some of the peaks, the change can be traced back due to a particular herb. Thus, the change in the fingerprinting pattern can be used to indicate the change in the composition of one of the ingredients of *Liu Wei Di Huang Wan* when the change is >15%. From the change in the pattern, one can identify which of the six herbs has changed its composition under favorable conditions.

12.5 Authenification of *Liu Wei Di Huang Wan* by MC-CE Device with Portable UV Detector

An MC-CE device offers several advantages over traditional CE: low sample consumption, portable analysis, and capability for integrating different functional units [41]. In the present work, an MC-CE device is developed to provide a portable device for fieldwork application based on previous CE work done as described in Section 12.4. Moreover, an MC-CE device provides a promising tool to assess the variability of each single pill for a given TCM preparation, with results reflecting the efficiency of the mixing and blending process for TCM manufacture. The methodology of using an MC-CE device for assessing the variability of *Liu Wei Di Huang Wan* by fingerprinting analysis is explored in the current work to verify the capability of an MC-CE device for quality assessment of TCM.

12.5.1 Instrumentation and Facilities

Computer-aided machines operated by CorelDRAW 10 (Corel Corporation, Ottawa, Ontario) software were used to control the CO_2 laser engraver (V-series, Pinnacle, Great Computer Corporation, Taipei) for ablating the desired channel pattern onto the bottom poly(methyl methacrylate) (PMMA) plate. A hot press (Guangju Machinery Company, Guangzhou, China) was used to bond the top and bottom PMMA plates together with an embedded separation capillary. The device fabricated was placed on a self-constructed sample workstation, with performance verified by a microscope during operation. An eight-channel, high-voltage power supply with each channel delivering 0 to +3,500 V was purchased (model MP-3500-FP, Major Science Co. Ltd., Taipei) and used to control the voltages and durations applied to microelectrodes placed in different vials of the MC-CE device. The analytes after MC-CE separation were detected by a portable UV-visible detector (model CE-P2, CE Resources).

12.5.2 Design and Fabrication of MC-CE Device

The PMMA plate (40 mm in length, 30 mm in width, and 0.15 mm in thickness), supplied by Ensinger Ltd. (Mid Glamorgan, UK), was used as the bottom plate for the ablation by a CO_2 laser under control by computer software to create a desired microchannel pattern with typical channel dimensions (100 μm in depth and 150 μm in width). Three holes (1.0 mm inside diameter [i.d.]) were drilled into the top PMMA plate to serve as three vials for sample reservoir, buffer reservoir, and sample waste reservoir (SR, BR, and SW, respectively) for application of required high voltages (Figure 12.9) [42].

Each of the vials is created with equal distance (5 mm) to the main channel. The 3-mm channel segment between SR and SW was designed to act

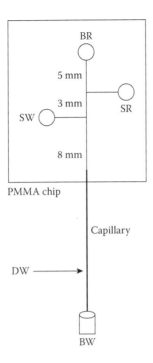

FIGURE 12.9
Schematic diagram showing a portable MC-CE device with double-T injectors and a bonded separation capillary. Vials with buffer reservoir (BR); sample reservoir (SR); sample waste reservoir (SW); buffer waste reservoir (BW); detection window (DW). (From Fung, Y.S. and Nie, Z. *Sep Sci* 2009, 1, 3–8.)

as a double-T injector to inject a large sample plug. An embedded capillary (50 μm i.d. and 13 cm in length) was bonded between two PMMA plates by a hot plate press machine to allow continued liquid contact from the capillary to the double-T injector via an 8-mm microchannel segment between SW and the end of the PMMA plate. The effective length of the separation capillary is ~9 cm, with the detection window created at the position of ~4 cm from the end of the separation capillary.

The two PMMA plates were bonded with the separation capillary by a hot bonding machine for 15 min under 0.6 MPa and 92°C. After hot bonding, the plate was cooled and cleaned in an ultrasonic bath (Figure 12.10). An air pump was used to remove residual water from the microchannels of the MC-CE device before use.

12.5.3 Procedures and Operation of MC-CE Device

Prior to each analysis, diluted NaOH solution, deionized water, and running buffer were used in sequence for 5 min each to condition the microchannels of the MC-CE device. A syringe pump was used to push the solution

FIGURE 12.10
Photograph of the MC-CE device incorporating an on-chip double-T injector and an embedded separation capillary (magnification 3×).

through the open end of the capillary to fill all vials. After conditioning, 10 µL of a well-mixed six-herb test solution containing five herbal ingredients according to expected formula and the remaining ingredient with 115% higher than expected value was pipetted into the SR vial. For sample injection, 2100 V was applied for 20 s to SR while SW was grounded and both BR and BW floated. Separation was carried out immediately after sample injection by applying 3500 V to BR while BW was grounded and both SW and SR were kept at 2100 V. Detection wavelength at 210 nm was used by portable UV-visible detector during all MC-CE runs.

12.5.4 Fingerprinting of *Liu Wei Di Huang Wan* by MC-CE Device

The working conditions of the MC-CE device were adopted on the optimized conditions of the CE procedure as described in Section 12.3, except that the sample was injected via the double-T injector instead of using hydrodynamic injection for CE injection. Due to the delay in time in the construction and testing of the MC-CE device and associated instruments, the samples of *Liu Wei Di Huang Wan* investigated previously for the CE work (old samples) had to be replaced after a 2-year storage period. To maintain samples continuity, the new samples were purchased from the same local pharmaceutical shop in Hong Kong.

The electropherogram of the new *Liu Wei Di Huang Wan* with six ingredients is shown in Figure 12.11 [42]. Compared to the electropherogram shown in Figure 12.7 for the old *Liu Wei Di Huang Wan* with six ingredients obtained using CE alone, significant differences were observed in the two electropherograms. The differences may be due to the following reasons. First, the capillary column used in MC-CE device was shorter (13 versus 45 cm) due to the voltage limitation of the multiple high-power supply for operation of the MC-CE device. This gave rise to shorter migration times, although the total number of components separated was still 24. Second, the use of double-T injector for sample injection was easier to realize in MC-CE device than the use of hydrodynamic injection previously used in the CE work. Third, the old and new herbs may have been from different regions or harvested in

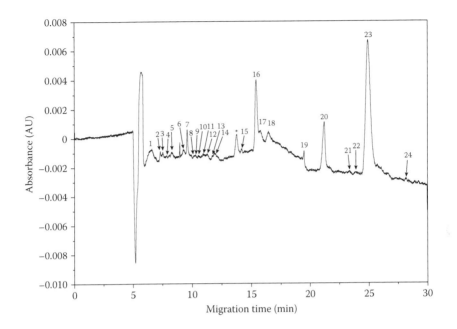

FIGURE 12.11

MC-CE electropherogram of the new *Liu Wei Di Huang Wan*. MC-CE conditions were buffer composed of 0.01 M borate, 0.05 M SDS, and methanol/water, 15% (v/v), with pH adjusted to 9.3; detection, 210 nm; capillary, 0.05 mm i.d. × 13 cm (window 9 cm); injection, 2100 V applied for 20 s across SW (grounded) and SR, while BW and BR floated; separation, 3500 V applied across BW (grounded) and BR, while SW and SR kept at 2100 V. * indicates internal standard peak (1.0 × 10^{-4} M benzoic acid). (From Fung, Y.S. and Nie, Z. *Sep Sci* 2009, 1, 3–8.)

different seasons, factors that are known to give rise to different compositions for herbs, even from the same botanical source.

The electropherograms for each of the six ingredients of the new *Liu Wei Di Huang Wan* are shown in Figure 12.12a–f. Based on the results, the six ingredients could also be divided into two groups. The first group included *C. Moutan* (Figure 12.12a), *R. Rehmanniae* (Figure 12.12b), and *F. Corni* (Figure 12.12c), with corresponding electropherograms. These three herbs were found to contribute to most of the peaks found in *Liu Wei Di Huang Wan*.

The electropherograms for the second group of herbs—*Poria* (Figure 12.12d), *R. Alismatis* (Figure 12.12e), and *R. Dioscorea* (Figure 12.12f)—showed no detectable peaks in the electropherograms of each herb alone. Although they did not contribute to peaks for identification, the three herbs could be detected indirectly via the interaction of their constituents on the detectable peaks from the other three herbs, leading to an observable shift in the relative migration time and a significant change in the corrected peak height.

Although the electropherograms for each of the six herbs of the old and new *Liu Wei Di Huang Wan* were different, both gave results that offer similar conclusions for the six individual herbs. For example, there were two groups of herbs.

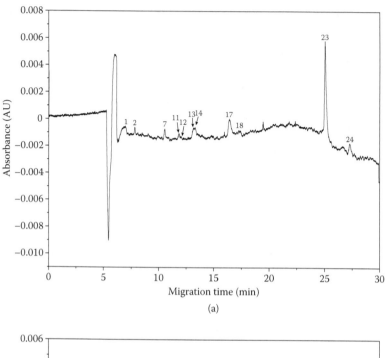

(a)

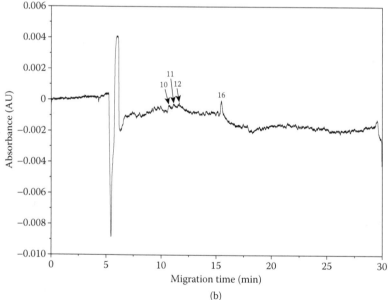

(b)

FIGURE 12.12

MC-CE electropherogram of a new batch of *C. Moutan* (a), *R. Rehmanniae* (b), *F. Corni* (c), *Poriae* (d), *R. Alismatis* (e), and *R. Dioscoreae* (f). CE conditions are the same as in Figure 12.11 (A) *C. Moutan* alone, (B) *R. Rehmanniae* alone; (C) *F. Corni* alone (D), *Poriae* alone, (E) *R. Alismatis* alone, (F) *R. Dioscoreae* alone.   *(Continued)*

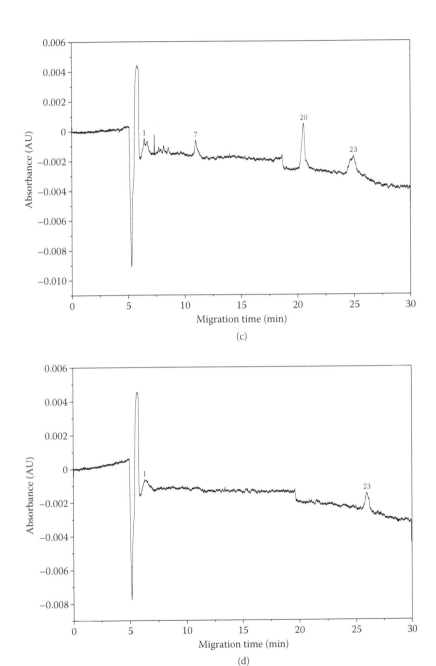

FIGURE 12.12 (Continued)
MC-CE electropherogram of a new batch of *C. Moutan* (a), *R. Rehmanniae* (b), *F. Corni* (c), *Poriae* (d), *R. Alismatis* (e), and *R. Dioscoreae* (f). CE conditions are the same as in Figure 12.11 (A) *C. Moutan* alone, (B) *R. Rehmanniae* alone; (C) *F. Corni* alone (D), *Poriae* alone, (E) *R. Alismatis* alone, (F) *R. Dioscoreae* alone. *(Continued)*

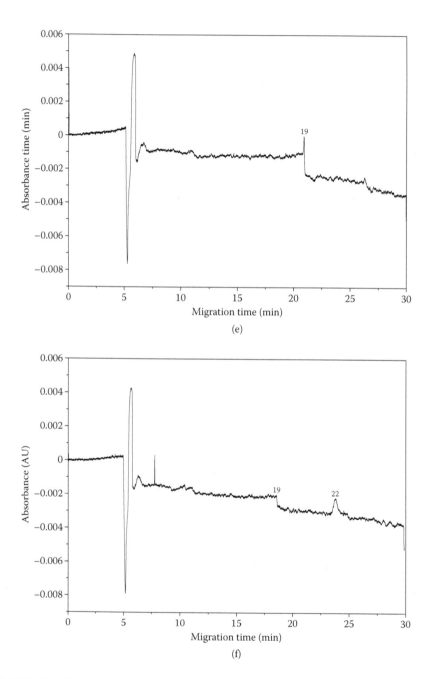

FIGURE 12.12 (Continued)
MC-CE electropherogram of a new batch of *C. Moutan* (a), *R. Rehmanniae* (b), *F. Corni* (c), *Poriae* (d), *R. Alismatis* (e), and *R. Dioscoreae* (f). CE conditions are the same as in Figure 12.11 (A) *C. Moutan* alone, (B) *R. Rehmanniae* alone; (C) *F. Corni* alone (D), *Poriae* alone, (E) *R. Alismatis* alone, (F) *R. Dioscoreae* alone.

The first group (*C. Moutan, R. Rehmanniae,* and *F. Corni*) was found to contribute to most of the peaks found in *Liu Wei Di Huang Wan.* The second group (*Poria, R. Alismatis,* and *R. Dioscorea*) had no peaks, but their presence can be detected via the change in the fingerprint pattern of the first group of herbs. Thus, for herbal medicine manufacturers, QC can be exercised for a given batch of the six herbs using the CE method developed in the present research.

12.5.5 Applicability Study

The same methodology was carried out by MC-CE device to assess variability of *Liu Wei Di Huang Wan* via addition of 15% of each ingredient, with subsequent observation of the change in the migration time and peak height of the electropherogram to indicate the variability of the target ingredient of *Liu Wei Di Huang Wan.* As *C. Moutan* was found to contribute to most of the peaks found in *Liu Wei Di Huang Wan,* and it was the major source for peaks 1, 2, 7, 11–14, 17, 18, 23, and 24 (Figure 12.12a), it was selected to add to the 100% sample to investigate the variability of *Liu Wei Di Huang Wan* due to *C. Moutan* (Figure 12.13).

Results from Table 12.4 showed that the migration times for most of the peaks from *C. Moutan* were shifted negatively, except peaks 16–24,

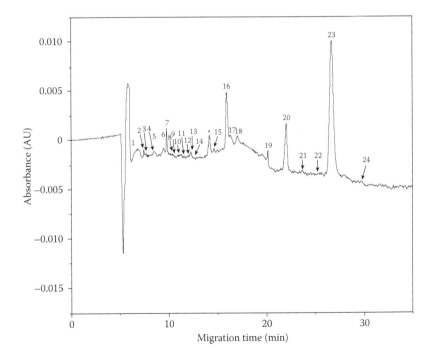

FIGURE 12.13
MC-CE electropherogram of a new batch of *Liu Wei Di Huang Wan* plus the addition of 15% A *C. Moutan.* CE conditions are the same as in Figure 12.11.

Microfluidic Chip-Capillary Electrophoresis Devices

TABLE 12.4

Variability of New *Liu Wei Di Huang Wan* plus 15% *C. Moutan* by Fingerprinting Analysis Using MC-CE Device

Peak		1	2	3	4	5	6	7	8	9	10	11	12	13	14	15	16	17	18	19	20	21	22	23	24
Sample A	t_n''	-	-	-	-	-	-	-	-	-	-	-	-	-	-		+	+	+	+	+	+	+	+	+
	H_n''	+	+					+									+	+	+	+	+	+	+	+	+

Note: Sample A *Liu Wei Di Huang Wan* plus 15% *C. Moutan*; t_n'' (relative migration time of peak n from sample A–F) = $|t_n^{115}$ (original migration time of peak n from sample A–F) – t_s (migration time of internal standard peak)$|$, where t_n^{115} refers to migration time for peak n at 115%. H_n'' (relative peak height of peak n from sample A–F) = H_n^{115} (original peak height of peak n from sample A–F)/H_s (peak height of internal standard peak), where H_n^{115} refers to peak height for peak n at 115%. The symbol + or – refers to deviation of t_n'' or H_n'' beyond the range of ±1.96 SD of t_n' or H_n' (100% *Liu Wei Di Huang Wan*). The space is left as blank when t_n'' or H_n'' was found to be within the range of ±1.96 SD of t_n' or H_n', respectively.

which were shifted positively. The peak heights of most of the peaks were more or less constant, except peaks 1, 2, 7, 16, 17, 20, and 22–24, which were shifted negatively. Thus, the addition of 15% *C. Moutan* was shown to produce a change that could be identified. As both the migration times and the peak heights of peaks 1, 2, 7, 17, 23, and 24 were changed and *C. Moutan* was the major source of these peaks (Figure 12.12a), peaks 1, 2, 7, 17, 23, and 24 can be used as the diagnostic peaks and the selective shift of the peaks can be traced back to *C. Moutan*.

Thus, the MC-CE device has been shown to be capable of indicating the variability of *Liu Wei Di Huang Wan* due to the change of 15% of each of the six ingredients of *Liu Wei Di Huang Wan*. From the change of the diagnostic pattern, one can identify which of the six ingredients had changed its composition under favorable conditions.

12.6 Summary

To meet the demand for QC of complex herbal drugs from established medicinal herb recipes, chromatographic fingerprinting had been advocated. To make use of the high separation efficiency and capability of the micellar electrokinetic chromatography (MEKC) method to separate both neutral and charged compounds extracted from herbal medicine, the analytical procedures on operations on extraction, separation, and quantitation of ingredients from a commercially available herbal drug were optimized. To demonstrate the application of MEKC for QC of a complex herbal preparation via fingerprinting of *Liu Wei Di Huang Wan*, a herbal preparation widely used in Asian countries comprising six ingredients at different levels, was selected for investigation in the present study.

The analytical procedures developed include the investigation of extraction methods and the development of an MEKC procedure for the separation of the extracted ingredients. Organic solvent systems commonly used in herbal extraction were investigated for their extraction efficiency, assisted by ultrasound, and the best extraction solvent for *Liu Wei Di Huang Wan* was the methanol:water system (1:1, v/v). For optimization of the MEKC separation, factors affecting the separation conditions and mode of detection for each of the six ingredients of *Liu Wei Di Huang Wan* were studied. The optimized conditions were hydrodynamic injection at 13 cm for 30 s; running voltage at +15 kV; and running buffer containing 10 mM borate, 50 mM SDS, and 15% methanol (v/v), with pH adjusted to 9.3.

Under the optimized conditions, 24 diagnostic peaks in total were identified, and a diagnostic pattern based on the change in the relative migration time and corrected peak height from the electropherograms was obtained for chromatographic fingerprinting of *Liu Wei Di Huang Wan*. The use of relative migration time and corrected peak height normalized to the internal standard peak of benzoic acid was found to improve the repeatability of the results and assist the fingerprinting process for identification of significant differences from the pattern of the diagnostic peaks. To provide a quantitative measure, each of the six ingredients of *Liu Wei Di Huang Wan* was lowered 15% during preparation. The CE fingerprinting methods were shown able to detect the change in herbal composition via a significant shift in the diagnostic fingerprinting pattern. Under favorable conditions, the source of each of the six herbs can be identified from information obtained from the fingerprinting pattern.

An MC-CE device designed to be portable for fingerprinting of herbal medicine is developed to enable on-site assessment of the quality of *Liu Wei Di Huang Wan*. The addition of 15% of each of the six ingredients to 100% *Liu Wei Di Huang Wan* can be identified by the MC-CE device. Under the optimized conditions, 24 diagnostic peaks in total were identified, and a diagnostic pattern based on the change in the relative migration time and corrected peak height was obtained. Based on the diagnostic pattern, peaks 1, 2, 7, 17, 23, and 24 can be used as diagnostic peaks, and the observed shift of the peaks can be traced back to *C. Moutan*. Thus, for a small change in the concentration of each of the six ingredients of *Liu Wei Di Huang Wan*, detectable features should be identified in the electropherogram of the final product, illustrating that an MC-CE device could be used for assessing the variability of *Liu Wei Di Huang Wan* in the field.

Compared to a laboratory procedure using CE alone, similar analytical performance and capability were found using the MC-CE device, as shown in the results and their associated interpretation. However, distinct variability of composition present in individual herb as raw material for *Liu Wei Di Huang Wan* is clearly shown by significant differences in the electropherograms for the two batches of herbs analyzed by CE and MC-CE device with a 2-year interval. Further work is needed to assess the variability of each ingredient herb from different sources via fingerprint pattern and to establish the relationship between the variability of herbal quality with the effectiveness of its medical treatment by collaboration with related medical users.

References

1. Harkey, M.R., Henderson, L., Gershwin, M.E., Stern, J.S. and Hackman, R.M. Variability in commercial ginseng products: An analysis of 25 preparations. *Am J Clin Nutr* 2001, 73, 1101–1106.
2. Bauer, R. and Tittle, G. Quality assessment of herbal preparations as a precondition of pharmacological and clinical studies. *Phytomedicine* 1996, 2, 193–198.

3. Mcgaw, L.J., Jager, A.K. and Staden, J. Variation in antibacterial activity of Schotia species. *S Afr J Bot* 2002, 68, 41–46.
4. Szeto, M.C., Zhong, Y.K., Tong, K.M., Kwong, B.N., Cheung, H.S., Chung, K.C. and Koo, Y.H. (Eds.). Quality control of herbal medicines in Hong Kong. The Hong Kong Society of Chinese Medicines, Hong Kong, 2003.
5. Fan, X.H., Cheng, Y.Y., Ye, Z.L., Lin, R.C. and Qian, C.C. Multiple chromatographic fingerprinting and its application to the quality control of herbal medicines. *Anal Chim Acta* 2006, 555, 217–224.
6. Dong, B., Sun, S.Q., Zhou, H.T. and Hu, S.L. Rapid and undamaged determination of Chishao by Fourier-transform infrared spectroscopy and clustering analysis. *Spectrosc Spect Anal* 2002, 22, 232–234.
7. Cheng, Y.Y., Yu, J. and Wu, Y.J. A new method for representing fingerprint feature: Visualization of analytical data of infrared spectra. *Chin J Anal Chem* 2002, 30, 1426–1430.
8. Yang, N.L., Cheng, Y.Y. and Qu, H.B. An approach to purifying process analysis of Chinese herbal extracts using NIRS. *Acta Chim Sinica* 2003, 61, 742–747.
9. Zou, H.B., Yang, G.S. and Qin, Z.R. Progress in quality control of herbal medicine with IR fingerprint spectra. *Anal Lett* 2005, 38, 1457–1475.
10. Zhan, D.Q. and Sun, S.Q. Discrimination of patchoulis of different geographical origins with two-dimensional IR correlation spectroscopy and wavelet transform. *Lect Notes Artif Int* 2005, 3801, 965–972.
11. Bilia, A.R., Bergonzi, M.C., Lazari, D. and Vincieri, F.F. Characterization of commercial Kava-Kava herbal drug and herbal drug preparations by means of nuclear magnetic resonance spectroscopy. *J Agr Food Chem* 2002, 50, 5016–5025.
12. Salvador, M.J., Lopes, G.N., Nascimento, V.F. and Zhcchi, O.L. Quality control of commercial tea by x-ray fluorescence. *X-Ray Spectrom* 2002, 31, 141–144.
13. Cheng, K.T., Chang, H.C., Huang, H. and Lin, C.T. RAPD analysis of Lycium barbarum medicine in Taiwan market. *Bot Bull Acad Sinica* 2000, 41, 11–14.
14. Kapteyn, J., Goldsbrough, P.B. and Simon, J.E. The use of RAPDs for assessment of identity, diversity, and quality of Echinacea. *Theor Appl Genet* 2002, 105, 369–376.
15. Meinl, W. and Pabel, U. Human sulphotransferases are involved in the activation of aristolochic acids and are expressed in renal target tissue. *Int J Cancer* 2006, 118, 1090–1097.
16. Fiehn, O., Kopka, J., Trethewey, R.N. and Willmtzer, L. Identification of uncommon plant metabolites based on calculation of elemental compositions using gas chromatography and quadrupole mass spectrometry. *Anal Chem* 2000, 72, 3573–3580.
17. Gong, F., Liang, Y.Z. and Chau, F.T. Combination of GC-MS with local resolution for determining volatile components in si-wu decoction. *J Sep Sci* 2003, 26, 112–122.
18. Gong, F., Liang, Y.Z. and Fung, Y.S. Analysis of volatile components from Cortex cinnamomi with hyphenated chromatography and chemometric resolution. *J Pharm Biomed Anal* 2004, 34, 1029–1047.
19. Yan, S.K., Xin, W.F. and Luo, G.A. Chemical fingerprinting of Gardenia jasminoides fruit using direct sample introduction and gas chromatography mass spectrometry detection. *J AOAC Int* 2006, 89, 40–45.

20. Rajani, M., Ravishankara, M.N., Shrivastava, N. and Padh, H. HPTLC-aided phytochemical fingerprinting analysis as a tool for evaluation of herbal drugs. A case study of Ushaq (Ammoniacum gum). *J Planar Chromatogr* 2001, 14, 34–41.
21. Smith, W.T. Identification of toxic herbs using TLC active constituent fingerprints. *J Planar Chromatogr* 1994, 7, 95–97.
22. Ma, K.W., Chau, F.T. and Wu, J.Y. Analysis of the nucleoside content of Cordyceps sinensis using the stepwise gradient elution technique of thin-layer chromatography. *Chin J Chem* 2004, 22, 85–91.
23. Babu, S.K. and Kumar, K.V. Estimation of trans-resveratrol in herbal extracts and dosage forms by high—Performance thin—Layer chromatography. *Chem Pharm Bull* 2005, 53, 691–693.
24. Lin, G., Li, P., Li, S.L. and Chan, S.W. Chromatographic analysis of Fritillaria isosteroidal alkaloids, the active ingredients of Beimu, the antitussive traditional Chinese medicinal herb. *J Chromatogr A* 2001, 935, 321–338.
25. Molgaar, P., Johnsen, S., Christensen, P. and Cornett, C. HPLC method validated for the simultaneous analysis of cichoric acid and alkamides in Echinacea purpurea plants and products. *J Agri Food Chem* 2003, 51, 6922–6933.
26. Ku, Y.R., Ho, Y.L., Chen, C.Y., Ho, L.K. and Chang, Y.S. Analysis of N-cis- and N-trans-feruloyl 3-methyldopamine in Achyranthes bidentata by HPLC. *J Liq Chromatogr Relat Technol* 2004, 27, 727–736.
27. Xie, Y., Jiang, Z.H. and Zhou, H. Simultaneous determination of six Aconitum alkaloids in proprietary Chinese medicines by high-performance liquid chromatography. *J Chromatogr A* 2005, 1093, 195–203.
28. Lin, I.H., Lee, M.C. and Chuang, W.C. Application of LC/MS and ICP/MS for establishing the fingerprint spectrum of the traditional Chinese medicinal preparation Gan-Lu-Yin. *J Sep Sci* 2006, 29, 172–179.
29. Liebich, H.M., Lehmann, R., Di Stefano, C., Haring, H.U., Kim, J.H. and Kim, J.R. Analysis of traditional Chinese anticancer drugs by capillary electrophoresis. *J Chromatogra A* 1998, 795, 388–393.
30. Fung, Y.S. and Tung, H.S. Application of capillary electrophoresis for organic acid analysis in herbal studies. *Electrophoresis* 2001, 22, 2242–2250.
31. Cheung, H.Y., Cheung, C.S. and Kong, C.K. Determination of bioactive diterpenoids from Andrographis paniculata by micellar electrokinetic chromatography. *J Chromatogr A* 2001, 930, 171–176.
32. Feng, H.T. and Li, S.F.Y. Determination of five toxic alkaloids in two common herbal medicines with capillary electrophoresis. *J Chromatogr A* 2002, 973, 243–247.
33. Qin, J.H., Leung, F.C. and Fung, Y.S. Rapid authentication of ginseng species using microchip electrophoresis with laser-induced fluorescence detection. *Anal Bioanal Chem* 2005, 381, 812–819.
34. Liu, L.H., Fan, L.Y. and Chen, H.L. Separation and determination of four active anthraquinones in Chinese herbal preparations by flow injection-capillary electrophoresis. *Electrophoresis* 2005, 26, 2999–3006.
35. Yu, L.J., Xu, Y. and Feng, H.T. Separation and determination of toxic pyrrolizidine alkaloids in traditional Chinese herbal medicines by micellar electrokinetic chromatography with organic modifier. *Electrophoresis* 2005, 26, 3397–3404.
36. Yu, L.J. and Li, S.F.Y. Dynamic pH junction-sweeping capillary electrophoresis for online preconcentration of toxic pyrrolizidine alkaloids in Chinese herbal medicine. *Electrophoresis* 2005, 26, 4360–4367.

37. Hsieh, S.C., Huang, M.F. and Lin, B.S. Determination of aristolochic acid in Chinese herbal medicine by capillary electrophoresis with laser-induced fluorescence detection. *J Chromatogr A* 2006, 1105, 127–134.
38. Meng, X.S. Analysis of Chinese Medicine (*Zhong Cheng Yao Fen Xi*). 2nd ed. People's Health Press, Beijing, China, 1998, pp. 211–229.
39. Shen, J.J., Lin, C.J., Huang, J.L., Hsieh, K.H. and Kuo, M.L. The effect of Liu-Wei-Di-Huang Wan on Cytokine gene expression from human peripheral blood lymphocytes. *Am J Chin Med* 2003, 31, 247–257.
40. Camilleri, P. (Ed.). *Capillary Electrophoresis Theory and Practice*. 2nd ed. CRC Press, Boca Baton, FL, 1997, pp. 199–200.
41. Manz, A., Harrison, D.J., Verpoorte, E.M.J., Fettinger, J.C., Paulus, A.L.H. and Widmer, H.M. Planar chips technology for miniaturization and integration for separation techniques into monitoring systems. *J Chromatogr* 1992, 593, 253–258.
42. Fung, Y.S. and Nie, Z. Microfluidic chip capillary electrophoresis for biomedical analysis. *Sep Sci* 2009, 1, 3–8.

13

On-Chip Binding Assay: MC-CE Device Integrated with Multisegment Circular-Ferrofluid-Driven Micromixing Injection for Assay of Free Bilirubin and Albumin Residual Binding Capacity for Bilirubin

Zhou Nie

Guangzhou Amway (China) Co., Ltd.
Guangzhou, People's Republic of China

Ying Sing Fung

The University of Hong Kong
Hong Kong SAR, People's Republic of China

CONTENTS

13.1 Introduction

Assessing the binding capability of biomolecules is significant in clinical
diagnosis, in particular for assessing the effect of diseases requiring urgent
treatment and intervention. In the present chapter, the presence of free and
protein-bound bilirubin in cord blood and its effect on the management of
bilirubinemia and associated Kernicterus disease is presented and discussed.

The fabrication and operation of the MC-CE device for on-site assessment
of free and albumin-bound bilirubin is presented and discussed. The information obtained to assist making clinical decision for intervention are discussed in light of the results presented.

13.1.1 Effect of Free and Protein-Bound Bilirubin on
          Bilirubinemia and Associated Kernicterus Disease

Hemoglobin is the major source of heme [1–4]; heme produces bilirubin
during catabolism. Total bilirubin content consists of two major fractions:
free bilirubin and protein-bound bilirubin; protein-bound bilirubin is
mainly albumin-bound bilirubin in plasma or serum [5,6]. A wealth of
experimental and clinical evidence indicates that the presence of toxic
unconjugated (free) bilirubin IXα in the blood can cause devastating brain
injury (*kernicterus*) for jaundiced newborns, affecting their IQ development.
The cause is attributed to the undesirable precipitation of bilirubin upon
strong interaction with unprotected brain tissues that are directly exposed
to the blood in the newborn before they are covered up by a blood–brain
tissue barrier during growth in the neonatal period.

Neonatal jaundice is a major clinical problem among Chinese and Asian
newborns. Most Chinese infants develop significant bilirubinemia ("yellow
baby") during the neonatal period well beyond the physiologic range. Many
reports have shown that kernicterus in newborns is related to the level of total
and free bilirubin in the serum. Some reports have indicated that kernicterus
appears in premature infants due to the low binding capacity of their serum

proteins for bilirubin, even though the total serum bilirubin level is considered to be "safe." Thus, the albumin residual binding capacity for free bilirubin gives rise to a mechanism to remove unbound bilirubin from the blood, providing a safety margin to protect against a sudden surge in free bilirubin in the blood of a neonate that could lead to damage to the growing, unprotected brain.

13.1.2 Albumin and Its Binding Interaction with Free Bilirubin

Serum albumin, as a carrier for bilirubin, can hold and transport bilirubin to other tissues. In skin, bone, muscle, and other tissues, the amount of extravascular bilirubin shows the same compartmental distribution as extravascular albumin [7,8]. To prevent undesirable bilirubin precipitation and associated toxicity, it is necessary to bind bilirubin to the major carrier, albumin, with high affinity so that the equilibrium concentration of free (unbound) bilirubin is kept very low. Thus, it is of utmost importance to measure the residual capacity of albumin to bind bilirubin in clinical diagnosis to assist in applying early medical intervention for individuals with low serum albumin binding capacity for bilirubin.

To prevent the occurrence of devastating brain injury (kernicterus) for jaundiced newborns, the time for exposure of unprotected brain tissue to free bilirubin in neonate blood has to be reduced to the shortest time possible. Thus, a reliable, quick, economic, and simple lab-on-a-chip device is needed for bedside monitoring of free bilirubin and the residual binding capacity of albumin for bilirubin to deliver results in time for making decisions for medical intervention. In addition, birth can occur at any time of the day, and the test should be done when evidence for bilirubinemia is noted. Thus, a simple lab-on-a-chip device such as a microfluidic chip-capillary electrophoresis (MC-CE) device that is capable of performing on-site assay for free bilirubin and its residual binding capacity by albumin is required. One of the critical demands for developing such a device is the capability for on-site titration at the microscale level, with suitable mixing of bilirubin to a given concentration of albumin.

13.1.3 Micromixing for MC-CE Device

A micromixer is an important component in MC-CE devices, because it determines the overall quality of the products formed. Conventional microfluidic mixers use chaotic mixing by either active mixing principles or passive mixing principles for the mixing operation. Due to the small size of the microchannels, the flow tends to be viscous, with low Reynolds number (<10). As the flow in the microfluidic device is essentially laminar, turbulent mixing is difficult to achieve in the microchannels and only diffusion mixing can occur. But diffusion alone is too slow to give rise to a good mix of two solutions in a given length of microchannel within a reasonable time.

To assist with good mixing, active microfluidic mixers have attracted recent attention because they can achieve better mixing effects than passive mixers for mixing large molecules, such as proteins.

Active microfluidic mixers use various external energy sources, such as magnetic, pressure, thermal, acoustic, or electrokinetic energy, to stir or agitate the flow for improving the mixing performance. Early types of active micromixers [9] used pressure field turbulence. Niu and Lee [10] generated an oscillating periodic flow using an actuator system to achieve better mixing efficiency. Lemoff and Lee [11] applied high voltages to a microchannel to assist in mixing. Lin et al. [12,13] used periodical direct current voltage switching to generate alternating electro-osmotic flow to mix channel fluids. Mao et al. [14] studied a linear temperature gradient across several parallel channels for micromixing applications. Acoustic streaming had been used to generate acoustic-induced flow for active mixing [15,16]. However, micromixers using active mixing schemes are, in general, more complex to fabricate and are often unsuitable for disposable applications due to considerations for their high fabrication cost and significant power consumption.

Nguyen et al. [17] studied the actuation of nanoferrofluid droplets driven by magnetic force in a closed-loop channel for micromixer application. The recent availability of strong micromagnets from rare-earth alloys makes it easy to generate a high magnetic field gradient at miniaturized areas. Hatch et al. [18] developed a ferrofluid-driven micropump using an external magnet. Yamahata et al. [19] also reported the use of similar magnetic actuation. The two devices could be used for continuously pumping liquids. The closed-loop circular MC-CE device uses a very simple design that can be easily fabricated at the microscale level. The magnet-driven micromixer has the advantages of low capital and running costs, minimal power consumption in operation, and highly effective and flexible mixing with adjustable number of cycles that can be activated directly within the MC device without additional fabrication requirements.

Based on a circular-ferrofluid-driven micromixer, a new multisegment incremental sample injection (MISI) procedure was developed, as described in this chapter, to prepare microscale samples with different albumin/bilirubin ratios for testing is a time-intensive operation. MISI can be realized by using a fixed concentration of bilirubin with the addition of different volumes of bilirubin to mix with the same volume of albumin. This approach greatly shortens the time for sample preparation prior to the use of CE/frontal analysis (FA) for free bilirubin determination. In addition, the mixing efficiency using a magnetic-driven micromixer can be easily optimized by controlling the rotating speed and cycle number of an external magnet. Different homogenized solutions with targeted bilirubin/albumin ratios can be prepared quickly on-site for assay of the binding capacity of albumin for free bilirubin using the MC-CE device integrated with MISI injection.

13.2 Materials, Standards, and Reagents

13.2.1 Preparation of Standards and Reagents

Bilirubin IXα and human serum albumin ([HSA] fraction V) were purchased from Sigma Chemical (Perth, WA, Australia), and ferrofluid APG S12n was provided from Ferrofluidic Inc. (Nashua, NH, USA). To avoid photodecomposition of bilirubin, all preparative procedures were performed in a dark room. All working standard solutions were diluted to the appropriate concentrations from stock solutions, centrifuged for 3 min at 13,200 rpm, and then filtered through a 0.45-μm membrane filter.

Running buffer (pH 7.4) was prepared by mixing 1 mmol/L ethylenediaminetetraacetic acid (EDTA) and 10 mmol/L sodium phosphate. Bilirubin stock solution was prepared by mixing 20 μL of 0.1 mol/L EDTA, 200 μL of 0.1 mol/L NaOH, and 2.5 mg of bilirubin with 780 μL of distilled water. Albumin stock solution was prepared by adding 0.048 g of HSA to 1 mL of phosphate-buffered saline buffer.

13.2.2 Magnetic-Driven Nanosized Ferrofluid

Ferrofluid is prepared by dispersing specially coated ferroparticles, with an average size of 10 nm, to a carrier oil with the addition of a dispersing agent to prevent deposition and maintain stability. Ferrofluid is always in colloidal state and thermally stable. One advantage of the oil-based ferrofluid is its immiscibility with the water-based bilirubin–albumin mixture, making it possible to drive the various segments of aqueous solutions without interfering interactions. The ferrofluid has a viscosity (η_{ff}) of 250 mPa·s, a density (ρ_{ff}) of 1.32 g/cm^3, and a surface tension (σ_{ff}) of 32 mN/m at 25°C. Thus, it can be considered a viscous and dense liquid that can be used to direct the movement of other liquid segments.

In the absence of an external magnetic field, ferroparticles are randomly diffused. However, in the presence of a magnetic field, the ferroparticles coaggregate along the magnetic field direction and repulse adjacent particles. In another words, the ferrofluid acts as magnet upon activation by an external magnetic field. Any change in the externally applied magnetic field leads to a quick response in changing the status of the ferrofluid from the nonmagnetic to the magnetic state, depending on the removal or introduction of an external magnetic field. An external magnetic field can be precisely positioned to control the movement of ferrofluid. The gradient of the external field and the magnetization value of the fluid are proportional to the forces holding the magnetic fluid. All of the above-mentioned actions showed that a ferrofluid plug can be used to drive the mixing liquids to move in a microchannel until a homogenized solution has been produced without the need for fabricating complicated microstructure or performing complex operations required in the current micromixers.

13.3 Design and Fabrication of MC-CE Device with MISI

13.3.1 Instrumentation and Facilities for Fabrication and Operation of MC-CE Device

Poly(methyl methacrylate) (PMMA) polymer chips with desired channel patterns were ablated by a CO_2 laser engraver controlled by computer software. A hot press bonding machine was used for PMMA MC-CE device fabrication. The channels of PMMA MC-CE device were observed on a home-made sample workstation during operation. The applied high voltage to the MC-CE device was controlled by an eight-channel, high-voltage power supply. A portable ultraviolet-visible detector was coupled with the separation capillary for detection after MC-CE separation.

Four magnets were mounted underneath the bottom of the MC-CE device at the region of the circular microchannel. The magnet was driven by a self-constructed electronic motor to generate a rotating magnetic field to control the movement of the ferrofluid for mixing of the bilirubin and albumin segments. When the magnet turned 360° in a circle, a full mixing cycle was counted for estimation of the mixing efficiency by the number of cycles performed during mixing.

13.3.2 Design and Layout of MC-CE Device with MISI

Compared to the rectangular serpentine devices currently used in micromixing, the design of the MC is very simple. A deep closed-loop circular channel is utilized to mix bilirubin and albumin. Two shallow long and straight channels were fabricated tangentially to the circular channel to inject and collect samples. The closed-loop circular mixing coil was ablated onto the PMMA plate by CO_2 laser. The MISI microchannel pattern was fabricated on-chip according to the computer-aided design manufacturing procedure under the control of commonly used software (CorelDRAW 10, Corel Corporation, Ottawa, Ontario, Canada) which delivered instruction to a CO_2 laser engraver for direct micromachining on a given PMMA chip. The access holes were also drilled as reservoir/vials on another PMMA slice as the cover. Finally, the two plates were bonded under a pressure of 0.6 MPa at 92°C for 15 min by a hot bonding machine.

The dimensions of the MC fabricated are 60 mm × 82.5 mm × 0.15 mm (width × height × thickness) (Figure 13.1) [20]. The circular channel was 200 μm in width, 250 μm in depth, and 62.8 mm in length, with a radius of 20 mm. Measured from the top of the channel, the channel was ablated typically to 100 μm in depth and 150 μm in width for all straight channels. The distance between each reservoir (see Figure 13.1, AR, WR, MR, and MW) and the double-T injector was 5 mm. The effective length of the left inlet straight channel for sample mixing was 48 mm. It consisted of three different lengths for buffer, sample, and albumin, respectively. As the channel length

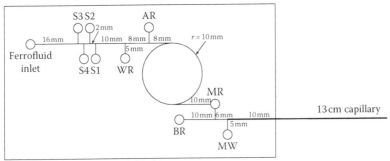

FIGURE 13.1
Schematic diagram showing the layout of MC-CE device with MISI and its coupling to the fused silica separation capillary. S1, S2, S3, and S4 are sample reservoirs containing the same concentration of bilirubin; AR, HSA reservoir; WR, waste reservoir for bilirubin and albumin; BR, buffer reservoir; MR, bilirubin–HSA mixture reservoir; MW, bilirubin–HSA mixture waste reservoir. (From Sun, H., Nie, Z. and Fung, Y.S. *Electrophoresis* 2010, 31, 3061–3069.)

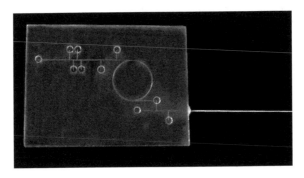

FIGURE 13.2
MC-CE device with MISI and the circular micromixer fabricated on PMMA plate by the CO_2 laser.

of albumin was kept constant at 8 mm, the sum of the variable channel length of the buffer and sample segment is 40 mm. To realize MISI injection on MC, different lengths of channel (S1–S4) were used to add different volumes of sample (bilirubin) for injection. For S1, for example, the injection length of S1 is 10 mm. Compared with the injection length of albumin at 8 mm, the ratio of the injection volume of bilirubin to albumin is 1.25. As the distance between each sample vial (S1–S4) is 2 mm, other injection ratios at 1.5, 1.75, and 2.0 of injection volume of bilirubin to albumin can be easily produced.

After a complete mixing of bilirubin and albumin inside the closed-loop circular channel, the mixture was then transferred to the right straight channel for detection. A large sample plug introduced by the 6-mm double-T injector and a 13-cm-long capillary (50 μm inside diameter) were used for CE separation (Figure 13.2).

13.4 Optimization of On-Chip Mixing and MISI Injection Procedures

13.4.1 Procedures and Operations

For each analysis, dilute NaOH solution, deionized water, and running buffer were used in sequence to condition the two straight microchannels of the MC-CE device for 5 min each. The whole operation procedure is divided into two phases: the before-mixing phase and the after-mixing phase. The before-mixing phase includes the procedure for MISI and mixing in the circular mixing zone; the after-mixing phase includes the injection of the well-mixed mixture for CE separation in the 13-cm fused silica capillary.

In the before-mixing phase, 10 μL of bilirubin solution was added to each of the four SRs and 10 μL of the HSA solution to the AR vial. For sample injection, 2100 V was applied to SR and AR while WR was kept at ground for 30 s. Immediately after sample injection, each vial was sealed using a PMMA membrane. The ferrofluid plug was then added to the ferrofluid inlet of the channel and moved to the circular channel by an external magnet. The bilirubin–albumin mixture in the inlet channel (Figure 13.3) was pushed and kept inside the circular channel for continuous circular mixing. The rotational speed of the magnet was adjusted, and the number of mixing cycles was controlled to vary different time durations for mixing bilirubin–albumin solution at specified ratios. Thus, the rotation speed and the different numbers of rotation cycles of the magnet could be optimized to obtain a fully mixed solution.

In the after-mixing phase, the fully mixed solution was pushed by an external magnet into MR for subsequent CE separation and detection. Injection of the well-mixed bilirubin–albumin solution for CE separation was then performed by applying 2100 V for 30 s to MR while MW was grounded. Immediately after the sample injection of a specified mixed

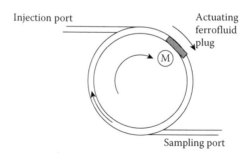

FIGURE 13.3

Schematic diagram showing the mixing operation carried out in the micromixer driven by a ferrofluid plug. The mixed bilirubin–albumin solution is driven by a ferrofluid plug within the circular channel. M, rotating magnet. (From Sun, H., Nie, Z. and Fung, Y.S. *Electrophoresis* 2010, 31, 3061–3069.)

bilirubin–albumin solution, high voltage (3500 V) was added to BR, grounded voltage to BW, and 2100 V to MW and MR at the same time.

13.4.2 Effect of Cycle Numbers

To prepare a given ratio of bilirubin to albumin at 3.5, loadings of 720 µmol/L albumin at AR and 1440 µmol/L bilirubin at S3 were added to the MC-CE device to test whether bilirubin and albumin can be well mixed using the magnetic-driven mixing concept. The effects of the rotation speed on the cycle number were investigated, and the rotation speed at 5 rpm was found to be capable of achieving satisfactory and well-controlled magnetic activation during rotation, which is equal to a cycle time of 12 s. The electropherograms obtained at different cycle numbers using a constant mixing ratio of bilirubin/albumin of 3.5 are shown in Figures 13.4–13.7.

Two sharp peaks were noted (Figure 13.4) after one complete cycle, indicating that free bilirubin was not at dynamic equilibrium after 1 cycle number and was dissociated from the bilirubin–albumin mixture. An uneven plateau and a peak were still visible at cycle number 2 (Figure 13.5). A small flat plateau and a peak were obtained at cycle number 3 (Figure 13.6), showing that the frontal condition due to a well-mixed mixture had been achieved. A small flat

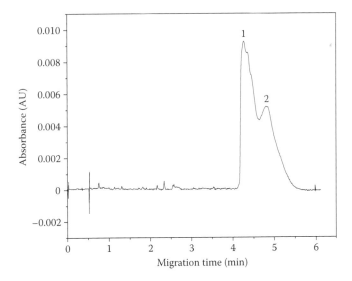

FIGURE 13.4
Electropherogram of a bilirubin–albumin mixture with a mixing ratio of 3.5 for a cycle number at 1. Rotation speed of the magnet is 5 rpm. Running buffer (pH 7.4): 1 mmol/L EDTA + 10 mmol/L sodium phosphate buffer; sample: 1440 µmol/L bilirubin + 720 µmol/L albumin. Detection at 440 nm. Before-cycle mixing: injection, SR and AR (2100 V) and WR (grounded) for 30 s; after-cycle mixing: injection, MR (2100 V) and MW (grounded) for 30 s. Separation: BR (3500 V), BW and MR (grounded), and MW (2100 V). BR, BW, SR, AR, MR, and MW are the same as in Figure 13.1. 1, albumin-bound bilirubin; 2, free bilirubin. (From Sun, H., Nie, Z. and Fung, Y.S. *Electrophoresis* 2010, 31, 3061–3069.)

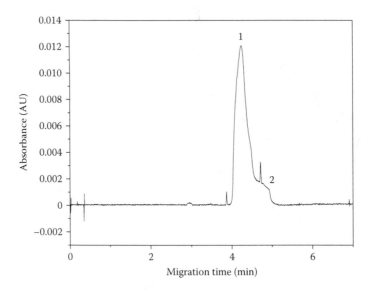

FIGURE 13.5
Electropherogram of a bilirubin–albumin mixture with a mixing ratio of 3.5 for a cycle number at 2. Other conditions are the same as in Figure 13.4.

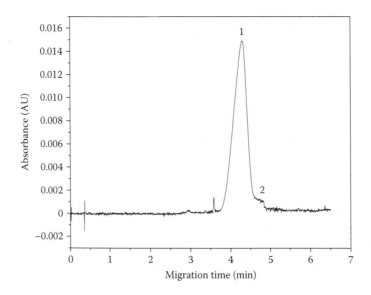

FIGURE 13.6
Electropherogram of a bilirubin–albumin mixture with a mixing ratio of 3.5 for a cycle number at 3. Other conditions are the same as in Figure 13.4.

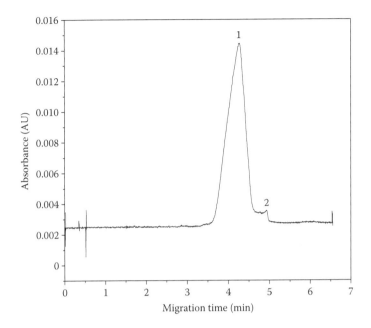

FIGURE 13.7
Electropherogram of a bilirubin–albumin mixture with a mixing ratio of 3.5 for a cycle number at 4. Other conditions are the same as in Figure 13.4. (From Sun, H., Nie, Z. and Fung, Y.S. *Electrophoresis* 2010, 31, 3061–3069.)

plateau and a peak were found to persist for increasing the cycle number to 4 (Figure 13.7). Thus, the cycle number was selected at 3 as it gave a sufficiently well-mixed bilirubin–albumin mixture within the shortest mixing time of 36 s.

As shown in Figure 13.8, the effect of the cycle number on the efficiency of mixing was demonstrated by plotting the absorbance of free bilirubin at different cycle numbers. The free bilirubin plateau expressed in absorbance units decreased exponentially from cycle number 1 to 2, followed by a gradual decline (Figure 13.8). The curve began to smooth at around cycle number 3, indicating that the mixing efficiency for the bilirubin–albumin mixture was improved with the cycle number due to a rapid reduction of the free bilirubin concentration.

13.4.3 Effect of the Velocity of the Driving Magnet

To reduce photodegradation of free bilirubin, the mixing time of the mixture should be kept as short as possible. This means that the velocity of the driving magnet should be as fast as possible. To investigate the effect of the velocity of the driving magnet on the efficiency of mixing, four different velocities of the driving magnet—5, 7.5, 10, and 15 rpm—were tested while the cycle number was kept at 3.

The free bilirubin plateau did not show a marked change with the increase in velocity of the driving magnet (Figure 13.9). However, for the high-viscosity

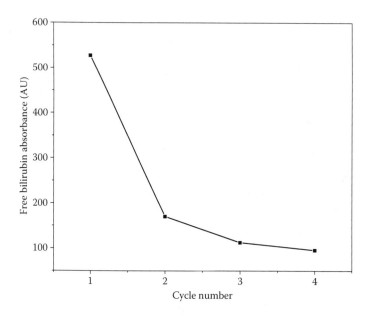

FIGURE 13.8
Effect of the cycle number of the circular micromixer on the mixing efficiency for a bilirubin–albumin mixture at a volume ratio of 3.5. Other conditions are the same as in Figure 13.4.

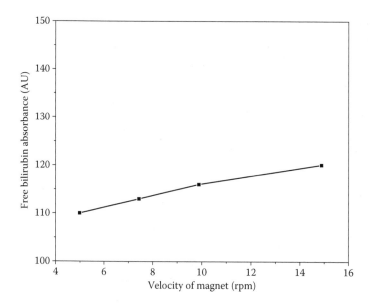

FIGURE 13.9
Effect of velocity of the driving magnet on the mixing efficiency of a bilirubin–albumin mixture at a volume ratio of 3.5 and a constant cycle number at 3. Other conditions are the same as in Figure 13.4.

ferrofluid plug, it tended to adhere to the wall of the circular microchannel due to its counteraction with the driving magnet. Moreover, the ferrofluid plug was easy to break down into several small segments under the high velocity of the driving magnet. This can affect the mixing efficiency for the mixture. Thus, the lowest velocity of driving magnet is adopted for the present mixing application. Limited by the control of the current electric motor used, the lowest velocity of 5 rpm was used.

13.5 Application of MC-CE Device for Assay of Free Bilirubin and Albumin Binding Capacity for Bilirubin

13.5.1 Integrating FA with MC-CE Device for Assay of Free or Unbound Bilirubin in HSA

Measurement under FA condition is essential for accurate assay of free or unbound bilirubin concentration as well as for the determination of albumin binding capacity for bilirubin. During the course of titration between bilirubin and albumin with fixed albumin and increasing bilirubin concentration as shown in Figure 13.10, the appearance of a flat plateau (peak 2) in the electropherogram at a given albumin/bilirubin ratio indicates that the frontal condition has been achieved with unbound bilirubin drawn out from the albumin peak (peak 1) at a fixed concentration, the same as the concentration of free or unbound bilirubin at equilibrium with albumin. Thus, the determination of the bilirubin concentration at the absorbance of the flat plateau represents the concentration of free or unbound bilirubin at equilibrium with the albumin–bilirubin mixture present in HSA (peak 1).

To achieve the frontal condition, a large and variable size of the sample plug for bilirubin has to be used for titration with a fixed albumin concentration. Compared to the injection length of albumin at 8 mm, the ratio of the injection volume of bilirubin to albumin is 1.25. As the distance between each sample vial (S1–S4) is 2 mm, other ratios at 1.5, 1.75, and 2.0 can be generated. This can be achieved by keeping the albumin concentration constant at 720 µmol/L while adding bilirubin (1440 µmol/L) from different sample reservoirs (S1–S4) to make up the desired ratios of albumin–bilirubin mixtures at 2.5, 3.0, 3.5, and 4.0.

13.5.2 Assay of Albumin Binding Capacity for Free Bilirubin with MC-CE/FA Device

An accurate assay of the binding capacity of HSA for free bilirubin is important, not only in the clinical application of the diagnostic test but also in establishment of the correct quantitative relationship between

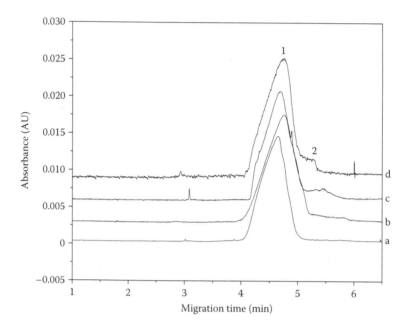

FIGURE 13.10
Electropherograms showing the change of peak profile after the addition of different amounts of bilirubin to samples with constant albumin concentration. Sample: AR, 720 μmol/L albumin and S1–S4, 1440 μmol/L bilirubin. MISI of different ratios of bilirubin to HSA. a, (S1) 2.5; b, (S2) 3.0; c, (S3) 3.5; and d, (S4) 4.0. 1, albumin-bound bilirubin; 2, free bilirubin. Other conditions are the same as in Figure 13.4.

bilirubin binding in serum and the uptake and toxicity of bilirubin in tissues. To investigate the binding capacity of HSA, different volume ratios of bilirubin–albumin mixture were added by MISI via the MC-CE device integrated with FA. To realize the MISI injection mode on the microchip, different lengths of the channel (S1–S4) were used to add different volumes of sample (bilirubin) for injection. For S1, for example, the injection length is 10 mm.

To assay the residual binding capacity of albumin for free bilirubin, the absorbance of the plateau due to free bilirubin was plotted against different amounts of bilirubin added with extrapolation to obtain the real amount of bilirubin needed to saturate the albumin before free bilirubin began to appear in the solution (Figure 13.11). The binding capacity of albumin for bilirubin is 17.16 ± 1.37 mg/100 mL ($n = 3$). For normal infants, taking 120 μmol/L as the average concentration of HSA, free bilirubin could be detected after accumulative addition of bilirubin exceeding 17.16 ± 1.37 mg in 100 mL of HSA under the physiologic condition. The results obtained match the normal range for newborns (6.5~20 mg/100 mL) [21].

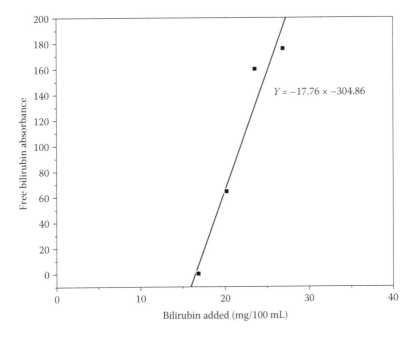

FIGURE 13.11
Determination of albumin binding capacity for bilirubin by MC-CE/FA device. Other conditions are the same as in Figure 13.4. (From Sun, H., Nie, Z. and Fung, Y.S. *Electrophoresis* 2010, 31, 3061–3069.)

13.6 Summary

To reduce the photodegradation of free bilirubin during sample preparation, a MISI method was successfully developed for on-site mixing using a circular micromixer driven by a ferrofluid plug. An MC-CE device integrated with MISI was fabricated to follow the binding interaction between bilirubin and HSA under FA conditions during the course of titration for the measurement of free bilirubin concentration and the residual binding capacity of albumin for bilirubin. The MISI procedure was optimized to inject bilirubin–albumin mixed segments at desirable ratios to the circular mixing coil to generate a well-mixed bilirubin–albumin solution for CE determination of the HSA binding capacity for unbound bilirubin under FA conditions. The binding capacity of normal albumin was found to be 17.16 ± 1.37 mg/100 mL ($n = 3$).

MISI can be realized by keeping a constant concentration of albumin with the addition of different volumes of bilirubin for mixing. This can greatly shorten the time for sample preparation. The actuation by ferrofluid plug is shown to be operable on microchannels for on-chip mixing of different microfluid segments as high magnetic field gradient can be generated and controlled for on-site application using an MC-CE device. In the present work, the integration of an MC-CE device with MISI with a ferrofluid-driven micromixer was shown to be suitable to determine free bilirubin concentration and the residual binding capacity of albumin for unbound bilirubin. The device developed was able to meet the clinical requirement for detecting an early rise of the free bilirubin concentration in HSA prior to the onset of jaundice.

Acknowledgments

We would like to acknowledge the contribution from Dr Ka Yin Wong, the Department of Paediatrics and Adolescent Medicine, The University of Hong Kong, Hong Kong SAR, People's Republic of China on the collaborated research reported in this chapter.

References

1. Lester, R. and Troxler, R.F. Recent advances in bile pigment metabolism. *Gastroenterology* 1969, 56, 143–169.
2. Vuchovich, D.M., Haimowitz, N., Bowers, N.D., Cosbey, J. and Hsia, D.Y.Y. The influence of serum bilirubin levels upon the ultimate development of low-birth-weight infants. *J. Mental Deficiency Res* 1965, 9, 51–60.
3. Arias, I.M. Inheritable and congenital hyperbilirubinemia. Models for the study of drug metabolism. *N Engl J Med* 1971, 285, 1416–1421.
4. Maisels, M.J. Bilirubin; on understanding and influencing its metabolism in the newborn infant. *Pediatr Clin North Am* 1972, 19, 447–501.
5. Lallff, J.J., Kasper, M.E., Wu, T.W. and Ambrose, R.T. Isolation and preliminary characterization of a fraction of bilirubin in serum that is firmly bound to protein. *Clin Chern* 1982, 28, 629–637.
6. Wu, T.W. Bilirubin analysis—The state of the art and future prospects. *Clin Biochem* 1984, 17, 221–229.
7. Schimid, R. and Hammaker, L. Metabolism and disposition of C^{14}-bilirubin in congenital nonhemolytic jaundice. *J Clin Invest* 1963, 42, 1720–1725.
8. Brodersen, R. Localization of bilirubin pools in the non-jaundiced rat, with a note on bilirubin dynamics in normal human adults and in Gilbert's syndrome. *Scand J Clin Lab Invest* 1972, 30, 95–98.
9. Glasgow, I. and Aubry, N. Enhancement of microfluidic mixing using time pulsing. *Lab Chip* 2003, 3, 114–120.
10. Niu, X. and Lee, Y.K. Efficient spatial-temporal chaotic mixing in microchannels. *J Micromech Microeng* 2003, 13, 454–462.
11. Lemoff, A.V. and Lee, A.P. AC magnetohydrodynamic micropump. *Sens Actuators B* 2000, 63, 178–185.

12. Lin, C.H., Fu, L.M. and Chien, Y.S. Microfluidic T-form mixer utilizing switching electroosmotic flow. *Anal Chem* 2004, 76, 5265–5272.
13. Fu, L.M., Yang, R.J., Lin, C.H. and Chien, Y.S. A novel microfluidic mixer utilizing electrokinetic driving forces under low switching frequency. *Electrophoresis* 2005, 26, 1814–1824.
14. Mao, H., Yang, T. and Cremer, P.S. A microfluidic device with a linear temperature gradient for parallel and combinatorial measurements. *J Am Chem Soc* 2002, 124, 4432–4435.
15. Yang, Z., Goto, H., Matsumoto, M. and Maeda, R. Active micromixer for microfluidic systems using lead-zirconate-titanate(PZT)-generated ultrasonic vibration. *Electrophoresis* 2000, 21, 116–119.
16. Yaralioglu, G.G., Wygant, I.O., Marentis, T.C. and Khuri-Yakub, B.T. Ultrasonic mixing in microfluidic channels using integrated transducers. *Anal Chem* 2004, 76, 3694–3698.
17. Nguyen, N.T., Ng, K.M. and Huang, X.Y. Manipulation of ferrofluid droplets using planar coils. *Appl Phys Lett* 2006, 89, 052509-1–052509-3.
18. Hatch, A., Kamholz, A.E., Holman, G., Yager, P. and Bohringer, K.F. A ferrofluidic magnetic micropump. *J Microelectromech Syst* 2001, 10, 215–223.
19. Yamahata, C., Chastellain, M., Parashar, V.K., Petri, A., Hofmann, H. and Gijs, M.A. Plastic micropump with ferrofluidic actuation. *J Microelectromech Syst* 2005, 14, 96–104.
20. Sun, H., Nie, Z. and Fung, Y.S. Determination of free bilirubin and its binding capacity by HSA using a microfluidic chip-capillary electrophoresis device with a multi-segment circular ferrofluid-driven micromixing injection. *Electrophoresis* 2010, 31, 3061–3069.
21. Michael, A.B. and Michael, C. *Oski's Essential Pediatrics*. Lippincott Williams & Wilkins, Philadelphia, PA, 2004.

14

Counting Organelles: Direct Counting of Mitochondrial Numbers

Wenfeng Zhao

Jiangsu Normal University
Xuzhou, People's Republic of China

Ying Sing Fung

The University of Hong Kong
Hong Kong SAR, People's Republic of China

CONTENTS

14.1 Introduction

Mitochondrial number is an important parameter to reflect the functions linked to muscular and dilated cardiomyopathy diseases, and the aging process. Thus, an accurate method for determination of mitochondrial number is essential to obtain important information for the assessment of the progression of the above-mentioned diseases. There are two major approaches for measurement of the mitochondrial number of a given biological sample, such as tissues and cells. The first, direct approach is based on the isolation of mitochondria from the biological sample for assessment of their activity and numbers. The methodology based on this approach is explored in this chapter. The second, indirect approach is based on measurement of a biomarker specifically linked to the mitochondrial number to obtain the average values of mitochondrial numbers present in a given sample. The methodology based on this indirect approach is discussed in Chapter 15.

In addition to mitochondria, many biological entities exist as particles, such as cells [1,2], bacteria [3], proteins [4–7], viruses [8], and organelles [9]. Lengthy sample pretreatments are often required to isolate mitochondria from other particles present in a biological sample. The extracts with enriched mitochondria are then separated, with the isolated mitochondria identified and counted. Capillary electrophoresis (CE) has become an important tool for separation of mitochondria due to its short analysis time, high separation efficiency, and consumption of microliters of samples and buffers.

In contrast to separation of molecular analytes, separation of particles is more complex as particle separation depends both on size [10] and surface chemistry [11]. In addition, organelles such as mitochondria are not rigid and nonconducting; they exhibit different behaviors compared to rigid spherical particles upon their electrophoretic separation. Thus, the theoretical aspects on factors affecting the migration of soft mitochondrial particles are presented, with issues discussed in Section 14.2.

Laser-induced fluorescence (LIF) offers a suitable mode for the detection of fluorescently labeled organelles due to its excellent high sensitivity and selectivity. To enable efficient collection of LIF light emitted from individual mitochondrial particles at a desirable spot at an inner capillary region, a microfluidic chip-capillary electrophoresis (MC-CE) device is fabricated. This device integrates microfluidic chip (MC) with LIF operated in fluorescence spike mode for highly efficient counting of individual mitochondrial particles passing through the detection zone for determination of the mitochondrial number, with results discussed in Section 14.3.

14.2 Theoretical Aspects of Electrophoretic Separation of Particles

14.2.1 Classical Model for Spherical Particles

Current theory developed to explain the electrophoretic behavior of particles is mostly based on the assumption that they are rigid, nonconducting spheres. It is known that a charged particle is surrounded by an ionic atmosphere in an electrolyte solution. The formation of a diffused ionic cloud during electrophoresis is attributed to the distortion of the electric double layer formed at the interface between the electrolytic solution and the particle surface. Based on the surface electrostatic [12] and electrokinetic theory [13], there are four forces acting on a charged particle: 1) positive electrophoretic forces and 2) negative forces due to the effect of relaxation of the electron cloud, 3) Stokes viscous drag, and 4) media friction. The effect of these four forces on a charged particle under an electric field is shown in Figure 14.1. The externally applied electric field gradient provides the driving force for electrophoresis. The relaxation effect arising from the deformation of the ionic atmosphere surrounding the migrating charged particle during electrophoresis provides the differential retardation force to separate analytes. Both the Stokes viscous drag and the media friction

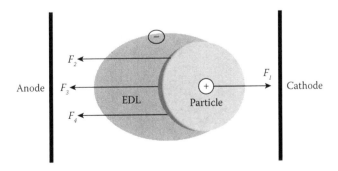

FIGURE 14.1
Schematic diagram showing the external forces acting on a charged particle from an electric field. F_1, force due to electric field gradient; F_2, force due to effect of relaxation; F_3, the retarding forces due to Stokes viscous drag; and F_4, the media friction.

force arising from the interaction of the particle with the surrounding media during its movement are opposing the movement of the particle.

Molecular electrophoretic theory assumes that the interaction between small analytes and the electrolytes in the solution is negligible. However, larger analytes with a charged surface can form an electrical double layer of counter ions, similar to the layer formed at the fused silica surface [14].

If the particle is assumed to be a rigid, nonconducting sphere with a small electrokinetic potential ($\varsigma \leq 25$ mV at 25°C) [15], its electrophoretic mobility (μ) should follow the equation

$$\mu = \frac{2\varepsilon\varsigma}{3\eta} f(\kappa r) \tag{14.1}$$

where ε is the dielectric constant of a solvent, ς is the zeta potential, η is the viscosity of the solvent, r is the radius of the particle, and κ is the Debye parameter. ς is defined as follows:

$$\varsigma = \frac{\sigma r}{\varepsilon(1+\kappa r)} \tag{14.2}$$

and σ is the net charge density of the particle [16], and σ is defined as follows:

$$\sigma = \frac{q}{4\pi r^2} \tag{14.3}$$

As the Hückel limit is approached, Equation 14.1 approaches the description of the electrophoretic mobility. Equation 14.2 suggests that the electrophoretic mobility of a particle is dependent on the zeta potential, particle diameter (r), and buffer compositions. Note that this equation is not universal and is based on several approximations, such as neglecting relaxation effect and idealizing particle surface [17]. Thus, Equations 14.1 and 14.2 are only applicable when the diffuse layer is negligible (e.g., $\kappa r \gg 1$, the Hückel limit; $\kappa r \ll 1$, and the Smoluchowski approximation or $|\varsigma| \leq 25$ mV).

14.2.2 Relaxation Effects

When the zeta potential of a particle is high, the description of the electrophoretic mobility of charged particles becomes very complex due to the relaxation effect arising from the distortion of the surrounding ionic atmosphere. Theories were first developed independently by Overbeek [18] and Booth [19] to predict the electrophoretic mobility of spherical nonconducting particles, and later were expanded by O'Brien and White [12]. As the particles are moving, the ionic atmosphere is lagging behind the particle in the shape of a comet-like tail, leading to an additional drag, termed as the "relaxation effect." The relaxation effect leads to a much stronger dependence of μ and κr than that according to Equation 14.2. The electrophoretic mobility goes to a maximum at a function of ζ for a specific κr ($\kappa r > 3$), but at the limit of $\kappa r \to 0$ and $\kappa r \to \infty$, the linear relationship between μ and ζ is reestablished. Thus, the mobility of a particle is decreased with an increasing zeta potential. Fortunately, particles with small potential ($|\zeta| \leq 25$ mV) [15] do not have significant relaxation effect. It is reported that the nonrespiring rat liver mitochondria show $\zeta = -21$ mV, with negligible relaxation effect [20]. In addition, the decrease in the dominance of the particle's electric field compared to κr is expected to cause a reduction in the relaxation effect. According to these factors, the relaxation effect can be ignored during the separation of mitochondrial particles.

14.2.3 Nonideal Particles

Based on the theories mentioned above, it is still difficult to interpret and describe how mitochondria migrate and separate under an electric field. The theory is developed based on the assumption that the particles are typically rigid and nonconducting spheres. However, mitochondria are nonspherical with deformable surfaces and mobile surface charges. Hayes' group used liposomes as a model to study their electrophoretic behavior under the effects of particle deformation [21,22]. Their theories were further modified by the Arriaga group using organelles as models [23–28]. Based on the electrokinetic theory and the surface characteristics of mitochondria, new theories are proposed to incorporate the electrophoretic behavior of deformable particles such as mitochondria, with discussions given in Section 14.2.4.

14.2.4 Nonspherical Particles

As the theory for the electrophoretic mobility of nonspherical particles is significantly more complex than the well-addressed theories for spherical particles, there are few systematic studies carried out on the electrophoresis of ellipsoid or rod-shaped particles [16]. The major force for electrophoretic mobility of nonspherical particles is due to the influence on the electric field-induced orientation. Tobacco mosaic virus (TMV) has been used as a model particle to demonstrate the dependence of the electrophoretic mobility on the orientation of the rod-shaped particles [29]. TMV can be considered as a rigid rod ~340 nm in length (r_2) and 15 nm in diameter (r_1). With the increase in the electric fields, the particle can approach $0°$ alignment [30]. A reduction of 20% of the frictional drag leads to a 25% increase in the electrophoretic mobility of TMV. Based on a similar calculation, the maximal change in the mitochondrial drag of an isolated mitochondrial particle ($r_1 = 200$ nm, $r_2 = 500$ nm) is only -3% (corresponding to a 3% increase in mobility). The 3% value is smaller than the standard error typically obtained from mitochondrial data [31]. As the consequence, change in 0.7% of the mitochondrial orientation angle is less significantly affected by the electrical fields. Combining these factors, the difference in mobility due to the orientation from electrical fields can be negligible compared to the standard error.

14.2.5 Uneven Surface Charge Distribution

One of the important factors affecting the electrophoretic migration of a particle is its surface charge [32]. An important assumption made about the electrophoretic behavior of a model particle is that the charge is evenly distributed on the surface of the particle. Although it can be assumed that the surface charges are negative as the mitochondrial surface is mainly composed of negatively charged lipids, it is difficult to approximate the net charge due to the uncertainty on the distribution of phospholipids between the inner and outer mitochondrial membrane surfaces, which can easily induce a dipole or more complex multipole moment on an individual mitochondrion. Mitochondria are known to contain hundreds of proteins, and the spatial distribution of membrane phospholipids is random. The outer membrane is highly permeable to enable the transportation of ions and small solutes into the intermembrane space [33]. The electron transport chain is located in the inner membrane of mitochondria, and a membrane potential and a pH gradient are generated across the membrane in which cations are accumulated in the matrix [34]. It has been shown that proteins and lipids are asymmetrically located in the mitochondrial membrane [35]. As the composition of the outer membrane of mitochondria determines the zeta potential of a particle, its surface characteristics contribute to a complicated electrophoretic behavior of mitochondrial particles during the separation process.

14.2.6 Deformable Particles

Under an electric field, the pressure and shear forces exerted on a "soft particle" are capable of deforming the normally spherical particle into an oblong ellipsoid with the long axis oriented with the electric field [36,37]. The viscous drag is reduced significantly with an increase in the electrophoretic mobility as the particle becomes more compressed. Adding nonionic lipid cholesterols can increase the rigidity of the membrane [21]. Although the mitochondrial membranes do contain some cholesterols, the typical levels in mitochondria are so low that they can only provide a slight stabilization. An additional factor is the effect of distribution of the surface charge on mitochondrial deformation in the presence of an applied electric field [22]. The difference between mitochondria and other "soft particles" is the presence of numerous cristae at the inner membrane which may act to improve the stability of the membrane. Despite the presence of cristae, it is still unavoidable that mitochondria may undergo some deformation in the presence of an electric field which can introduce variability on the migration time.

14.2.7 Separation of Individual Mitochondrial Particles

How particles migrate and separate under an electric field is an interesting topic for study. As discussed above, when the Hückel limit is approached, the analytes are considered as charges in electrophoresis. Above the Hückel limit [38] but below the Smoluchowski approximation [39], electrophoretic migration becomes dependent on the size of the particle through both the Henry function and ζ via Equation 14.2. Separation of particles in this range is just size dependent, assuming that they have the same σ [10,15,17,40,41]. It is not a nontrivial problem to separate particles that contain differences in both size and surface chemistry to relate the difference in electrophoretic behavior directly to particle characteristics such as size or σ. Therefore, the ionic strength of the buffer should be adjusted to achieve a suitable κ to eliminate these issues so that the system can be operated either in Hückel limit or within the Smoluchowski approximation. Particles in the microrange satisfy the Smoluchowski approximation, meaning that Equation 14.2 can be refined as

$$\varsigma = \frac{\sigma}{\varepsilon \kappa} \qquad (14.4)$$

and Equation 14.1 becomes

$$\mu = \frac{\sigma}{\eta \kappa} \qquad (14.5)$$

Under these conditions, the electrophoretic behavior is without any direct dependence on the size but instead on the charge density at the particle surface. In another words, particles of different sizes with the same surface charge density will comigrate. As for mitochondrial particles, the κr is fairly large. According to Equation 14.1, for large κr values, the ζ potential (and particles mobility with identical or similar surface charge density) is independent of particle size. An intact mitochondrion is typically in the 200–500 nm range [42–44], and the buffers used in the separation system typically have an ionic strength of 4.15 mM KCl, which corresponds to a $\kappa = 0.208$ nm^{-1}. Even using a smaller size of mitochondria for calculation, κr is approximately equal to 40, a value that is still larger than the threshold of the Smoluchowski approximation, considered typically to be $\kappa r \approx 20$ [15]. Therefore, migration of mitochondrial particles in electric field is no longer directly dependent on their size and mainly governed by a function of their surface charge density.

14.3 Separation of Mitochondria by CE and MC Electrophoresis

14.3.1 CE Separation of Individual Organelles

The major problem for CE separation of organelles is the strong interaction between the organelles and the capillary wall, which can lead to adsorption of analytes, such as large peptides and proteins, at the inner surface of the capillary. Whiting and Arriaga [45] compared a separation of individual mitochondria from rat and mouse livers using uncoated and dynamically coated capillaries. It is possible to separate mitochondria from mouse liver with uncoated capillary in 30-min runs without a significant amount of carryover. However, rat liver mitochondria exhibit strong interaction with the inner surface of uncoated capillary, making it impossible to obtain reliable data. Therefore, the inner surface of the capillary should be modified to minimize the adsorption of the mitochondria.

Generally, there are two common approaches used for capillary modification. First, by adding a component to the buffer that will preferentially adsorb to the wall (dynamic coating); and second, by irreversible covalent immobilization of appropriate molecules [46–48]. The former approach is simpler to perform, but it requires the maintenance of a constant concentration of a surface-modifying chemical (e.g., polyvinyl alcohol [49–51]) in the buffer system during the entire separation process, which is not always acceptable.

Capillaries with a permanent coating at the inner surface of the capillary wall are now available from commercial sources. Typically, permanent coatings are available based on strong adsorption [52], chemical modification [53,54], or polymerization [55–59] at the inner capillary surface. Polyethylene glycol and polyacrylamide (PAA) are two of the most commonly used polymers. The surface characteristics of mitochondrial outer membrane can lead to a strong interaction between mitochondria and the surface of a bare fused silica capillary. Acrylamide permanently coated capillaries are frequently used and considered as one of the best permanent coatings available due to their long-term stability and repeatable performance [59–61].

14.3.2 MC Separation of Individual Organelles

Reports on procedures using MC for biological analysis and clinical diagnosis are increasing in recent years. Compared to CE, MC offers faster separation, higher sample throughput, and the capability to integrate with other analytical platforms. Duffy et al. [62] had successfully separated mitochondria isolated from bovine liver tissue and selectively labeled with 10-nonyl acridine orange (NAO) using a commercially available double T glass microchip. The results showed a marked reduction (five times) in separation time in comparison with the CE-based methods (4 min versus 20 min), a small volume for sample consumption (80 pL versus 1 nL), and an equal sensitivity. Allen et al. [63] demonstrated a high-throughput CE-LIF method for separation of individual mitochondria in a microfluidic device. Mitochondria labeled with membrane-permeable Oregon green diacetate succinimidyl ester, an amine-reactive dye, were injected into an MC and broken down rapidly under a pulse laser. After photolysis, a rapid CE separation could be carried out to determine the amine portion of each mitochondrion. Microchip-free flow electrophoresis (μ-FFE) with on-line detection had been used by Kostal et al. [64] to separate fluorescently labeled mitochondria extracted from rat skeletal muscle myoblasts. This study suggested that the determination of mitochondria using μ-FFE could be accelerated compared to CE-LIF as reported previously.

Although MC has shown many advantages compared to CE for separation of mitochondria, its low resolution due to the use of a very short separation path (1–2 cm) has prevented its application to real samples with complex sample matrixes. A high resolution is needed for counting mitochondria to avoid overlapping of particles arriving at the same time to the detection zone. In addition, extremely high detection sensitivity is required for counting a single mitochondrion as a fluorescent spike. Thus, in the present work, an MC-CE device is developed using CE with high separation efficiency to reduce particle overlap at the detection zone. The integration of LIF with MC also enables counting mitochondrial particles at the most desirable position inside the capillary column to enhance fluorescence collection to increase the detection sensitivity. Details on the design and fabrication of the MC-CE device are given in the following sections.

14.4 MC-CE Device for Counting of Mitochondrial Number

14.4.1 Counting of Particles by LIF

The fluorescence spikes or bursts from CE-LIF have been attributed to mitochondrial particles by the Arriaga group and others [61,65–69] and have been used to count mitochondrial events for characterization of the particle profile. Arriaga's group [65–67] had used a postcolumn LIF detector (Figure 14.2) with an externally applied sheath flow in the form of a cuvette to enclose the eluates migrated out from the separation capillary. As the exit of the capillary is placed vertically, the exiting analyte stream was drawn by gravity into a pear-shaped droplet (Figure 14.2b) that was carried down by the sheath flow. The exiting analyte stream diameter depends on the relative flow rates between the sheath flow and the analyte flow. Thus, the sheath volumetric flow can be increased to a suitable rate to suppress the analyte stream to the desired width. Thus, analytes were concentrated prior to LIF detection. As the organelles migrate out from the capillary, they can be focused hydrodynamically to a narrow stream where the laser was impacted to excite the exited organelles with minimal scattered light from the capillary wall. The 488 nm excitation line from an argon ion laser was used for excitation, and the fluorescence from each organelle was sampled for spike measurement by the microscope objective.

To obtain size information via measurement of the scattered light, a light-scattering LIF detector was further developed from the postcolumn LIF detector to measure the scattered laser light intensity at a specific angle [42]. The fluorescence scattering plot was then used to make the intrinsic correction for the variations in the refraction index of mitochondria to obtain both size and refraction index of mitochondria.

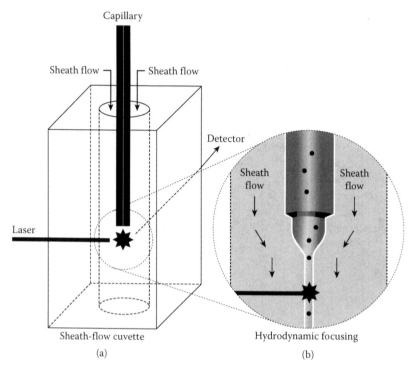

FIGURE 14.2
Postcolumn LIF detector with sheath curvette (a) and hydrodynamic focusing (b).

Although the postcolumn LIF detector with sheath curvette and hydro-dynamic focusing provides satisfactory detection sensitivity, it uses self-constructed equipment and requires careful optical alignment after replacement or installation of new capillary to optimize laser excitation and fluorescence collection. The operation is tedious; thus, a different optical arrangement is adopted for LIF measurement in the MC and MC-CE device, with details given in the next section.

14.4.2 Integration of MC-CE Device with LIF for Counting of Mitochondrial Number

The conditions for counting particles via fluorescent spikes by LIF are as follows:

1. The dye should be uniquely binding with the target organelles and excess nonbinding dye should not show any fluorescence that is the same as the bounded dye to avoid false-positive and false-negative errors.

2. Particles with different mass-to-charge (m/z) ratios should be separated from each other without overlap at the time when they are passing through the detection zone.

3. The fluorescence intensity should be strong enough to enable a single particle's detection.

4. The size of each individual particle represented by the fluorescence spike width should be smaller than the length of the detection zone.

5. The height of the fluorescent spikes should represent a single or multiple numbers of particles with the same m/z ratios.

To satisfy the above-mentioned conditions, the instrumentation for LIF measurement using MC and MC-CE device requires a highly sensitive photon-counting photomultiplier tube (PMT) together with a drastic reduction in stray and scattered light. Due to the need for operation of MC, laser excitation and fluorescence collection have to be positioned on the underside of the MC to assist operation from above (Figure 14.3). NAO was selected as

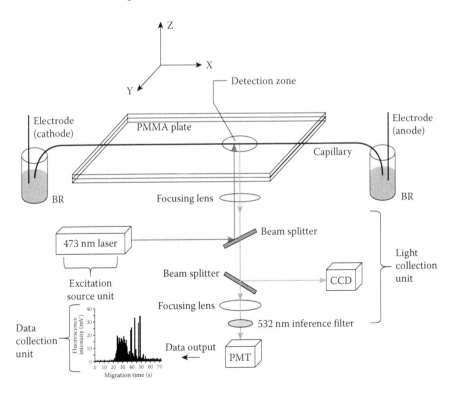

FIGURE 14.3
Schematic diagram showing the layout for integration of LIF detection mode with the MC-CE device. BR, buffer reservoir; PMMA, poly(methyl methacrylate).

the dye for LIF detection due to its specific interaction with mitochondria with excitation at 473 nm and emission at 525 nm. In MC work as described in Section 14.3.2, the resolution is limited by the short separation length of the microchannel. Thus, a 60-cm-long capillary column was used in the MC-CE device to resolve different subgroups of mitochondria from cell extracts.

The MC-CE device integrated with LIF was fabricated and shown in Figure 14.2. A confocal single-point optical system is used to eliminate stray light interference. An air-cooled 20-mW 473 nm laser diode-pumped solid laser was used for excitation and a bandpass filter as a stop for scattered light. An optically on-line charge-couple device (CCD) camera was used to assist focusing the laser onto the inner diameter of the capillary to reduce scattered light. The fluorescence was spectrally filtered by a 530 nm bandpass filter prior to detection by a PMT (Hamamatsu Photonics, Hamamatsu Japan). The optical alignment for a new and replaced capillary is much easier to perform compared to the procedures using the postcolumn LIF detector with a sheath curvette for hydrodynamic focusing.

14.4.3 Operations and Procedures

14.4.3.1 Selection of Detection Zone

To make full use of the laser irradiation and to avoid light scattering, a CCD imaging system was used to allow a precise focusing of the laser beam onto a desired spot at the detection zone of the separation capillary. Proper focusing of the laser beam is important as the spot size can increase due to beam divergence, giving rise to the reduction in the detection sensitivity [70]. As shown in Figure 14.4, the laser beam is focused onto an area smaller than the internal diameter of the capillary walls, with results showing a drastic reduction of background noise due to laser light scattered from the capillary walls.

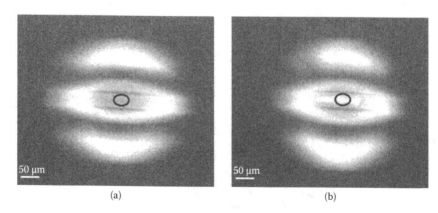

(a)   (b)

FIGURE 14.4
CCD images of the separation capillary before (a) and after (b) careful focusing by a laser beam.

14.4.3.2 Sample Preparation and Extraction of Mitochondria

Mitochondria from HepG2 cells were extracted by the differential centrifugation protocol [42]. In brief, HepG2 cells ready to be harvested during growth were collected for 5 min by centrifugation at 900 × g. After washing twice by buffer at 4°C containing 210 mM D-mannitol, 70 mM sucrose, 5 mM 4-(2-hydroxyethyl)-1-piperazineethanesulfonic acid (HEPES), and 5 mM ethylenediaminetetraacetic acid (EDTA), with pH adjusted to 7.4 by potassium hydroxide, the number of cells was counted by a hemocytometer and resuspended in 1 mL of buffer at ~4 × 10^6 cells/mL. A glass homogenizer was then used to disrupt the cells, with cell breakage assessed by a light microscope for an aliquot treated with Trypan blue. Cell debris was removed for 5 min by centrifugation at 1,400 × g. The processes were repeated three times. The final supernatant was pelleted for 10 min by centrifugation at 14,000 × g. The pellet was resuspended by repeating pipette in 1 mL of CE buffer. For NAO treatment, various concentrations from 0 to 100 µM NAO were added to the suspension of mitochondria. After 10-min ice incubation in the dark, the mitochondrial suspension was washed by the CE buffer, pelletizing twice at 16,000 × g for 10 min, and resuspending in the buffer. The mitochondrial suspension was kept on ice for storage and analyzed by direct injection into MC-CE device after warming to ambient temperature.

14.5 Results and Discussion

14.5.1 MC-CE Electropherogram for Separation of Mitochondria

Although centrifugation has been used extensively to fractionate organelles for further study, it is still difficult to resolve particles with similar or overlapping densities. For example, the densities of lysosomes and mitochondria from human liver are 1.15–1.20 and 1.20 g/mL, respectively [10]. Other strategies to separate organelle fractions that cannot ordinarily be resolved by density are needed. CE has been shown as an attractive alternative to separate organelles based on their surface charges instead of organelle densities. The ability to separate individual mitochondria by capillary zone electrophoresis arises from the differences in their electrophoretic mobility. It is preferred to separate mitochondria in their natural state (i.e., pH 7.4). Thus, the fundamental properties of organelles can be studied under their natural conditions at pH 7.4.

A PAA-coated fused silica capillary (60 cm in length and 50 µm inside diameter) precoated with PAA was used for CE separation. The repeatability of fluorescence produced by 6-µm fluorescently labeled polystyrene beads was used to assess LIF detector response. For conditioning, the capillary was rinsed sequentially for 5 min each with water, methanol, and CE buffer (pH 7.4) containing sucrose (250 mM) and HEPES (10 mM) to prevent aggregation of organelles. Sample was injected by pressure at 4 psi for 5 s, followed by separation at −400 V/cm. The electropherograms obtained are given in Figure 14.5, showing a single peak with migration time at 800 s and a peak profile with a large tailing by injecting extract from HepG2 cells with no NAO added (Figure 14.5B) and a flat featureless background fluorescence for NAO alone with no cell extract (Figure 14.5A).

With the addition of NAO to cell extract for interaction with mitochondria, fluorescence spikes appear in the electropherogram (Figure 14.6); these spikes are attributed to the mitochondrial particles. Note that the fluorescence spikes started to appear at 400 s and were complete after 1000 s, with a profile completely different from the tailing peak shown in Figure 14.5, at which only one tailing peak is shown at 800 s. The results indicate that the mobility of mitochondria is changing upon interaction with NAO and that different amounts of NAO are interacting with different subgroups or segments of mitochondria to give rise to a large spread of fluorescence spikes from 400 to 1000 s.

The widths of the fluorescence spikes are fairly constant at ~90 ms, indicating that the passing time of each mitochondrial event over the detection zone is more or less the same. Assuming that the charge density of the NAO–cardiolipin complexes over the surface of the mitochondrial membrane is uniform and that each mitochondrial particle is moving freely with equal passing time over the detection zone, the height of the fluorescence spike represents the size of the mitochondrial particle. The distribution of fluorescent

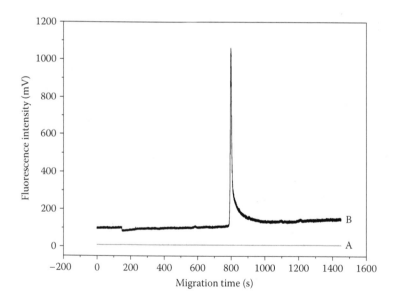

FIGURE 14.5
Electrophoregram of mitochondria extracted from HepG2 cells and separated directly without NAO label (A) and the electropherogram of 10^{-5} M NAO alone (B).

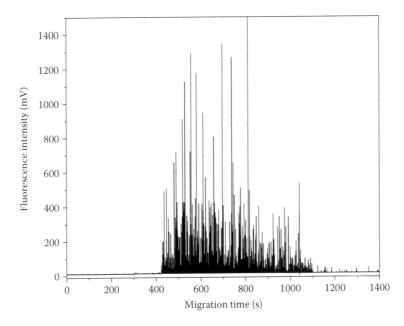

FIGURE 14.6

Electropherogram of mitochondria extracted from HepG2 cells and staining with saturated NAO (5 μM).

spikes observed in Figure 14.6 represents the separation of individual mito-chondrial particles by the MC-CE device with elution order according to the difference of electrophoretic mobility attributed to the variation of the charge density of NAO–cardiolipin complexes formed at the membrane surface of different mitochondrial subgroups.

14.5.2 Counting Mitochondrial Number from HepG2 Cell Extracts

The measuring time of the PMT detector to capture the fluorescence emitted from NAO has to be shorter than the travel time of individual mitochondria to pass through the detection volume. The typical travel time of individual mitochondrion passing through the detection volume was ~50 ms [70]. The longer passing time recorded in our experiment is attributed to the follow-ing reasons. First, we used a different CE buffer, which gives rise to a lower velocity of mitochondria passing through the detector volume. Second, the on-column detection window of ~50 μm used in our instrument setup is longer than that reported in the literature.

A typical electropherogram after the addition of 5 μM NAO to the HepG2 cells extract shows numerous fluorescent spikes (Figure 14.6), whereas the electropherogram without NAO treatments shows only a flat background (Figure 14.5). The results indicate that NAO treatment is necessary for the detection of individual mitochondria and the presence of other organelles

TABLE 14.1

Number of Mitochondrial Events Counted from
HepG2 Cell Extract

	HepG2 Cell Extract		
	1	2	3
Mean ($n = 3$)	1,702	1,788	1,912
SD ($n = 3$)	168	220	115
RSD% ($n = 3$)	15.0	11.0	10.3

Note: For each extract, three injections of 60 nL to the
MC-CE device were made. Based on results given,
the average number of mitochondrial events per
HepG2 cell is estimated as 1,107.

in the cell extraction do not lead to light scattering and autofluorescence.
The fluorescence spikes are thus attributed to individual mitochondrial particles containing NAO–cardiolipin complexes at the membrane surface.

As the signal-to-noise ratio for mitochondrial fluorescent spikes is obtained
under conditions close to the detection limit of the instrument, a "peak-free"
area is used to determine the background noise, which is estimated to be
<14% relative standard deviation (RSD). The threshold value used to define
the appearance of a fluorescent spike is taken as those with more than five
times that of the RSD of the background noise. The outcome for the evaluation of the entire electropherogram of mitochondria produces sequences of
numbers, describing positions (i.e., migration times or electrophoretic mobility) and maximum intensities for all peaks, corresponding to the analyzed
particles. From these data, some statistical conclusions may be drawn from
the distribution of peak intensities and the mobility of the detected event.
For the results given in Figure 14.6, the number of mitochondrial events
counted from HepG2 cell extracts is shown in Table 14.1 for triple sets of data.

Note that comigrating mitochondria can show up as a single event with
fluorescence from all comigrating mitochondria. As this can affect their
quantitation, a statistical overlap theory (SOT) [71,72] was used to estimate
the probability to detect multiple mitochondria as a single event. As SOT
has been applied to electrophoretic separations for assessing the temporal
distribution of particles for CE-LIF [27,73,74], it can be used for the present
calculation. The results based on SOT [71,72] indicate that the chances for
doublets and triplets at the most crowded regions as shown in Figure 14.6
are <5%, which means that the majority of events observed are attributed to
individual mitochondrial particles.

14.5.3 Characterization of Mitochondria for Zeta Potential and Electrophoretic Mobility

In addition to mitochondrial number, characterization of mitochondrial
physical properties, such as zeta potential and electrophoretic mobility, is

important for comparison between mitochondrial subgroups extracted from HepG2 cells. As reviewed in the theory section, the electrophoretic mobility of a particle is dependent on its zeta potential, size, and buffer composition; the observed variations in the electrophoretic mobility of individual organelles shown in Figure 14.6 can be explained by the variation in the first two parameters. Due to the fact that the buffer composition is constant in a given separation, the observed individual electrophoretic mobility is expected to be a function of zeta potential and size. Based on the electron micrographs of mitochondria extracted from HepG2 cells, the mitochondria were slightly ellipsoid and had an average radius (R) of 500 nm. Taking into account the mobility obtained in our experiments under the Booth-Overbeek theory [18,75] and conditions described in this chapter, the buffers used during separation typically have an ionic strength of 4.15 mM KCl, which corresponds to a κ value of 0.208 nm^{-1}. The κr value obtained strongly suggested that the electrophoretic mobility of mitochondria is mainly dependent on the ζ potential and not size. Actually, the ζ potential is determined by the outer membrane composition of mitochondria in a given buffer system. Thus, the variation in the electrophoretic mobility of mitochondria probably reflects the variation in their surface composition. However, the statement is a simplified description under several assumptions such as the geometry of mitochondria are considered as spherical and nondeforming particles and they do not interact with the capillary wall. Based on above-mentioned theory and the results in Table 14.2 and Figure 14.6, the median value of zeta potential in the sample tested are 16–44 mV. The results on the distribution of electrophoretic mobility of mitochondria extracted from HepG2 cell are summarized in Table 14.2. The large spread of electrophoretic mobility from −1.2 to −4.2 × 10^{-4} (cm^{-1} V^{-1} s^{-1}) leads to the appearance of a distribution of mitochondrial events as shown in Figure 14.6.

TABLE 14.2

Electrophoretic Mobility and Zeta Potential for Mitochondria Extracted from HepG2 Cells

	HepG2 Cell Extract		
	1	2	3
Electrophoretic Mobility (10^{-4} cm^{-1} V^{-1} s^{-1})			
Range	−1.78 ± 0.31	−2.10 ± 0.19	−2.07 ± 0.21
Mean ($n = 3$)	−1.78	−2.10	−2.07
RSD% ($n = 3$)	17.4	9.04	10.1
Zeta Potential (mV)			
Range	16–44		
Mean ($n = 3$)	27		
RSD% ($n = 3$)	11		

Note: For each extract, three injections of 60 nL to the MC-CE device were made.

14.6 Summary

Direct method based on fluorescent spike counting to measure individual mitochondrion for computing the overall mitochondrial number from cell extracts has been used in the present study as it produces results independent of the different techniques used. The advantages of the direct method compared to the indirect method (see Chapter 15) are its extremely high detection sensitivity for assessing mitochondrial number and the provision of additional chemical information, such as zeta potential, electrophoretic mobility, and number and size of individual mitochondrion. The heterogeneities of the organelles can be shown through these measurements to reflect the conditions of mitochondrial subgroups in the cell extracts. Important information to indicate the variability of mitochondrial subgroups such as cardiolipin content per mitochondrion, profile of cardiolipin among individual mitochondrion, and distribution of mitochondria size as well as m/z ratios have been shown to be retrieved from the experimental results from the present study.

The MC-CE device developed in the present study was shown to exhibit the following advantages for mitochondrial measurement: 1) easy operation and alignment of optical path for LIF measurement; 2) a suitable spot can be chosen inside the separation capillary for measurement of mitochondrial event with reduction of scattered light; and 3) as mitochondria particles are measured at the middle of the capillary, they are measured in a free flow without disturbance from the capillary wall. The advantages of MC-CE measurement are reflected in the measured results in the present study showing less scattered results, repeatable width of LIF fluorescent spikes, better repeatability in measurement, and less spread of mitochondrial number measured among different cell extracts.

Acknowledgments

We would like to acknowledge the contribution from Dr Wai-Sum O, the Department of Anatomy, the University of Hong Kong, Hong Kong SAR, People's Republic of China on the collaborated research reported in this chapter.

References

1. Armstrong, D.W. and He, L. Determination of cell viability in single or mixed samples using capillary electrophoresis laser-induced fluorescence microfluidic systems. *Anal Chem* 2001, 73, 4551–4557.

2. Korohoda, W. and Wilk, A. Cell electrophoresis—A method for cell separation and research into cell surface properties. *Cell Mol Biol Lett* 2007, 13, 312–326.
3. Buszewski, B., Szumski, M., Kłodzińska, E. and Dahm, H. Separation of bacteria by capillary electrophoresis. *J Sep Sci* 2003, 26, 1045–1049.
4. Witos, J., Cilpa, G., Yohannes, G., Öörni, K., Kovanen, P., Jauhiainen, M. and Riekkola, M.L. Sugar treatment of human lipoprotein particles and their separation by capillary electrophoresis. *J Sep Sci* 2010, 33, 2528–2535.
5. Liu, M.Y., McNeal, C.J. and Macfarlane, R.D. Charge density profiling of circulating human low-density lipoprotein particles by capillary zone electrophoresis. *Electrophoresis* 2004, 25, 2985–2995.
6. Kim, Y.H. and Kim, Y.S. Effect of nanoparticles in protein separation by capillary electrophoresis. *Bull Korean Chem Soc* 2010, 31, 479–482.
7. Valaskovic, G.A., Kelleher, N.L. and McLafferty, F.W. Attomole protein characterization by capillary electrophoresis-mass spectrometry. *Science* 1996, 273, 1199–1202.
8. Okun, V.M., Ronacher, B., Blaas, D. and Kenndler, E. Analysis of common cold virus (human rhinovirus serotype 2) by capillary zone electrophoresis: The problem of peak identification. *Anal Chem* 1999, 71, 2028–2032.
9. Gunasekera, N., Musier-Forsyth, K. and Arriaga, E.A. Electrophoretic behavior of individual nuclear species as determined by capillary electrophoresis with laser-induced fluorescence detection. *Electrophoresis* 2002, 23, 2110–2116.
10. Radko, S.P., Stastna, M. and Chrambach, A. Size-dependent electrophoretic migration and separation of liposomes by capillary zone electrophoresis in electrolyte solutions of various ionic strengths. *Anal Chem* 2000, 72, 5955–5960.
11. Petersen, S.L. and Ballou, N.E. Effects of capillary temperature control and electrophoretic heterogeneity on parameters characterizing separations of particles by capillary zone electrophoresis. *Anal Chem* 1992, 62, 1676–1681.
12. O'Brien, R.W. and White, L.R. Electrophoretic mobility of a spherical colloidal particle. *J Chem Soc Faraday Trans* 1978, 77, 1607–1626.
13. Hunter, R.J. *Zeta Potential in Colloid Science Principles and Applications.* Transaction Publishers, London, 1981.
14. Baker, D.R. *Capillary Electrophoresis.* Transaction Publishers, John Wiley & Sons, Inc., New York, 1995.
15. Radko, S.P., Stastna, M. and Chrambach, A. Capillary zone electrophoresis of sub-µm-sized particles in electrolyte solutions of various ionic strengths: Size-dependent electrophoretic migration and separation efficiency. *Electrophoresis* 2000, 21, 3583–3592.
16. Ho, C.C., Ottewill, R.H. and Yu, L. Examination of ellipsoidal polystyrene particles by electrophoresis. *Langmuir* 1997, 13, 1925–1930.
17. Radko, S.P. and Chrambach, A. Separation and characterization of sub-µm- and µm-sized particles by capillary zone electrophoresis. *Electrophoresis* 2002, 23, 1957–1972.
18. Overbeek, J.T.G. Theory of the relaxation effect in electrophoresis. *Kolloidchem Beihefte* 1943, 54, 287–364.
19. Booth, F. The cataphoresis of spherical, solid non-conducting particles in a symmetrical electrolyte. *Proc Roy Soc A* 1950, 203, 514–533.
20. Aiuchi, T., Kamo, N., Kurihara, K. and Kobatake, Y. Significance of surface potential in interaction of 8-anilino-1-naphthalenesulfonate with mitochondria fluorescence intensity and zeta potential. *Biochemistry* 1977, 16, 1626–1630.

21. Pysher, M.D. and Hayes, M.A. Examination of the electrophoretic behavior of liposomes. *Langmuir* 2004, 20, 4369–4375.

22. Pysher, M.D. and Hayes, M.A. Effects of deformability, uneven surface charge distributions, and multipole moments on biocolloid electrophoretic migration. *Langmuir* 2005, 21, 3572–3577.

23. Navratil, M.G., Mabbott, G.A. and Arriaga, E.A. Chemical microscopy applied to biological systems. *Anal Chem* 2006, 78, 4005–4019.

24. Strack, A., Duffy, C.F., Malvey, M. and Arriaga, E.A. Individual mitochondrion characterization: A comparison of classical assays to capillary electrophoresis with laser-induced fluorescence detection. *Anal Biochem* 2001, 294, 141–147.

25. Duffy, C.F., Fuller, K.M., Malvey, M., O'Kennedy, R. and Arriaga, E.A. Determination of electrophoretic mobility distributions through the analysis of individual mitochondrial events by capillary electrophoresis with laser-induced fluorescence detection. *Anal Chem* 2002, 74, 171–176.

26. Poe, B.G., Navratil, M.G. and Arriaga, E.A. Analysis of subcellular sized particles—Capillary electrophoresis with post-column laser-induced fluorescence detection versus flow cytometry. *J Chromatogr* A 2006, 1137, 249–255.

27. Johnson, R.D., Navratil, M., Poe, B.G., Xiong, G.H., Olson, K.J., Ahmadzadeh, H., Andreyev, D., Duffy, C.F. and Arriaga, E.A. Analysis of mitochondria isolated from single cells. *Anal Bioanal Chem* 2007, 387, 107–118.

28. Chen, Y., Xiong, G.H. and Arriaga, E.A. CE analysis of the acidic organelles of a single cell. *Electrophoresis* 2007, 28, 2406–2415.

29. Grossman, P.D. and Soane, D.S. Orientation effects on the electrophoretic mobility of rod-shaped molecules in free solution. *Anal Chem* 1990, 62, 1592–1596.

30. O'Konski, C.T. and Haltner, A.J. Characterization of the monomer and dimer of tobacco mosaic virus by transient electric birefringence. *J Am Chem Soc* 1956, 78, 3604–3610.

31. Fuller, K.M. and Arriaga, E.A. Capillary electrophoresis monitors changes in the electrophoretic behavior of mitochondrial preparations. *J Chromatogr B* 2004, 806, 151–159.

32. Jones, H.K. and Ballou, N.E. Separations of chemically different particles by capillary electrophoresis. *Anal Chem* 1990, 62, 2484–2490.

33. Nicholls, D.G. and Ferguson, S.J. *Bioenergetics.* 3rd ed. Academic Press, New York, 2002.

34. Saraste, M. Oxidative phosphorylation at the *fin de siècle. Science* 1999, 283, 1488–1493.

35. Zammit, V.A., Corstorphine, C.G., Kolodziej, M.P. and Fraser, F. Lipid molecular order in liver mitochondrial outer membranes, and sensitivity of carnitine palmitoyltransferase I to malonyl-CoA. *Lipids* 1998, 33, 371–376.

36. Levine, S., Levine, M., Sharp, K.A. and Brooks, D.E. Theory of the electrokinetic behavior of human erythrocytes. *Biophys J* 1983, 42, 127–135.

37. Donath, E., Walther, D., Shilov, N.V., Knippel, E., Budde, A., Lowack, K., Helm, C.A. and Mohwald, H. Nonlinear hairy layer theory of electrophoretic fingerprinting applied to consecutive layer by layer polyelectrolyte adsorption onto charged polystyrene latex particles. *Langmuir* 1997, 13, 5294–5305.

38. Zhou, L., Chomyn, A., Attardi, G. and Miller, C.A. Myoclonic epilepsy and ragged red fibers (MERRF) syndrome: Selective vulnerability of CNS neurons does not correlate with the level of mitochondrial tRNAlys mutation in individual neuronal isolates. *J Neurosci* 1997, 17, 7746–7753.

39. D'Aurelio, M., Pallotti, F., Barrientos, A., Gajewski, C.D., Kwong, J.Q., Bruno, C., Beal, M.F. and Manfredi, G. In vivo regulation of oxidative phosphorylation in cells harboring a stop-codon mutation in mitochondrial DNA-encoded cytochrome c oxidase subunit I. *J Biol Chem* 2001, 276, 46925–46932.

40. VanOrman, B.B. and McIntire, G.L. Analytical separation of polystyrene nanospheres by capillary electrophoresis. *J Microcolumn Sep* 1989, 1, 289–293.

41. Huff, B.V. and McIntire, G.L. Determination of the electrophoretic mobility of polystyrene particles by capillary electrophoresis. *J Microcolumn Sep* 1994, 6, 591–594.

42. Andreyev, D. and Arriaga, E.A. Simultaneous laser-induced fluorescence and scattering detection of individual particles separated by capillary electrophoresis. *Anal Chem* 2007, 79, 5474–5478.

43. Unverferth, D.V., Leier, C.V., Magorien, R.D., Croskery, R., Svirbely, J.R., Kolibash, A.J., Dick, M.R., Meacham, J.A. and Baba, N. Improvement of human myocardial mitochondria after dobutamine: A quantitative ultrastructural study. *J Pharmacol Exp Ther* 1980, 215, 527–532.

44. Unverferth, D.V., Leier, C.V., Brierley, G.P., Magorien, R.D. and Baba, N. Human myocardial mitochondria: Size differences in parts of the cell. *Exp Mol Pathol* 1981, 35, 184–188.

45. Whiting, C.E. and Arriaga, E.A. CE-LIF analysis of mitochondria using uncoated and dynamically coated capillaries. *Electrophoresis* 2006, 27, 4523–4531.

46. Yang, R.M., Liu, Y.H. and Wang, Y.M. Hydroxyethylcellulose-graft-poly (4-vinylpyridine) as a novel, adsorbed coating for protein separation by CE. *Electrophoresis* 2009, 30, 2321–2327.

47. Swedberg, S.A. Characterization of protein behavior in high-performance capillary electrophoresis using a novel capillary system. *Anal Biochem* 1990, 185, 51–56.

48. Russo, M.V. Ethylbenzene-modified fused-silica columns for capillary electrophoresis. *Chromatographia* 2002, 56, 175–180.

49. De Nobili, M., Bragato, G. and Mori, A. Capillary electrophoretic behaviour of humic substances in physical gels. *J Chromatogr A* 1999, 863, 195–204.

50. Gilges, M., Kleemiss, M.H. and Schomburg, G. Capillary zone electrophoresis separations of basic and acidic proteins using poly(vinyl alcohol) coatings in fused silica capillaries. *Anal Chem* 1994, 66, 2038–2046.

51. Law, W.S., Zhao, J.H. and Li, S.F.Y. On-line sample enrichment for the determination of proteins by capillary zone electrophoresis with poly(vinyl alcohol)-coated bubble cell capillaries. *Electrophoresis* 2005, 26, 3486–3494.

52. Wang, Y. and Dubin, P.L. Capillary modification by noncovalent polycation adsorption: Effects of polymer molecular weight and adsorption ionic strength. *Anal Chem* 1999, 71, 3463–3468.

53. Oda, R.P. and Landers, J.P. Application of capillary electrophoresis to the analysis of proteins. *Bioseparations* 1995, 5, 315–328.

54. Jorgenson, J.W. and Lukacs, K.D. Capillary zone electrophoresis. *Science* 1983, 222, 226–272.

55. Liu, Q., Lin, F. and Hartwick, R.A. Poly(diallyldimethylammonium chloride) as a cationic coating for capillary electrophoresis. *J Chromatogr Sci* 1997, 36, 126–130.

56. Strelec, I., Pacáková, V., Bosáková, Z., Coufal, P., Guryca, V. and Stulík, K. Modification of capillary electrophoresis capillaries by poly(hydroxyethyl methacrylate), poly(diethylene glycol monomethacrylate) and poly(triethylene glycol monomethacrylate). *Electrophoresis* 2002, 4, 528–535.

57. Liu, J.K., Pan, T., Woolley, A.T. and Lee, M.L. Surface-modified poly(methyl methacrylate) capillary electrophoresis microchips for protein and peptide analysis. *Anal Chem* 2004, 76, 6948–6955.

58. Sun, X.F., Li, D. and Lee, M.L. Poly(ethylene glycol)-functionalized polymeric microchips for capillary electrophoresis. *Anal Chem* 2009, 81, 6278–6284.

59. Engelhardt, H. and Cuñat-Walter, M.A. Preparation and stability tests for polyacrylamide-coated capillaries for capillary electrophoresis. *J Chromatogr A* 1995, 716, 27–33.

60. Barron, A., Sunada, W.M. and Blanch, H.W. The use of coated and uncoated capillaries for the electrophoretic separation of DNA in dilute polymer solutions. *Electrophoresis* 1995, 16, 64–74.

61. Zhao, W.F., Chen, Q.D., Wu, R.G., Wu, H., Fung, Y.S. and O, W.S. Capillary electrophoresis with LIF detection for assessment of mitochondrial number based on the cardiolipin content. *Electrophoresis* 2011, 32, 3025–3033.

62. Duffy, C.F., MacCraith, B., Diamond, D., O'Kennedy, R. and Arriaga, E.A. Fast electrophoretic analysis of individual mitochondria using microchip capillary electrophoresis with laser induced fluorescence detection. *Lab Chip* 2006, 6, 1007–1011.

63. Allen, P.B., Doepker, B.R., Chiu, D.T. High-throughput capillary-electrophoresis analysis of the contents of a single mitochondria. *Anal Chem* 2009, 81, 3784–3791.

64. Kostal, V., Fonslow, B.R., Arriaga, E.A. and Bowser, M.T. Fast determination of mitochondria electrophoretic mobility using micro free-flow electrophoresis. *Anal Chem* 2009, 81, 9267–9273.

65. Duffy, C.F., McEathron, A.A. and Arriaga, E.A. Determination of individual microsphere properties by capillary electrophoresis with laser-induced fluorescence detection. *Electrophoresis* 2002, 23, 2040–2047.

66. Chen, Y. and Arriaga, E.A. Individual electrophoretic mobilities of liposomes and acidic organelles displaying pH gradients across their membranes. *Langmuir* 2007, 23, 5584–5590.

67. Chen, Y. and Arriaga, E.A. Individual acidic organelle pH measurements by capillary electrophoresis. *Anal Chem* 2006, 78, 820–826.

68. Wolken, G.G. and Arriaga, E.A. Simultaneous measurement of individual mitochondrial membrane potential and electrophoretic mobility by capillary electrophoresis. *Anal Chem* 2014, 86, 4217–4226.

69. Satori, C.P., Kostal, V. and Arriaga, E.A. Review on recent advances in the analysis of isolated organelles. *Anal Chim Acta* 2012, 753, 8–18.

70. Kok, S.J., Isberg, K., Gooijer, C., Brinkman, U.A.T. and Velthorst, N.H. Ultraviolet laser-induced fluorescence detection strategies in capillary electrophoresis: Determination of naphthalene sulphonates in river water. *Anal Chim Acta* 1998, 360, 109–118.

71. Davis, J.M. Extension of statistical overlap theory to poorly resolved separations. *Anal Chem* 1997, 69, 3796–3805.

72. Navratil, M., Poe, B.G. and Arriaga, E.A. Quantitation of DNA copy number in individual mitochondrial particles by capillary electrophoresis. *Anal Chem* 2007, 79, 7691–7699.

73. Whiting, C.E. and Arriaga, E.A. Evaluation of individual particle capillary electrophoresis experiments via quantile analysis. *J Chromatogr A* 2007, 1157, 446–453.

74. Zeng, H.T., Yeung, W.S.B., Cheung, M.P.L., Ho, P.C., Lee, C.K.F., Zhuang, G.L., Liang, X.Y. and O, W.S. In vitro-matured rat oocytes have low mitochondrial deoxyribonucleic acid and adenosine triphosphate contents and have abnormal mitochondrial redistribution. *Fertil Steril* 2009, 91, 900–907.

75. Duffy, C.F., Gafoor, S., Richards, D., Admadzadeh, H., O'Kennedy, R. and Arriaga, E.A. Determination of properties of individual liposomes by capillary electrophoresis with postcolumn laser-induced fluorescence detection. *Anal Chem* 2001, 73, 1855–1861.

15

Indirect Assessment of Organelles in Cell Extract: Determination of Mitochondrial Number by Cardiolipin Content Using MC-CE Device Integrated with Laser-Induced Fluorescence

Wenfeng Zhao

Jiangsu Normal University
Xuzhou, People's Republic of China

Ying Sing Fung

The University of Hong Kong
Hong Kong SAR, People's Republic of China

CONTENTS

15.1 Introduction

The monitoring of the number of mitochondria, the organelle responsible for energy generation in cells, is obvious important, as it reflects essential cellular functions of target cells under investigation. However, it is a difficult analytical task to determine mitochrondrial number of target cells without the interference from other organelles and cellular debris.

In the present work, the specific interaction between the dye NAO and cadiolipid, a lipid uniquely present at the surface of mitochondria membrane, is utilized to enable indirect determination of mitochondrial number using MC-CE device integrated with laser-induced fluorescence detection. The limitations of current method are presented and advantages using MC-CE device discussed. The analytical performance of the MC-CE device fabricated is given and discussed for its applicability in assessing mitochrondrial numbers in harvested cells.

15.1.1 Needs for Mitochondrial Number Determination

Mitochondria are organelles that are present in most eukaryotic cells. They are recognized as "cellular power plants," because they are primary sites for energy generation, providing cells with a source of chemical energy called adenosine triphosphate or ATP [1–3]. As mitochondria exhibit strong effects on many important cellular functions, methods to assess their numbers are needed to understand their functions and related complex biological processes. The number of mitochondria per cell is an important parameter for consideration of mitochondrial biogenesis. Mitochondria are complex organelles composed mainly of lipids and proteins. Some lipids are unique to mitochondria, and they can be used as biomarkers to assess mitochondrial number by counting the number of particles containing these unique lipids.

15.1.2 Requirements and Analytical Issues

Mitochondrial number is an important parameter to reflect the functions linked to muscular and dilated cardiomyopathy diseases, and also to the aging process. Direct counting of the number of mitochondria based on the presence of unique lipids in the particles is currently the primary method for assessing mitochondrial numbers. Details on the methodology for direct counting of mitochondrial numbers are given in Chapter 14.

Transmission electron microscopy (TEM) is the current method for assessing mitochondrial number, but it suffers the following disadvantages. First, the instrumentation is expensive and requires special skill for operation; second, a large number of particles needs to be counted to produce statistically significant results; and third, sample preparation for a direct counting method requires high operation skill for procedures that are tedious, labor-intensive, and time-consuming. Thus, an alternative approach based on the assay of a unique biomarker for indirect determination of mitochondrial number is explored in the present work. This approach uses quick and easy operational procedures to produce results representing an average estimate of mitochondrial number from the concentration of a molecular biomarker uniquely produced by mitochondria.

15.2 Current Methods for Mitochondrial Number Determination

Due to the vital function of mitochondria, the development of methods to assess mitochondrial number is very important. Microscopy techniques [4,5], patch clamping [6], flow cytometry [7], and optical trapping [8] have been applied to measure functions attributed to mitochondria, such as mitochondrial membrane potential, metabolism, apoptosis, and mitochondrial calcium homeostasis. However, study of their relationship to mitochondrial number, a critical index of mitochondrial function, is rare due to the lack of suitable methods for assessing mitochondrial number.

The best-established method for assessing mitochondrial number is TEM; it is used to confirm the identity of mitochondria via their membrane structure. Conventional procedures involve the use of glutaraldehyde fixation, osmication, dehydration, infiltration by the plastic monomer and catalyst, and polymerization. Estimation of mitochondrial number is obtained based on the TEM tomography involving a three-dimensional reconstruction of selected specimens from numerous electron images representing different angular views [5]. The main drawback of this method is that the sample preparation is laborious and time consuming. Operating the diamond knife is also a great challenge, even for an experienced technician, to cut successive ultrathin sections within a cell. In addition, TEM operation requires placing the sample in a high-vacuum chamber and employing high-energy electron beams for bombardment of the observed sample; this process can disrupt the function and integrity of the mitochondria under investigation.

Arriaga's group has successfully coupled capillary electrophoresis (CE) with laser-induced fluorescence (LIF) detection (CE-LIF) to separate and detect mitochondrial particles [9–13]. A spike is detected when each fluorescently labeled mitochondrial particle has reached the detection zone [9]. Thus, the number of mitochondria isolated from culture cells or tissues can be estimated based on the collection and analysis of detected fluorescent spikes. The main advantage offered by the CE technique is that it is generally simple to use a fluorescent label to tag specific organelles within a short time frame. This technique has permitted the determination of hundreds to thousands of mitochondrial particles within 20 min. In addition, the intensities of fluorescent signal from an appropriate dye provide a quantitative measurement for a given component of mitochondria, such as cardiolipin content, or other functional parameters, including mitochondrial membrane potential or endocytic capacity.

15.3 Cardiolipin as Biomarker for Mitochondrial Number

One of the most noteworthy aspects of the inner mitochondrial membrane is that it contains a unique lipid named cardiolipin that can be used as a biomarker for estimating mitochondrial number. The distribution of various phospholipids from different sources is summarized in Table 15.1. The results indicate that the cardiolipin content of mitochondria varies from different sources. However, it is fairly constant from cells or tissues obtained from a given source [14].

Although the cardiolipin content of mitochondria varies from different sources, it is fairly constant from cells or tissues obtained from a given source. Thus, cardiolipin has become the "signature lipid of mitochondria" [14]. It is suggested to be important for the assembly of complexes at the membrane to maintain functional conformation [15,16]. As shown in Figure 15.1, cardiolipin has a dimeric structure, with each structure carrying two negative charges

TABLE 15.1

Phospholipid Content of Mitochondria from Different Sources

Source of Mitochondria	% of Total Phospholipids (w/w)								
	CL	PC	PE	PI	PG	PS	Sph	Lyso	Phac
Rat liver	12–17	34–55	22–36	5–8	1	1	1–3	1–2	1–2
Bovine liver	17	43	35	5	—	—	—	—	—
Bovine heart	16–20	38–45	30–37	6–10	11	1	3	6	
Human heart	18	43	34	—	—	—	—	—	—
Rat kidney	9–20	36–41	30–37	3	3	1	1	1	—
Potato	17–19	33–44	26–33	7–13	5	—	—	3	—

Note: CL, cardiolipin; PC, phosphatidylcholine; PE, phosphatidylethanolamine; PI, phosphatidylinositol; PG, phosphatidylglycerol; PS, phosphatidylserine; Sph, sphingomyelin; Lyso, lysophosphoglycerides; Phac, phosphatidic acid.

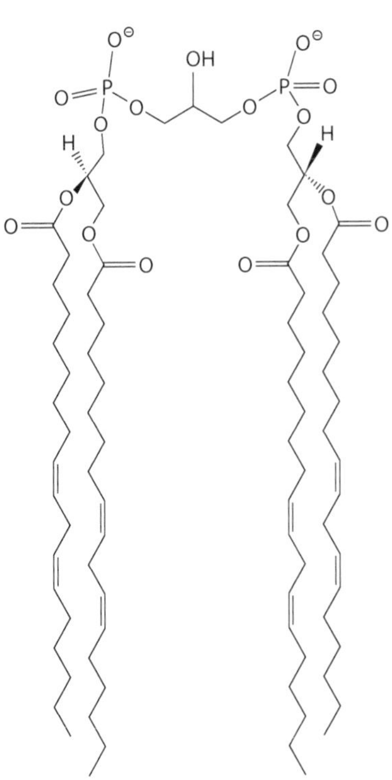

FIGURE 15.1
Typical molecular structure of cardiolipin.

and four fatty acid chains linked by a glycerol backbone. It plays a key role in all basic mitochondrial functions, including mitochondrial electron transport chain stabilization, ATP synthesis, and substrate transport [17,18]. Accumulating evidence suggests that cardiolipin also plays active roles in several mitochondria-dependent steps of apoptosis [19].

Quantitation of cardiolipin provides important data to assist in understanding mitochondrial metabolism at the molecular level. Moreover, this unique lipid is present at a relatively constant amount in the inner membrane. Thus, it has a high potential to serve as a biomarker for mitochondria. The development of methods for quantitation of the amount of cardiolipin could provide a simple method for the determination of mitochondrial number if the cardiolipin content for each mitochondrion is fairly constant.

For any substance X that is exclusively present in mitochondria, the number of mitochondria per cell (N) can be calculated using the following equation:

$$N = \frac{C}{C_m} \tag{15.1}$$

where C is the measured total amount of X per cell and C_m is the average amount of X per mitochondrion. The only requirement is to determine the cardiolipin content of mitochondria isolated from a given number of cells for the calculation of C. Based on the average amount of X per mitochondrion information (C_m) [20], the mitochondrial number (N) can then be calculated from Equation 15.1. Thus, the mitochondrial number per cell can be determined by other methods in addition to TEM.

15.4 Analytical Methods for Cardiolipin Determination

Due to the lack of chromophores in its structure, cardiolipin is difficult to detect. Techniques that have been used for cardiolipin determination include (1) imaging techniques; fluorescence [21,22] and confocal laser scanning [23] microscopy; (2) fluorescence spectrometry [24,25]; and (3) separation techniques, such as high-performance liquid chromatography (HPLC) [26,27], flow cytometry [28], and CE [29]. Microscopic techniques are usually used together with other means, such as fluorescent probes, to visualize cardiolipin for quantitative measurement.

15.4.1 Fluorescence Probe for Cardiolipin

The most commonly used fluorescence probe for cardiolipin is 10-*N*-nonyl acridine orange (NAO) (Figure 15.2). It is extensively used to identify the exact location of cardiolipin and to perform quantitative assays of cardiolipin in living cells due to its high specificity for cardiolipin [30]. NAO can bind to cardiolipin to form NAO–cardiolipin complexes (2:1 or 1:1 mole ratios), emitting red and green fluorescence, respectively. The affinity constant is 2×10^6 M^{-1} for NAO–cardiolipin conjugation (1:1 complex) and only 7×10^4 M^{-1} for those from phosphatidylserine and phosphatidylinositol [31]. The strong electrostatic attraction between the phosphate groups of cardiolipin and the quaternary ammonium of NAO leads to the formation of the cardiolipin–NAO complex. Quantitative studies making use of the strong interaction between NAO and cardiolipin have been reported previously [24,25,29]. However, most of these techniques are used to determine the cardiolipin content in bulk mitochondria isolated from a given cell population.

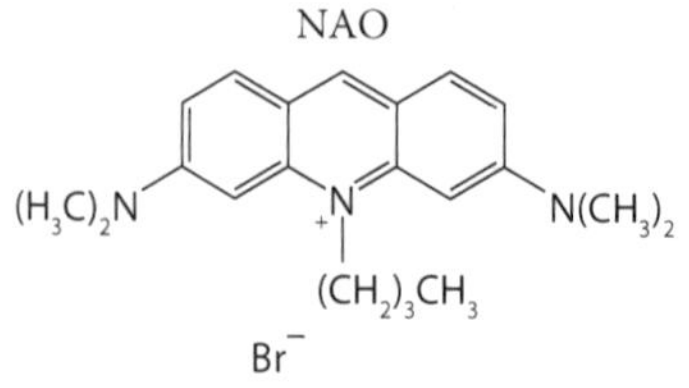

FIGURE 15.2
Molecular structure of NAO.

TABLE 15.2

Comparison of Current Analytical Methods for Cardiolipin Determination

Technique	In Vivo	Single Cell	Individual Mitochondrion	Separation Process
Confocal microscopy	√	√	√	√
Fluorescence spectrometry	×	×	×	×
Flow cytometry	×	×	√	×
HPLC	×	×	×	×
CE	√	√	√	√

Note: HPLC, high performance liquid chromatography; CE, capillary electrophoresis; √, yes; ×, no.

15.4.2 Comparison of Methods for Cardiolipin Determination

From the application of existing methods reported in the literature (Table 15.2), CE offers the best overall method for the determination of cardiolipin content for individual mitochondria and for bulk mitochondria. In addition, CE offers other advantages, such as the ability to analyze a very small sample volume, low consumption of solvents, rapid analysis, high resolution power, and low operational cost.

The integration of microfluidic chip (MC)-CE with LIF detection provides an extremely high detection sensitivity to enable achieving mass detection limits down to the zeptomole (10^{-21}) range [32]. The above-mentioned advantages make CE almost an ideal technique for analyzing complex biological samples for their mitochondrial content.

15.5 Integration of CE Methodology with LIF Operated in Indirect Enhanced Fluorescence Detection Mode

15.5.1 Current CE Methods for Cardiolipin Determination

A CE-ultraviolet (UV) method was reported by Qi et al. [33] for cardiolipin determination in bulk mitochondria through on-line dye–cardiolipin interaction. A 0.1 mM solution of NAO was added to the running electrolyte to enable absorbance detection at 497 nm for complexes formed by NAO–cardiolipin interaction. However, the linear relationship did not apply at submicromolar cardiolipin concentrations, thus narrowing its scope of application. In addition, the sensitivity is not sufficient for many studies with a limited supply of mitochondria.

Making use of the high sensitivity of LIF detection, a CE-LIF method was developed by Fuller et al. [32] to analyze cultured NS1 cells for cardiolipin content from an individual mitochondrion. The isolated mitochondria were

TABLE 15.3

Comparison of Methods for Cardiolipin Determination

Analytical Technique	Working Range	LOD	Time of Migration	Reference
Direct fluorescence quenching				
NAO mediation	0.2–10 µM	0.2 µM	—	[21,22]
L-Cysteine–capped CdTe quantum dots mediation	0.13–1.04 µM	18.5 nM	—	[23]
LC/CE separation				
LC/UV	0.2–3.2 mM	5 µM	~13 min	[24]
LC/LIF	0.5–4 µM	0.05 µM	~10 min	[25]
CE/indirect photometric detection	19–100 mg/L	1.3 µM	Within 10 min	[26]
CE/UV	0.5–100 µM	0.05 µM	2.4 min	[33]
CE/LIF	0.1–200 µM	9 nM	1.75 min	[32]

stained with NAO to form stoichiometrically a 2:1 or 1:1 NAO–cardiolipin complex at different dye concentrations. The green fluorescent 1:1 complex was selected because the fluorescence intensity is much stronger than that from the red fluorescent 2:1 complex.

Although the detection sensitivity is high using CE-LIF, it suffers the interference from constituents of the complex sample matrix that could quench the fluorescence intensity and hence affect the determination of mitochondrial number. With an aim to reduce sample matrix interference and enhance detection sensitivity, Zhao et al. [34] had developed a new CE-LIF procedure based on indirect enhanced fluorescence detection (IEFD). Details on IEFD are given in the following sections.

For comparison of direct and separation methods, the performance of existing separation methods and direct fluorimetric methods for cardiolipin determination are given in Table 15.3. For direct methods, the working range is, in general, less than that of the separation method, which is attributed to sample matrix interference and the capability of the separation method to remove interferents before detection. For limit of detection (LOD), the separation method is slightly better than the direct method.

For comparing HPLC and CE methods, both the working range and LOD for CE are better than those of HPLC. This may be attributed to the use of a largely organic background electrolyte (BGE) as the mobile phase which improve both sensitivity and linearity of the working range. A detailed explanation is given in Section 15.5.2.

15.5.2 IEFD Mode to Enhance Detection Selectivity and Sensitivity

The fluorescence of NAO with excitation at 473 nm and emission at 525 nm was found to be enhanced in the presence of cardiolipins in a solvent mixture containing MeOH, acetonitrile (ACN), and H_2O at 40, 50, and 10% (v/v),

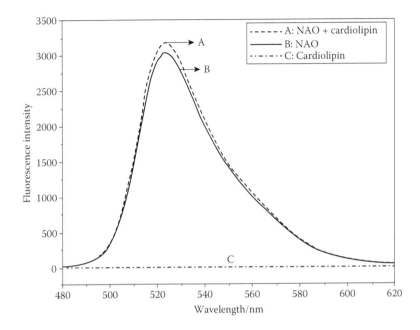

FIGURE 15.3

Fluorescence spectra of NAO, cardiolipin, and mixed standard solution. (A) NAO + cardiolipin (20 and 10%, respectively). (B) NAO (20 µM). (C) Cardiolipin (20 µM). Solvent: MeOH, ACN, and H_2O [40, 50, and 10% (v/v), respectively]. Excitation/emission, 473/525 nm, respectively.

respectively, as shown in Figure 15.3. Thus, an indirect detection mode can be used to detect cardiolipin. As other phospholipids coextracted from the biological sample matrix do not show significant fluorescence at 525 nm, sample matrix interference can be greatly reduced. In addition to the improvement of detection selectivity due to unique fluorescence enhancement, the IEFD mode can also increase detection sensitivity due to the differential measurement of the fluorescent peak against the background fluorescence from the running buffer and allows the use of a higher amplification of the photomultiplier tube (PMT) for light measurement.

15.6 MC-CE Device Integrated with LIF Operated in IEFD Mode

15.6.1 Design and Layout of the MC-CE Device

The MC-CE device integrated with LIF operated in IEFD mode is shown in Figure 15.4. It consists of two separate units: the optical unit and the MC-CE

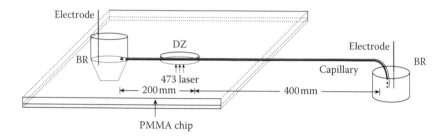

FIGURE 15.4
Layout of the MC-CE/LIF Device. BR, buffer reservoir; DZ, detection zone; PMMA, poly(methyl methacrylate).

separation unit. The optical unit consists of a charge-coupled device camera (WATEC America CORP, New York, USA) for positioning of LIF excitation, a 473 nm laser (Beijing, China), a 532 nm filter (Edmund Optics, Singapore) to cut off scattered light, and a highly sensitive PMT (model R212, Hamamatsu, Japan) for measuring fluorescence light. All optical elements are aligned in confocal arrangement to reduce stray and scattered light to enter PMT for detection.

The second unit is an MC-CE device that is set on top of the optical unit in a horizontal position to facilitate sample introduction via the capillary end to allow viewing of the LIF excitation area situated in the inner diameter of the separation capillary at the detection zone in confocal path to allow fluorescence light to enter PMT for light measurement. The MC-CE device could be adjusted in three directions to locate the optimized position for LIF irradiation and collection of fluorescent light. The software developed by the Chinese Academy of Sciences (Dalian, China) was used for acquisition and calculation of data obtained from the experiment.

15.6.2 Procedures and Operations

Mitochondria from HepG2 cells were isolated by differential centrifugation; ~10,000 cells collected were homogenized. The supernatants were centrifuged three times prior to obtaining the mitochondrial pellets. The procedure developed by Fuller et al. [32] was used for lipids extraction by suspending the harvested pellets in a series of buffers for vortex mixing and extraction. The final extract was subjected to centrifugation at $7,000 \times g$ for 15 min. The bottom chloroform layer was withdrawn and evaporated to dryness at 40°C under vacuum before reconstitution in 100 μL of CE buffer as the test sample.

The 100-μL test sample for CE injection is sufficient as each injection requires 50–100 nL. The operations of the MC-CE device include using CE buffer containing 20 mM NAO in a mixed solvent [MeOH–ACN–H_2O at 40, 50, and 10% (v/v), respectively]. Laser power (6 mW) is used to avoid saturation of photoemission from fluorophores. The use of a largely organic BGE as the mobile phase improves both sensitivity and linearity of working range for the following reasons. First, the use of a mixed organic solvent increases the viscosity and reduces the polarity of the solvent, hence stabilizing the NAO–cardiolipin complex and lowering the quenching by oxygen present in the mixed solvent. Thus, the fluorescence intensity is enhanced in the mixed organic solvent. Second, the electrophoretic mobility of cardiolipin increases rapidly when ACN was <40% (v/v). A larger amount of ACN leads to an increase in electro-osmotic flow (EOF) and a reduction in the effective charge at the inner capillary surface. However, an increase in ion pair formation was found at a higher ACN concentration, which gives rise to the opposite effect on EOF.

Thus, the actual EOF in a mixed solvent varies, depending on the balance of the various factors described. When ACN exceeds 50%, the resolution becomes poor. As a compromise, the optimized ratios for the mixed solvent [H_2O–MeOH–ACN at 10, 40, and 50% (v/v), respectively] were used. In summary, the MC-CE device offers a rapid method with high separation efficiency and low consumption of solvents and reagents for cardiolipin determination.

15.7 Application of the MC-CE Device Integrated with LIF Operated in IEFD Mode for Cardiolipin Determination

15.7.1 Optimization of Working Conditions

The electropherogram of a standard cardiolipin in BGE is shown in Figure 15.5. The negative peak was attributed as the EOF marker and the positive peak as the cardiolipin–NAO complex. As running buffer with high organic content was used, the wall adsorption problem of NAO was removed as well as the undesirable reversal of EOF. Thus, the peak after EOF marker was assigned as cardiolipin. A linear working range for cardiolipin determination was found to vary from 0.1 to 200 μM ($r = 0.9955$) and a detection limit (signal-to-noise ratio [S/N] = 3) at 9 nM was achieved within a 2-min CE run.

The injection of other phospholipids, such as phosphatidylserine (PS), phosphatidylcholine (PC), phosphatidylglycerol (PG), phosphatidic acid (Phac), and phosphatidylethanolamine (PE), into the MC-CE device does not give rise to observable peaks. As the magnitude of EOF in unbuffered organic solvents follows the order ACN > H_2O > MeOH [34], the use of a largely organic BGE reduces the CE separation time to within 2 min.

The optimized amount of NAO added to sample was studied by injecting three standard NAO solutions (200, 20, and 5 μM) into MC-CE device for separation, showing matching peak profiles. From the results, 5 μM

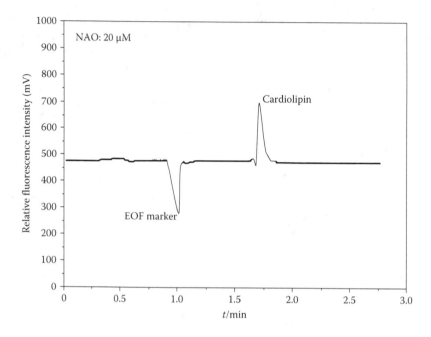

FIGURE 15.5
Electropherogram of a standard cardiolipin mixture in background electrolyte solution. Cardiolipin, 20 μM cardiolipin; BGE, 20 μM NAO in a mixed solvent [MeOH–ACN–H$_2$O of 40, 50, and 10% (v/v), respectively]. Effective length of separation capillary, 35 cm. Total capillary length, 60 cm. CE run voltage, 18 kV.

NAO was not selected because of the low background signal and adequate change in peak area for cardiolipin determination. The relative peak area was found to increase with increasing NAO. However, the background noise was also increased at high NAO concentration (200 μM). Moreover, a plateau instead of a peak appeared using cardiolipin concentration >150 μM and was attributed to the saturation effect, creating difficulties for measuring the fluorescence signal. As both a high background noise and a reduction in S/N occurred at increased dye concentration, 20 μM NAO was selected as the operating dye concentration.

15.7.2 Analytical Performance

The analytical performance of the developed MC-CE device for determination of cardiolipin is summarized in Table 15.4. Using 20 μM NAO in BGE [MeOH–ACN–H$_2$O at 40, 50, and 10% (v/v), respectively], a satisfactory linear working curve was found from 0.1 to 200 μM, with a correlation coefficient (γ^2) of 0.9955, detection limit (S/N = 3) at 9 nM, and precision for peak area and migration time measurement ($n = 5$) at 2.47 and 1.08%, respectively.

TABLE 15.4

Analytical Performance of the MC-CE Device Integrated with LIF Operated in IFED Mode for Cardiolipin Determination

Analytical Performance	Cardiolipin
Linear range	0.1–200 μM
Detection limit[a]	9 nM
Regression equation	Y = 4.86X + 128
Correlation coefficient for linear range	0.9955
Peak area[b] (RSD%)	2.47
Migration time[b] (RSD%)	1.08

[a] The detection limit is calculated based on S/N = 3.
[b] $n = 5$ for mean measurement of peak area and migration time.

Based on the measured cardiolipin content and assumption of a 100% extraction efficiency of cardiolipin from given cell line samples, <30 HepG2 cells is estimated to be measurable. Based on a constant cardiolipin content of 15.45 amol per mitochondrion (see Chapter 14 for details), the mitochondrial number using cardiolipin as a biomarker is estimated to be 1,428 per HepG2 cell. The result obtained is slightly different from the results obtained in Chapter 14. The slight variation may due to the natural heterogeneity among cells. Note that both results are in agreement with literature data [1] that human liver cells usually possess 1,000–2,000 mitochondria per cell.

15.8 Summary

Using the MC-CE/LIF device fabricated, an analytical procedure has developed for the evaluation of the number of mitochondria by an indirect approach based on cardiolipin as the biomarker for mitochondria. The MC-CE/LIF device is simple and easy to operate in a normal laboratory environment. The use of the indirect approach for cardiolipin determination could lower the detection limit compared to direct methods [21–23]. Samples down to 30 HepG2 cells could be assessed for their mitochondrial counts. However, the results obtained can only indicate the average mitochondrial number for samples extracted from a given cell or tissue based on the cardiolipin content. The assumption of constant cardiolipin content per mitochondrion is needed to be valid for a given biological sample using direct counting procedures as described in Chapter 14, in particular for comparison of mitochondrial number between different cell types.

The direct method as described in Chapter 14 is based on counting a fluorescent spike as a measure of individual mitochondrion to compute the overall mitochondrial number. The heterogeneities of the organelles can also be measured at the same time. The statistics from the fluorescent spikes can be used to calculate the distribution profile of cardiolipin among individual mitochondrion, important information to indicate the variability of cardiolipin content per mitochondrion in different mitochondria under different biomedical conditions. However, it requires much longer measurement time and expensive instrumentation for operation.

In contrast, the indirect approach using cardiolipin for assessing mitochondrial number as described in this chapter gives the average value of mitochondrial number using a fast procedure and affordable instrumentation for operation in a normal laboratory environment. The availability of a simple, fast, and sensitive procedure based on MC-CE/LIF device developed in the present work could provide an affordable alternative for assessing mitochondrial number, other than the tedious TEM procedure for a given number of cells and tissues, making the important mitochondrial number measurement available to researchers in related biomedical areas.

Acknowledgments

We would like to acknowledge the contribution from Dr Wai-Sum O, the Department of Anatomy, the University of Hong Kong, Hong Kong SAR, People's Republic of China on the collaborated research reported in this chapter.

References

1. Saraste, M. Oxidative phosphorylation at the *fin de siècle*. *Science* 1999, 283, 1488–1493.
2. Brenner, C. and Kroemer, G. Mitochondria—The death signal integrators. *Science* 2000, 289, 1150–1151.
3. Scheffler, I.E. *Mitochondria*. Transaction Publishers, Wiley-Liss, New York, 2008.
4. Boustany, N.N., Drezek, R. and Thakor, N.V. Calcium-induced alterations in mitochondrial morphology quantified in situ with optical scatter imaging. *Biophys J* 2002, 83, 1691–1700.
5. Navratil, M.G, Mabbott, G.A. and Arriaga, E.A. Chemical microscopy applied to biological systems. *Anal Chem* 2006, 78, 4005–4019.
6. Cavelier, L., Johannisson, A. and Gyllensten, U. Analysis of mtDNA copy number and composition of single mitochondrial particles using flow cytometry and PCR. *Exp Cell Res* 2000, 259, 79–85.

7. Jonas, E.A., Knox, R.J. and Kaczmarek, L.K. Giga-ohm seals on intracellular membranes: A technique for studying intracellular ion channels in intact cells. *Neuron* 1997, 19, 7–13.

8. Ashkin, A., Schütze, K., Dziedzic, J.M., Euteneuer, U. and Schliwa, M. Force generation of organelle transport measured in vivo by an infrared laser trap. *Nature* 1990, 348, 346–348.

9. Strack, A., Duffy, C.F., Malvey, M. and Arriaga, E.A. Individual mitochondrion characterization: A comparison of classical assays to capillary electrophoresis with laser-induced fluorescence detection. *Anal Biochem* 2001, 294, 141–147.

10. Duffy, C.F., Fuller, K.M., Malvey, M., O'Kennedy, R. and Arriaga, E.A. Determination of electrophoretic mobility distributions through the analysis of individual mitochondrial events by capillary electrophoresis with laser-induced fluorescence detection. *Anal Chem* 2002, 74, 171–176.

11. Poe, B.G., Navratil, M.G. and Arriaga, E.A. Capillary electrophoresis with post-column laser-induced fluorescence detection versus flow cytometry. *J Chromatogr A* 2006, 1137, 249–255.

12. Johnson, R.D., Navratil, M., Poe, B.G., Xiong, G.H., Olson, K.J., Ahmadzadeh, H., Andreyev, D., Duffy, C.F. and Arriaga, E.A. Analysis of mitochondria isolated from single cells. *Anal Bioanal Chem* 2007, 387, 107–118.

13. Chen, Y., Xiong, G.H. and Arriaga, E.A. CE analysis of the acidic organelles of a single cell. *Electrophoresis* 2007, 28, 2406–2415.

14. Günther, D. Lipids of mitochondria. *Biochim Biophys Acta* 1985, 822, 1–42.

15. Robison, N.C., Zborowski, J. and Talbert, L.H. Cardiolipin-depleted bovine heart cytochrome-c-oxidase—Binding stoichiometry and affinity for cardiolipin derivatives. *Biochemistry* 1990, 29, 8962–8969.

16. Yue, W.H., Zou, Y.P., Lu, L. and Yu, C.A. Crystallization of mitochondrial ubiquinol-cytochrome c reductase. *Biochemistry* 1991, 30, 2303–2306.

17. Hainesa, T.H. and Dencher, N.A. Cardiolipin: A proton trap for oxidative phosphorylation. *FEBS Lett* 2002, 528, 35–39.

18. Schlame, M., Rua, D. and Greenberg, M.L. The biosynthesis and functional role of cardiolipin. *Prog Lipid Res* 2000, 39, 257–288.

19. Gonzalvez, F. and Gottlieb, E. Cardiolipin: Setting the beat of apoptosis. *Apoptosis* 2007, 12, 877–885.

20. Kaewsuya, P., Danielson, N.D. and Ekhterae, D. Fluorescent determination of cardiolipin using 10-*N*-nonyl acridine orange. *Anal Bioanal Chem* 2007, 387, 2775–2782.

21. Kaewsuya, P., Miller, J.D., Danielson, N.D., Sanjeevi, J. and James, P.F. Comparison of N-alkyl acridine orange dyes as fluorescence probes for the determination of cardiolipin. *Anal Chim Acta* 2008, 626, 111–118.

22. Zhao, W.F., Fung, Y.S., O, W.S. and Cheung, M.P.L. L-Cysteine-capped CdTe quantum dots as a fluorescence probe for determination of cardiolipin. *Anal Sci* 2010, 26, 879–883.

23. Barceló-Coblijin, G. and Murphy, E. An improved method for separating cardiolipin by HPLC. *J Lipids* 2008, 43, 971–976.

24. Zhao, W.F., O, W.S., Fung, Y.S. and Cheung, M.L. Analysis of mitochondria by capillary electrophoresis: Cardiolipin levels decrease in response to carbonyl cyanide 4-(trifluoromethoxy) phenylhydrazone. *Eur. J. Lipid Sci. Tech.* 2010, 112, 1058–1066.

25. Haddadian, F., Shamsi, S.A., Schaeper, J.P. and Danielson, N.D. Capillary electrophoresis of phospholipids with indirect photome detection. *J Chromatogr Sci* 1998, 36, 395–400.

26. Gu, Z.Y., Zou, L., Fang, Z., Zhu, W.H. and Zhong, X.H. One-pot synthesis of highly luminescent CdTe/CdS core/shell nanocrystals in aqueous phase. *Nanotechnology* 2008, 19, 135604.
27. Murray, C.B., Kagan, C.R. and Bawendi, M.G. Synthesis and characterization of monodisperse nanocrystals and close-packed nanocrystal assemblies. *Annu Rev Mater Sci* 2000, 30, 545–610.
28. Murray, C.B., Norris, D.J. and Bawendi, M.G. Synthesis and characterization of nearly monodisperse CdE (E = sulfur, selenium, tellurium) semiconductor nanocrystallites. *J Am Chem Soc* 1993, 115, 8706–8715.
29. Li, H., Shih, W.Y. and Shih, W.H. Stable aqueous ZnS quantum dots obtained using (3-mercaptopropyl)trimethoxysilane as a capping molecule. *Nanotechnology* 2007, 18, 495–605.
30. Li, Z. and Du, Y.M. Biomimic synthesis of CdS nanoparticles with enhanced luminescence. *Mater Lett* 2003, 57, 2480–2484.
31. Mao, H., Yao, J.N., Wang, L.N. and Liu, W.S. Easily prepared high-quantum-yield CdS quantum dots in water using hyperbranched polyethylenimine as modifier. *J Colloid Interface Sci* 2008, 319, 353–356.
32. Fuller, K.M., Duffy, C.F. and Arriaga, E.A. Determination of the cardiolipin content of individual mitochondria by capillary electrophoresis with laser-induced fluorescence detection. *Electrophoresis* 2002, 23, 1571–1576.
33. Qi, L.N., Danielson, N.D., Dai, Q. and Lee, R.M. Capillary electrophoresis of cardiolipin with on-line dye interaction and spectrophotometric detection. *Electrophoresis* 2003, 24, 1680–1686.
34. Zhao, W.F., Chen, Q.D., Wu, R.G., Wu, H., Fung, Y.S. and O, W. S. Capillary electrophoresis with LIF detection for assessment of mitochondrial number based on the cardiolipin content. *Electrophoresis* 2011, 32, 3025–3033.

16

Protein Characterization and Quantitation: Integrating MC-CE Device with Gas-Phase Electrophoretic Mobility Molecular Analyzer

Tongmei Ma

South China University of Technology
Guangzhou, People's Republic of China

Gloria Kwan-lok SZE

Ho Man Tin Government Offices, Kowloon
Hong Kong SAR, People's Republic of China

Ying Sing Fung

The University of Hong Kong
Hong Kong SAR, People's Republic of China

CONTENTS

16.1 Introduction

The recent rapid development of protein biomarkers and protein drugs give rise to an analytical challenge for the characterization and measurement of proteins, in particular at trace levels in complex body fluids with the presence of other proteins such as human serum plasma. Tedious sample pretreatment is often employed to clean up samples and enrich analyte proteins prior to protein determination. It is often not possible to determine newly discovered protein biomarkers such as those present at ultralow level in body fluid.

In the present chapter, two important areas for protein determination and characterization by GEMMA have been assessed. First, its capability for protein counting and determination is evaluated using milk protein as the analyte protein. Second, a novel setup using MC-CE device to cleanup urine samples prior to counting of analyte proteins by GEMMA is investigated. Its analytical performance for determining urinary proteins at ultra-trace levels is given and discussed in light of the results presented.

16.1.1 Needs and Requirements for Protein Characterization and Measurement

Proteomics study became important after the completion of the Human Genome Project in 2003. Genetic defects can be detected by the measurement of new proteins or the absence of essential proteins. In the course of proteomics study, new proteins have been identified as biomarkers for diseases and also for use as protein drugs. This led to the development of new methods for quantitation of protein biomarkers for diagnosis of diseases, prognosis for development of medical conditions at different stages, and assay of drugs and metabolites.

The traditional procedures for protein assay are based on gel or two-dimensional electrophoresis, a time-consuming operation, and one with limited resolution to deal with complex samples, such as urine. Thus, new procedures based on mass spectrometry (MS) have been developed for protein identification and quantitation. However, the high cost of MS instrumentation prevents its use for routine analysis, and the requirement of a high vacuum during MS operation excludes its application to nonvolatile high-molecular-weight proteins.

Recent advances in the gas-phase electrophoretic mobility molecular analyzer (GEMMA) that operates under ambient atmospheric pressure enable characterization of nonvolatile high-molecular-weight proteins. Its high detection sensitivity based on particle counting is extremely useful in the determination of proteins at trace levels in urine. The recent development of microfluidic chip-capillary electrophoresis (MC-CE) device provides much potential to develop procedures to meet the requirements for bedside clinical analysis and home-care monitoring needs, as it is small, portable, and can deliver results on-site. The integration of MC-CE with GEMMA to develop a device for characterization of proteins in biofluids is discussed in this chapter, with preliminary results presented.

16.1.2 Development and Application of GEMMA

Since the methodology of GEMMA was first introduced by Kaufman and co-workers [1], the development of the GEMMA technique over the past 20 years has provided an analytical tool with great promise in characterizing nanometer-sized protein molecules in analytical biochemistry [1–9].

GEMMA possesses the capability to measure and detect all kinds of globular proteins, including large noncovalent protein–ligand and protein–protein complexes. Globular proteins, viewed as nearly spherical particles, have shown excellent correlation between the electrophoretic mobility diameters (EMDs) and the relative molecular masses according to the Millikan equation, ranging from several kilodaltons to megadaltons and corresponding to the nanometer-range size in the EMDs [2,6]. Thus, the first focused application area for GEMMA is to act as a particle analyzer, in particular, for spherical particles.

Very recently, the GEMMA method has been demonstrated to not only be applicable to simple globular proteins but also to intact, weakly bound, noncovalent, biospecific protein complexes; viruses and virus fragments; DNA; and bacteriophages [3,6–12]. GEMMA not only delivers information on the size and molecular mass of proteins but also provides details about the structures of large protein molecules [7,13,14] and monitors chemical reactions such as the reduction process of immunoglobulin G [6] and partial dissociation of human rhinoviruses [10].

As a relatively new technique, GEMMA can be compared to conventional methodologies used for protein identification such as capillary electrophoresis (CE), X-ray crystallography, and MS. The advantages of GEMMA over these methods include its outstanding sensitivity across the full mass range without decomposing the particle into atoms; the simple spectral structure obtained which can give detailed information, such as the concentration and size of the particle; and the fast procedure and its operation at atmospheric pressure, especially for noncovalent proteins [6]. Thus, GEMMA is viewed as a complementary technique to MS for investigating protein molecules [7,8,15]. In all, the successful application of GEMMA explored in previous work has established its important role for characterizing intact protein molecules.

16.2 Principle and Methodology for GEMMA Measurement

16.2.1 Instrumentation and Operations

The GEMMA system utilizes three major components developed in aerosol physics: a nanoelectrospray source with a charge reduction device; a differential mobility analyzer (DMA) that is capable of resolving particles based on their different mobility in air; and an ultrafine condensation particle counter (UCPC), which detects individual particles by means of light scattering. Details on the GEMMA (TSI Inc., Shoreview, MN) instrumentation and operation have been described thoroughly in the operations manual [16–19] and previous research [1,2,5–7]. A schematic block diagram representing

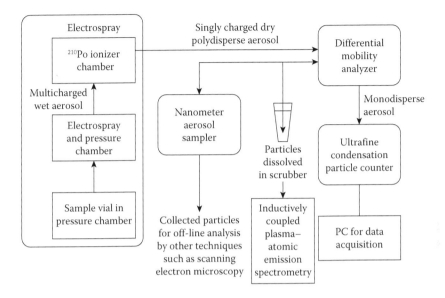

FIGURE 16.1
Schematic diagram showing the various functional components of the GEMMA.

the three components of the GEMMA system is shown in Figure 16.1: a TSI model 3480 electrospray (ES) aerosol generator (electrospray ionization [ESI] unit), a TSI model 3080/3085 DMA, and a TSI model 3025A UCPC.

The operations of different components of the GEMMA system are as follows:

1. An ESI unit to produce ultrafine, multicharged droplets from an aqueous protein-containing solution. A hot ionized airstream is used to dry the wet multidisperse and multicharged droplets generated by the ESI unit, and then they are neutralized by a radioactive α source to singly charged nanoparticles, which are then transferred from the ESI unit to the DMA.

2. The singly charged particles are separated by a DMA at atmospheric pressure based on their differential electrophoretic mobility in air. The electrophoretic mobilities of several biomolecules in air have been studied and are shown to correlate well with their molecular masses [2,5,20]. The DMA device was originally developed in aerosol physics for the separation of submicrometer-to-micrometer particle aerosols; recently, its working range was extended to the nanometer region, or in terms of molecular mass from several kilodaltons to megadaltons, which is suitable for analysis of protein molecules and thus closing the gap between classic aerosol particle technology (micrometer to millimeter range) and MS (subnanometer to 10 nm) [21–23].

3. The size-separated monodisperse protein ions are measured using the UCPC, a sensitive detector capable of counting, with almost 100% efficiency, the individual particles with diameters >2.5 nm. Alternatively, the size-separated monodisperse protein ions can be collected by a nanometer aerosol sampler (NAS) for subsequent off-line characterization by other analytical techniques.

16.2.2 Charge-Neutralized ES for Generation of Singly Charged Polydisperse Protein Aerosols

In GEMMA operation, the aerosol droplets are generated by the ES from the protein-containing sample solution initially as ultrafine, multicharged droplets. A schematic diagram showing the model 3480 ES aerosol generator (TSI Inc.) is given in Figure 16.2 [16]. ES is assisted by a regular air supply together with argon as a carrier gas at adjustable pressure to deliver a constant feed of sample to the ES for generation of multicharged polydisperse particles. CO_2 is used as a blanket gas during ion generation to depress

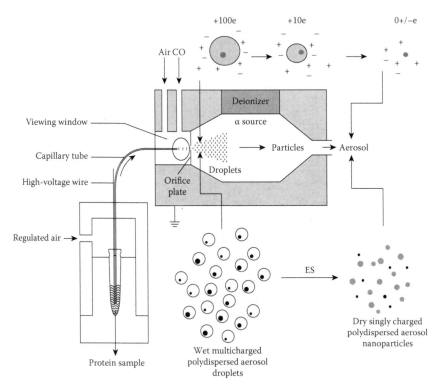

FIGURE 16.2
Schematic diagram showing the charge-neutralized electrospray (model 3480 aerosol generator, TSI Inc.) and associated regions for charge neutralization and generation of dried singly charged polydisperse nanoparticles from wet multicharged polydisperse aerosol droplets.

unwanted discharge. After aerosol generation, an α radioactive source is used to neutralize the wet multicharged aerosol droplets into dried singly charged nanoparticles at the exit of the ESI unit.

The protein sample solution is prepared and transferred to a vial that is directly placed inside an enclosed cylindrical pressure chamber. The pressure chamber accommodates a capillary (polyimide coated, 25 cm in length, 25 μm inside diameter, and 150 μm outside diameter) and a high-voltage Pt wire, both of which are immersed in the solution. A differential pressure of 3.7 psi causes the solution to be pushed through the capillary at a flowrate of 66 nL/min into the ES chamber. The flowrate (Q) through the capillary is governed by the following equation, based on a Poiseuille flow in circular tube:

$$Q = c \cdot \pi \left(\frac{D_c}{2} \right) \cdot \left(\frac{1}{8\mu} \right) \cdot \left(\frac{\Delta P}{L} \right) \tag{16.1}$$

where D_c is the capillary inner diameter (micrometers), μ is the liquid viscosity (poise), ΔP is the pressure across the capillary (psi), L is the length of the capillary (centimeters), and c is a constant for unit conversion (4.14×10^4).

The ES transfers the sample to an aerosol by charging the liquid, pushing it through the capillary, and exerting an electric field at the capillary tip that draws the charged solution out of the capillary. The shape of the droplet at the end of the capillary tip could be inspected from the viewing window and controlled to be the "cone-jet" mode by adjusting the ES voltage. The typical voltages range from 2.0 to 2.5 kV and the corresponding currents range from 200 to 300 nA. Immediately after the multi-charged aerosol droplets are generated at the capillary tip, they are transported into the ionization chamber by filtered air and CO_2 mixture. The air and CO_2 flowrates are typically set to be 1.0 and 0.1 L min^{-1}, respectively, and CO_2 is used to stabilize the ES against corona discharge. The transfers of particles must be done quickly because after the primary droplet is generated, the liquid begins to evaporate, which decreases the surface area of the droplet. This increases the density of the surface charge and produce less monodisperse aerosols.

Using ionized air, the multicharged droplets are then evaporated and reduced in size to yield particles with a charge distribution centered on neutral for most of the particles (~99% of all aerosolized particles), with a small fraction (~1% of all aerosolized particles) carrying a single positive or negative charge. The multicharged aerosol droplets inside the ionization chamber are subsequently neutralized by a ^{210}Po α-radiation source. Charge reduction occurs from the interaction of the multicharged particles with ionized primary gas ions. This neutralization process and evaporation produce predominantly neutral particles with a small fraction of the particles possessing a single positive charge. The equilibrium charge distribution generated by the ionizer of the ESI unit can be represented by a theoretical model developed by Wiedensohler [24]. A schematic diagram showing the formation of singly charged protein aerosol by the charge-neutralized ES is given in Figure 16.2.

With the right choice of concentration (less than about one analyte molecule per droplet), singly charged particles in the nanosize range can be generated. The solution must be completely evaporated in the ES chamber. The typical buffer solution used extensively in the ES is 20 mM ammonium acetate (NH_3Ac) in ultrapure water or high-quality deionized distilled water (DI).

16.2.3 DMA for Separation and Isolation of Singly Charged Monodisperse Protein Aerosols

DMA is the heart of the GEMMA system. It belongs to a subsystem of electrostatic classifiers (model 3080/3085, TSI Inc.) that contains a control platform for all instrument controls, including airflow, high voltage, and operation of DMA. It offers high sensitivity, high resolution, and low sample consumption in its operation as an electrostatic classifier for nanoparticles with different mass-to-charge ratios (m/z).

In the GEMMA procedure, proteins in the sample solution are sprayed by the ESI unit into singly charged polydisperse protein aerosols. DMA is then used to separate the charged proteins based on the flow trajectory directed by an applied electric field (Figure 16.3). The polydisperse particles entering DMA are subjected to two forces: a downward force directed by the gas

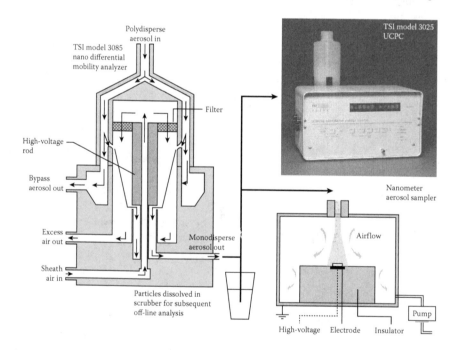

FIGURE 16.3
Schematic diagram showing the integration of the differential mobility analyzer (DMA) with ultrafine condensation particle counter (UCPC), nanometer aerosol sampler (NAS), and scrubber for characterization of singly charged monodisperse aerosols exited from the DMA.

flow velocity and a sideward attractive force directed by the electric field in a direction perpendicular to the gas flow direction. Only aerosols with a specific m/z can exit the DMA under a given electric field gradient and gas flowrate. Thus, by control of the electric field and gas flowrate, protein aerosols of specified m/z can be isolated and sampled at the exit of DMA for online counting of their number by UCPC, as well as for accumulative collection by NAS for subsequent off-line scanning electron microscopy (SEM) investigation. For chemical characterization, aerosols exiting from the DMA can also be directly scrubbed into a given solution for subsequent off-line analysis by other chemical analysis techniques, such as inductively coupled plasma-atomic emission spectrometry (ICP-AES).

The polydisperse particles entering the DMA are separated by their electrical mobility according to the following equation [17]:

$$Z_p = \frac{neC}{3\pi\mu d_p} \tag{16.2}$$

where Z_p is the electrical mobility, n is the number of elementary charges on the particle, e is the elementary charge (1.6×10^{-19} C), C is the Cunningham slip correction factor, μ is the gas viscosity [(dyne seconds/square centimeter) poise], and d_p is the particle diameter (centimeters).

The mobility is inversely related to the particle size. The electric field inside the DMA influences the flow trajectory of the charged particles. The DMA contains an inner cylinder that is connected to a negative power supply (0 to −10 kV) with precisely controlled voltage. Particles with negative charge strike the outer electrode, and neutral particles are removed by the excess flow. Particles with a positive charge are attracted toward the center electrode due to the electric field gradient. Only particles within a narrow range of electrical mobility are able to follow the correct trajectory to pass through an open slit at the DMA exit.

The number of the detected particles is directly plotted against the inverse electrophoretic mobility of the particular protein. By means of the Millikan equation [2,7], which correlates the molecular diameter to the singly charged particle mobility in air, the plot displayed can be transformed to a spectrum of particle concentration distribution versus an equivalent EMD with computer software, resulting in a size distribution spectrum. As the EMD is a function of the molecular mass and effective density of each protein molecule, the spectrum can also be presented as a particle concentration versus protein molecular mass instead of its size [5], resulting in a mass spectrum. As a result, the approach provides a method to obtain the molecular weights of proteins. The typical analysis time using GEMMA is only ~2–3 min for one scan, making it a fast and sensitive technique for protein characterization.

There are two selections of DMAs in the 3080 GEMMA series: the long DMA (3080 DMA) and nano-DMA (3085 DMA). The long DMA has a broad working range from 5 to 1000 nm. The nano-DMA is designed for particles up to 20 nm, but it can analyze particles from 2 to 150 nm. The model 3080/3085 DMA is capable of resolving particles whose mobility differs by only several percent, which is improved compared with the older version of DMA. The DMA consists of two cylindrical electrodes, and the electrostatic field is created by the potential difference between the central electrode connected to a negative power supply and the electrically grounded outer cylindrical electrode, thus separating positively single-charged particles based on their electrophoretic mobility in the sheath airflow. The coaxial flow of the filtered air is orthogonal to the electric field and its flowrate is set at 15–20 L·min⁻¹. The size of the particle exiting the DMA is primarily controlled by the scanning voltage between 0 V and −10 kV.

Compared to the long DMA, the aerosol flows into the nano-DMA through a short tube that quickly widens in the conical section to reach a narrow annular channel. This could reduce the distortions of the flow field. The sheath flow is routed though the center electrode from the bottom before it turns 180° and passes through the same filter as the long DMA to accommodate the axial aerosol inlet. The extra concentric cylinder below the inlet slit increases the polydisperse flow, which reduces the particle transport time and therefore reduces particle loss due to diffusion.

16.2.4 UCPC for Online Counting of Protein Aerosol Numbers

After the separation of the dried polydisperse singly charged protein aerosols at atmospheric pressure exiting from the charge-neutralized ESI unit by DMA into monodisperse singly charged aerosols according to their different electrophoretic mobility in air, the monodisperse singly charged protein aerosols are subjected to online counting of their numbers and subsequent off-line characterization using other analytical techniques upon their accumulative collection on filters.

For online nanoparticle counting, the fractions isolated from DMA with a specific range of m/z are transferred to the UCPC for quantitative counting. UCPC enables growth of ions in a supersaturated cloud chamber to a size detectable by laser light scattering. The UCPC consists of three major parts: the sensor, the microprocessor-based signal processing electronics, and the flow system [18]. The sensor part of the condensation particle counter (CPC) itself is made up of a saturator, condenser, and an optical detector. The sheath flow enters the saturator section where it passes over a heated, liquid-soaked wick. The wick continually draws liquid from the isopropanal reservoir bottle. The isopropanal evaporates and saturates this sheath flow with vapor. At this point, the aerosol stream containing the particle is recombined with the vapor-saturated sheath flow as a laminar flow. These streams are then cooled using a thermoelectric device installed in the condenser. The vapor cools, becomes supersaturated, and begins to condense onto the particle nuclei carried by the aerosol stream to form larger detectable droplets.

A laser-diode light source is placed above the aerosol exit for illumination. The scattered light from the droplets at 90° side-scatter angle is focused onto a low-noise photodiode for detection. The scattered light pulses collected by the photodetector are converted into electrical pulses. The electrical pulses are then counted to produce results for measurement of the particle concentration. The particles are injected into a flow of saturated n-butanol vapor at 37°C. A supersaturated vapor condenses onto the particles to form droplets by cooling to 10°C. With the laser-diode light source in the UCPC, the micrometer-sized droplets produced can be detected and counted individually by means of light scattering. The flow ratio between the aerosol flow into the UCPC and sheath flow for the DMA is kept constant at 0.1, thus the aerosol flowrate is ~1.5–1.8 L·min^{-1}. The model 3025A UCPC uses this vapor sheath technique to make it possible to measure the number concentration of airborne particles that are >2.5 nm in diameter.

16.2.5 NAS for Accumulative Collection and Characterization of Protein Aerosols

In addition to the UCPC, the selected protein particles from the DMA can be sampled by a NAS that collects particles on a collection plate mounted on an electrode surface for SEM study. A NAS consists of a sampling chamber (Figure 16.3) with an electrode on which the collecting plate is mounted to collect particles [19]. A potential up to −10,000 V is applied to the electrode to produce a negative potential to attract positive particles. Typically, a transmission electron microscope grid is mounted to the electrode to collect particles for microscopic analysis. The electrode comes in two sizes, 9.5 and 25 mm, to control the spot size of deposited particles. The voltage controls the mass of the positively charged particles to be drawn to it. As the voltage increases, the degree of deflection increases, and larger m/z particles are deposited onto the surface. All surfaces have to be wiped free of oil and residues with isopropanol. The total amount of particles deposited onto the substrate is governed by the flowrate and electrode voltage. Upon accumulative collection of aerosols on filters placed on the collection plate, the particles can be examined for their morphology and chemical composition by electron microscopy and microprobe X-ray analysis.

16.3 Advantages and Limitations of GEMMA for Protein Characterization and Measurement

The GEMMA measurement provides a refined separation method for characterization of nanosized protein particles. The charge-neutralized ES isolates dried and intact protein from the sample solution without molecular breakdown in the form of singly charged polydisperse protein aerosols. With the use of DMA for gas-phase electrophoretic mobility separation, targeted singly charged monodisperse protein aerosols can be isolated for UCPC counting by laser light scattering in the cloud chamber upon particle growth and visualization. Based on the Millikan equation, this method provided a way to obtain the molecular weights of nonvolatile and large proteins. The typical analysis time using GEMMA is only ~2–3 min for one scan, making it a fast and sensitive technique for protein characterization.

The exiting particles from the DMA can be collected by a NAS for further chemical analysis and electron microscopy investigation. As the GEMMA system works under ambient atmospheric pressure with operations done at room temperature, the proteins isolated can be characterized in their native forms with minimal distortion. In a typical experimental run, the DMA voltage is scanned for 135 s (120 s for increasing voltage and 15 s for the return) to cover the sample diameter range of 2–1000 nm. No average and no smoothing algorithm were necessary for each GEMMA spectrum of the urine sample under study.

As a particle analyzer working at the molecular size range, its measurement time should be very fast, its detection sensitivity extremely high, and its working range wide and adequate. The detectable particle concentration ranges vary from <0.01 to 10,000 particles/cm^3, with a dynamic range of six orders of magnitudes. The capability for off-line integration with SEM and other advanced analytical techniques greatly expands the scope for characterization of proteins collected by a NAS.

The limitations include the high capital and running costs of GEMMA instrumentation, which exclude its use for routine monitoring. However, it can provide a tool as a reference to validate the determination of difficult protein samples. The need for high-purity solvent and volatile buffer constituents for GEMMA requires expensive solvent and a lengthy and tedious cleanup procedure for complex samples with nonvolatile components. The interference from nonvolatile low-molecular-weight impurities and the limited resolution of GEMMA for high-molecular-weight macromolecules have restricted GEMMA's scope of application. To provide a more practical tool for characterization of proteins from real samples that often consist of complex sample matrixes, integration with an MC-CE device of high separation efficiency is needed.

Here, two real sample applicability studies have been investigated using GEMMA for protein characterization and measurement. The first applicability study involves the characterization of proteins in fresh and imitation milk to illustrate the integration of GEMMA with ICP-AES to distinguish the difference in milk proteins from different sources. The second applicability study involves the characterization and quantitative measurement of urinary proteins by the integration of an MC-CE device with GEMMA to clean up complex urine samples prior to GEMMA characterization and to address the various issues for quantitative protein measurement using the GEMMA system. Both studies are discussed in detail in the sections to follow.

16.4 Characterization and Quantitative Measurement of Proteins by GEMMA with Integrated Techniques

Two applicability studies are presented in this chapter for characterization and quantitative measurement of proteins by GEMMA and integrated methods. The first applicability study covers the application of GEMMA and integrated methods for characterization of proteins from fresh and imitation milk to illustrate the enhancement for protein characterization by integrating different techniques. The second study concentrates on the integration of MC-CE devices with GEMMA to enhance quantitative measurement of trace proteins in urine, a complex sample matrix.

16.4.1 Characterization of Proteins from Fresh and Imitation Milk by GEMMA and Integrated Methods

The aim of this applicability study is to explore the analytical capability for coupling GEMMA with ICP-AES to provide information on the association of Ca with different casein micelles at nanometric sizes for fresh milk and imitation milk with high Ca content. The coupled technique combines the advantages of the GEMMA system for high-resolution particle sizing from 2 to 833 nm (corresponding to a molecular weight of ~8 kDa to 100 MDa, respectively) and an elemental selective and sensitive ICP system for the determination of Ca upon collection of targeted size fractions separated by the GEMMA system. The coupling of the two techniques can provide useful sizing and chemical information to assist in interpretation of the nature of the high-Ca imitation milk compared to fresh milk.

16.4.2 Instrumentation and Related Operations

The GEMMA system (TSI Inc.) consists of an ES to generate aerosol containing charged nanometric particles, which are subsequently discharged to singly charged particles by a ^{210}Po radioactive source. The particles are then carried to the scanning mobility particles sizer where polydisperse particles are separated based on their differential mobility in a flow trajectory under an applied electric field. The exiting monodisperse particles are counted by the CPC or by a NAS, which collects the monodisperse particles for subsequent quantitative analysis by ICP-AES. The ICP-AES (model Liberty 110, Varian Australia Pty. Inc., Mulgrave, Victoria, Australia) equipped with v-groove nebulizer was used for all Ca determinations. See Figure 16.1 for a schematic diagram of the equipment setup.

The monodisperse particles separated by GEMMA were collected by a NAS onto an aluminum plate and subsequently extracted into solution by placing the plate in 4 mL of 2% HNO_3 and then ultrasonicating for 45 min. Alternatively, the outlet stream from the DMA was directly bubbled into 4 mL of 2% HNO_3. The solutions were subsequently analyzed by ICP-AES for their Ca content. All glassware and collection plates were washed with 10% (v/v) HNO_3 prior to rinsing with distilled water, following by air-drying or oven drying in enclosed environment to avoid contamination.

16.4.3 Standards and Buffer Solutions

The stock solution of 1000 mg/L Ca was prepared by weighing 2.5 g of $CaCO_3$ (extra pure, Merck, Darmstadt, Germany), dissolving it in 2 mL of concentrated HNO_3, and making up to 100 mL using distilled water. The standard Ca solutions for calibration were made up to five concentrations (2.0, 1.0, 0.6, 0.4, and 0.2 mg/L) by dilution with distilled water. Standard polystyrene (PS) nanobeads were purchased from Polysciences (Warrington, PA) with a certified particle size of 55 nm and an SD of 1.9 nm.

The buffer for an ES must be volatile so that it can be evaporated in the spray chamber. The buffer conductivities must be within a range within which the ES can be operated to produce small particles, as the primary droplet size produced by the ES is inversely proportional to the cube root of the conductivity [1]. NH_3Ac at 20 mM was extensively used for the ES system as it is a volatile and weak buffer with easily adjustable pH and conductivity. The buffer solution was used to clean the capillary before, after, and between samples. It was also used as the solvent to prepare samples. NH_3Ac (analytical grade reagent, Aldrich Chemical Co. Inc., Milwaukee, WI) solution at 20 mM was prepared by dissolving 0.77 g of NH_3Ac in 500 mL of DI water. The pH of the buffer was adjusted using ammonium hydroxide (25%, Peking Chemicals Co. Ltd., Beijing) or acetic acid (100%, BDH, Poole, Dorset). High-purity gases, CO_2 ($\geq$99.9%, v/v), NO_2 ($\geq$99.995%, v/v), compressed air, and Ar ($\geq$99.995%, v/v), were purchased locally (Hong Kong Oxygen Co. Ltd., Hong Kong).

16.4.4 Sample Preparations and Analytical Procedures

The liquid milk samples were purchased from local food store. The sample descriptions are listed in Table 16.1. The milk sample was directly used after dilution to 0.1% (w/w) with distilled water for total Ca determination using ICP-AES or defatted by centrifugation before dilution with the spray

TABLE 16.1

Information on Milk Sample from Package Label

Sample	Label Description	Ca Content (mg/mL)
M1	Fresh milk, whole-fat milk	>1.10
M2	High-Ca skimmed milk, imitation milk	>1.75

buffer for particle-sizing analysis using GEMMA. For ICP-AES operations, the instrument was allowed to warm up for 20 min after the plasma was ignited. Sample contamination was avoided by rinsing with 2% HNO_3 for 2 min followed by pure distilled water for 1 min between samples, before sample aspiration for 30 s. The integration time for each peak scan was 3 s, and three replicates were taken for each scan.

16.4.5 Optimization of GEMMA Working Conditions

To achieve the conditions for delivering a single molecule in each aerosol droplet generated and the isolation of required singly charged monodisperse protein from the exit of GEMMA for measurement by other integrated methods, the working conditions for the charge-neutralized ES, DMA, sample, and buffer pH were optimized per details given in the following sections.

16.4.5.1 Charge-Neutralized ES

To achieve the formation of a stable and repeatable ES, the major criterion is the formation of a stable Taylor cone in the form of a cone-jet mode. This was achieved when the ES voltage was in the range 1.8–2.5 kV and the current was approximately −500 nA. The spray voltage and air flowrate have to be optimized [16]. The ES airflow and CO_2 flow were kept at 2.0 and 0.2 L·min^{-1}, respectively. The flowrate of the CO_2 adjusted to 0.2 L·min^{-1} (in 1/10 ratio to airflow) gave a minimal background count of particles when aspirated with the buffer as registered by the CPC, indicating that the polydisperse wet aerosols generated by the ES were sufficiently dried and neutralized prior to separation by the DMA.

16.4.5.2 DMA

DMA operated at a sheath flow ranging from 3.0 to 15 L·min^{-1}. At the minimum sheath flow of 3.0 L·min^{-1}, the DMA measured particle sizes ranging from 13.7 to 710 nm, and at the maximum sheath flow of 15 L·min^{-1}, it measured particle sizes ranging from 5.7 to 225 nm. Our preliminary results showed that the particle sizes were ~10–50 nm; therefore, the sheath flow that could achieve this size range was selected for optimization. The ratio of sheath air to polydisperse aerosol flowrate is normally set to 10:1. To achieve the best resolution, the flowrate of the incoming polydisperse aerosol must be equal to that of the exiting monodisperse aerosol, under the control of the CPC drawing flowrate. The exiting monodisperse particles drawn to the CPC were limited to a rate of either 0.3 or 1.5 L·min^{-1} proposed by the CPC. As higher flowrates produce better resolutions to measure smaller particles (Figure 16.4), the flowrate of sheath flow varied while the aerosol flowrate was kept at 1.5 L·min^{-1}. Optimization for the separation of sample

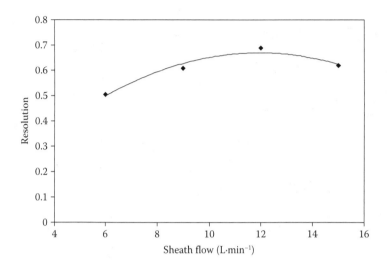

FIGURE 16.4
Effect of DMA sheath flow on the resolution of two particle fractions. Buffer: 20 mM NH_3Ac (pH 8.0). Sample: M2 diluted to 1% (v/v) in buffer.

M2 was conducted. The results show that the best resolution of the two different sizes of particles was achieved when the sheath flow and aerosol flow were 11.5 and 1.5, respectively. These flows were adopted for use in subsequent studies.

16.4.5.3 Sample and Buffer pH

When making a protein sample, the pH of the solution must be greater than the isoelectric point (pI) for the protein or the sample will stick to the capillary walls [16]. As the pI value of milk protein is <pH 5.8 [25], the buffer must be adjusted to pH 7.0 or above to avoid precipitation. The buffer pH was adjusted to alkali condition to avoid acid dissolution of Ca from casein micelles. To study the effect of pH at the alkali side with regard to Ca dissolution from casein micelles, a skimmed milk sample (M2), which had an original pH of ~6.7, was diluted to 0.1% (w/w) using NH_3Ac buffer. The pH is adjusted from pH 6.0 to 10.0 using ammonium hydroxide or acetic acid. The solution was then centrifuged at 13,200 rpm for 60 min to remove solids and fats until a clear solution was obtained. The solution was then filtered through a 0.1-μm membrane filter. The Ca content in the permeate was then determined by ICP-AES.

The Ca concentration of diluted M2 was fairly constant from pH 6.0 to pH 8.0 at ~0.33 μg/L, and it dropped rapidly at increasing pH. The decrease in Ca in the permeate at high pH may due to the formation of $Ca(OH)_2$ that precipitates from the solution and is retained in the filtrate. Based on a dilution factor of 1000, the soluble Ca content of the original milk is 0.33 mg/L.

TABLE 16.2

Optimized GEMMA Operation Parameters for Sizing Milk Proteins

Parameter	Optimized Condition
Buffer pH	7.8
DMA sheath flowrate (L·min⁻¹)	11.5
DMA aerosol flowrate (L·min⁻¹)	1.5
Particle scan size range (nm)	6.6–259
CPC aerosol inlet flowrate (L·min⁻¹)	1.5
NAS gas flowrate (L·min⁻¹)	1.5
NAS sampling time (min)	60

Note: DMA, differential mobility analyzer; CPC, condensation particle counter; NAS, nanometer aerosol sampler.

In comparison to the total Ca in skimmed milk (1.72 mg/g), the Ca content of the milk sample shows negligible Ca dissolution from casein micelles and aggregates, as ~19% Ca is in soluble form and 81% is in the solid particulate fraction. The pH of the buffer should be kept between 6.0 and 8.0 to avoid Ca dissolution from casein micelles and aggregates. Thus, pH 7.8 was selected for the subsequent study.

In summary, the optimized working conditions for GEMMA separation of milk samples are listed in Table 16.2. To achieve the best separation, the following adjustments have to be made: adjust ES voltage to maintain the formation of Taylor cone; keep carrier gas flowrate for air and CO_2 at 2.0 and 0.2 L·min⁻¹, respectively; keep ES buffer at pH 7.8; and maintain DMA sheath flow and aerosol flow at 11.5 and 1.5 L·min⁻¹, respectively.

16.4.6 Characterization of Proteins from Fresh and Imitation Milk by GEMMA and Integrated Techniques

16.4.6.1 Sizing of Proteins from Fresh and Imitation Milk

The preliminary results on sizing of samples M1 and M2 by GEMMA showed two particle sizes within ~10–50 nm, and they are designated as D_A for smaller particles with diameter of ~15 nm and as D_B for larger particles with diameter of ~30 nm. To assess separation efficiency for the GEMMA system, the resolution used in a chromatogram was adopted by replacing retention time by particle diameter [26]:

$$\text{Resolution} = \frac{2 \times (D_B - D_A)}{w_A + w_B} \tag{16.3}$$

where D_A and D_B refer to particle diameters and w_A and w_B are peak widths for fine and large particles, respectively. Peak width measurement was used on the assumption that the peaks were symmetrical, and w was determined by extension of the side arms of the peak to reach the baseline.

To reduce the probability of catching two particles within a single aerosol generated by the ES, sample dilution was progressively increased until a stable particle size distribution was accomplished. The results of the effect of sample dilution on the diameter of two groups of micelles particles generated from skimmed milk (M2) are shown in Figure 16.5. The lowering of milk concentration leads to improved resolution from 4 to 0.5%. However, further lowering the concentration to 0.25% led to disappearance of the large particle peak. Only one peak was found at 0.1% concentration.

The two groups of particles at ~15 and 30 nm followed the same pattern on size reduction upon dilution, and both become stable at concentration below 1% (w/w). The improvement in peak separation upon sufficient dilution may be due to a significant reduction of probability for incorporation of two types of particles in the same droplet generated by the ES. This leads to a clearer boundary between the two groups of particles and hence improved resolution. The results indicate the two groups of micelles particles are dissociated until a stable core structure with a narrow size distribution was found

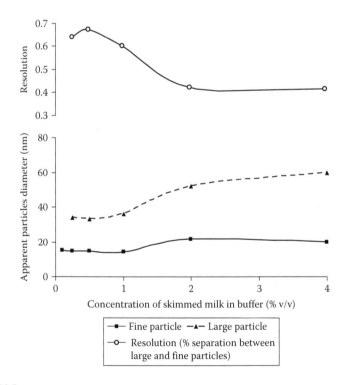

FIGURE 16.5
Effect of sample concentration on resolution and apparent diameter of fine and large particles. Sample: skimmed milk (M2). Buffer: 20 mM NH_3Ac (pH 7.8).

at 0.5% sample concentration. Further lowering of the sample concentration leads to breakdown of the micelle structure into submicelle units with the presence of fine particles only. Thus, 0.75% (w/w) is selected as it gave the best peak separation with stable peaks.

To validate particle sizing by GEMMA, standard 50-nm PS nanobeads with certified mean particle size at 55 nm and SD of 1.9 nm were used. The particle size as measured by GEMMA under the optimized conditions was mean particle size of 51.8 nm and SD of 1.19 nm based on the repeated experiment, both of which are agreeable with the certified particles sizes of the standard PS beads. However, the size distribution shows a slightly tailing at particle sizes <40 nm, likely due to the presence of nonvolatile surfactants, stabilizers, and other compounds in the PS standards that are included in the ES droplets, occurring after the evaporation of the volatile buffer.

To compare the size distribution of fresh milk and high-Ca imitation milk, size profiles were obtained using GEMMA (Figure 16.6). Under the optimized working conditions, particles with two different sizes (fine and large particles) were found within the detection range. A typical peak profile for the two milk samples is shown in Figure 16.6. Results for peak size, size ratio, and number ratio of the two peaks (large and fine particles) are shown in Table 16.3. From the ratio of the EMD of fresh milk and imitation milk, the large peak (casein micelles) is too big as a dimer for the fine peak (submicelles). The model of a nonrigid structure of casein micelles with a core casein–micelle structure surrounded by dissociable submicelles [27] agrees with the experimental results, as the large particle may represent the casein–micelles and fine particles the dissociated submicelles.

From the numerical results tabulated in Table 16.3, the concentration of fine particles is much more than that of large particles for both M1 and M2 samples. In general, the size of fine and large particles for M2 (imitation milk) is larger than those for sample M1 (fresh milk). However, the number counts (1822 and 800) and the ratio for number counts (2.3) between fine and large particles are more or less the same for both samples. The results indicate that the addition of Ca for M2 led to larger particle formation for both fine and large particles. However, the ratio of casein micelles (large particles) to submicelles (fine particles) is kept constant.

16.4.6.2 Determination of Protein-Associated Ca from Fresh and Imitation Milk

Two methods for off-line coupling of ICP with particles separated by the DMA were investigated. The first method was to connect the exit of the monodisperse particles from the DMA to the NAS for particle collection and subsequent elemental analysis by SEM with microprobe X-ray analysis. The second method was to insert the exit tubing from the DMA into a 2% nitric acid reservoir for scrubbing Ca particles into solution for subsequent ICP-AES analysis. For particle collection, an aluminum plate (10 mm × 10 mm)

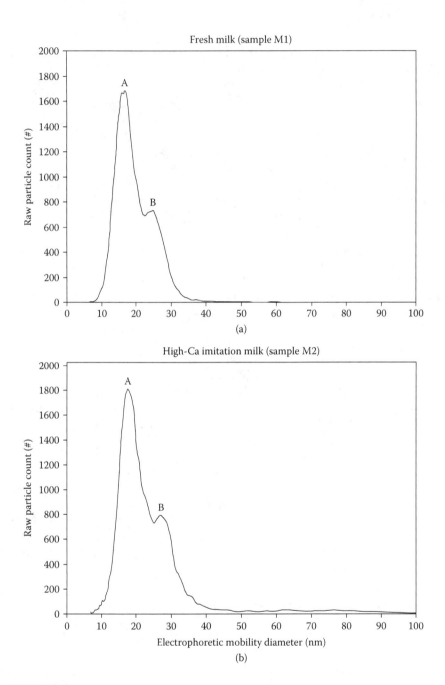

FIGURE 16.6
(a) Peak profile of fresh milk (M1) and (b) high-Ca imitation milk (M2). Two particle sizes were found in both samples. (A) Fine particles. (B) Large particles.

TABLE 16.3

Size Distributions of Proteins from Fresh and Imitation Milk by GEMMA

Sample	Peak	EMD (nm)	EMD Ratio (Peaks A/B)	Particle No. Count	No. Ratio (Peaks A/B)
M1	A	15.8	0.63	1700	2.3
	B	25		729	
M2	A	16.4	0.59	1822	2.3
	B	27.6		800	

Note: Peak heights and particle numbers for peaks A and B are estimated from
   Figure 16.6.
EMD = electrophoretic mobility diameter.

was mounted to the 25-mm electrode in the NAS sampling chamber. The electrode was kept at −7.0 kV under a gas flowrate of 1.5 L·min⁻¹ to match the flowrate of the CPC.

For effective particle collection, the harmonization of flowrates of ES, GEMMA, and NAS are required. For the ES capillary sample flowrate, Q (nanoliters per minute), is calculated using the following equation [1]:

$$Q = c.\pi \left(\frac{D_c}{2} \right)^4 \cdot \left(\frac{1}{8\mu} \right) \cdot \left(\frac{\Delta P}{L} \right) \tag{16.4}$$

where D_c is the capillary inner diameter in micrometers, μ is the liquid viscosity in poise, ΔP the pressure across the capillary in psi, L is the length of the capillary in centimeters, and c is a constant for unit conversion (4.14×10^{-4}).

The flowrate of sample under the optimized flowrate using a capillary of 40-µm-diameter, 25-cm-length at 3.7 psi was calculated to be 432 nL/min using Equation 16.4. As the initial sample concentration in the ES buffer was 0.75% (w/w) and its liquid viscosity was assumed to be 0.89×10^{-2} poise (for water at 20°C) and the lowest total Ca concentration in milk sample was 1.2 mg/g and the sample was diluted to 0.75%, a collection time of 50 min (assuming 100% collection) would give a Ca concentration >0.3 ng/mL (practical detection limit of ICP-AES [28]). Therefore, allowing 60 min to collect particles by a NAS would be sufficient to determine the Ca concentration.

The voltage to obtain peak maximum (the mean particle diameters) upon repeated scanning was used for collection, and the sample was sprayed for 60 min in both the methods. The aluminum plate with collected particles was placed in a sealed bottle, and 4 mL of 2% HNO_3 was then added, followed by ultrasonication for 45 min. The solution obtained was used for quantitative analysis by ICP-AES. Results from both off-line sampling methods are shown in Table 16.4. Collections of particles on the aluminum plate are more scattered as shown by large relative standard deviation (RSD), which may be due to the rebound of particles from surface during impact leading to sample loss. The method is time-consuming, involving a total

TABLE 16.4

Distribution of Ca-Associated Proteins in Fine and Large Particles

Sample[a]	Peak[b]	NAS[c]			Acid Scrubber[d]		
		[Ca] (μg/L)	RSD% (n = 6)	Peak Ratio (A/B)	[Ca] (μg/L)	RSD% (n = 6)	Peak Ratio (A/B)
M1	A	0.81	78.7	3.24	1.55	15.3	2.54
	B	0.25	108		0.61	18.0	
M2	A	0.30	20.5	0.59	1.12	12.5	0.57
	B	0.51	42.8		1.98	11.7	

Note: NAS, nanometer aerosol sampler; RSD, relative standard deviation.
[a] M1, fresh milk, whole-fat milk; M2, high-Ca skimmed milk, imitation milk.
[b] A, fine particles; B, large particles.
[c] NAS, collecting plate (10 mm × 10 mm) mounted on top of a 25-mm-diameter NAS electrode.
[d] Acid scrubber, direct scrubbing of monodisperse particles into 4 mL of 2% HNO_3 solution.

time of ~1.5 h for sampling collection and ultrasonication of the collection plate. In contrast, the scrubbing method using acid bath showed relatively less RSD in repeatability and was more direct. The tube must be inserted to the reservoir in an angle to ensure a stable stream of bubbles. The latter method is preferred for the present study.

The results indicate that small-sized particles have higher Ca concentration for both M1 and M2 samples. The use of NAS and acid scrubber shows more or less the same peak ratio (A/B) for M1 and M2 samples. However, the concentration of Ca in solution collected by the scrubber method is always higher and the RSD is always better than with NAS particle collection for both fine and large particles in M1 and M2 samples. This provides further support for sample loss due to particle rebound from the surface during impact leading in NAS collection as shown in a previous particle-sizing study given in Section 16.4.6.1.

For sample M1, the fine particles have three times higher Ca concentration compared to large particles. However, the reverse is observed for sample M2. The peak ratio (A/B) clearly indicates this difference, with nearly 3 for M1 and <1 (~0.58) for M2. The distributions of Ca between the two particle sizes were relatively constant for both M1 and M2. The results show that the addition of Ca to the imitation milk leads to a stronger association with the large particles to such an extent that the peak ratio is reversed. In fresh milk, most Ca are associated with organic matters and bound to casein micelle and submicelle structures. Ca added to imitation milk is likely to be inorganic in nature with a stronger association to the large particles at which electrical repulsion is less due to a larger surface area.

In summary, the integration of GEMMA with ICP-AES has been shown to expand the scope of GEMMA with both sizing and chemical information available to understand the partition of Ca among milk proteins with different sizes and structures. Among the two methods investigated for the collection of protein particles separated by GEMMA for subsequent Ca determination, the scrubbing method is found to be better than the NAS method, as shown by a better RSD, higher collection efficiency, and much shorter collection time. The results obtained have provided experimental support for the casein micelle/submicelle model proposed to explain the distribution of Ca among caseins and other milk proteins. The integration with MC-CE device for sample cleanup and enhancement of detection sensitivity is discussed in Section 16.5.

16.5 Integrating MC-CE Device with GEMMA for Characterization and Quantitation of Urinary Proteins

16.5.1 Urinary Protein Determination

The widespread availability and simple collection procedure for human urine makes it a reasonable source for protein analysis in clinical laboratories. Recent advances in understanding the proteome of the human genome during the progression of genetic diseases has lead to the discovery of many important proteins for treatment and early diagnosis requiring a nonintrusive means to detect them. As a result, the analysis of urinary protein concentration and size distribution is of great diagnostic importance. For example, excretion of urinary proteins is one of the most common abnormalities, reflecting the medical conditions of the kidney or urinary tract, and a detailed, accurate analytical assay of urinary proteins is essential for confirmation of certain disease diagnoses or physiological states. The concentrations of urinary proteins, such as human serum albumin (HSA) and β2-microglobulin (β2-MG), in urine can provide information for specific and accurate clinical analysis for diabetes mellitus and diabetic nephropathy [23,29–32]. In addition, the concentration of human chorionic gonadotropin (hCG) in urine from pregnant women can be used to assess different stages during pregnancy, especially for the early pregnancy test.

For quantitative measurement of urinary proteins, techniques such as CE, high-performance liquid chromatography, and MS, have been developed. However, with the increasing demand for detecting trace levels of urinary proteins, these techniques are limited by low detection sensitivity and insufficient selectivity to handle urine samples with complex sample matrixes [16–19,22–24,33]. As GEMMA offers extremely high detection sensitivity and requires high-purity volatile solvent and electrolyte in operation, its integration with an MC-CE device designed for analyte isolation, sample cleanup, and solvent switching is investigated in the present study.

16.5.2　Requirements for GEMMA Measurement of Urinary Proteins

For quantitative measurement of protein concentration sprayed from sample solution, the following conditions are required:

1. Every protein molecule present in the sample solution must have an equal chance of being sprayed into droplets for entering the GEMMA system;
2. Each droplet generated from the sample solution should contain none or one of the target protein molecules. The presence of two or more protein molecules should be kept to a minimum to increase the detection sensitivity and to avoid interference from aggregates for GEMMA measurement;
3. The sprayed aerosol droplets should be completely dried to remove all solvent and volatile electrolytes and leave behind the analyte protein as a dried nanosized aerosol;
4. The dried nanosized protein should be singly charged to enable selective collection by its mass-to-charge value;
5. Other constituents from samples and buffer solution should be volatile and removed during evaporation of the droplet;
6. The singly charged aerosol at a desired m/z value should be isolated from other multicharged aerosols for number counting or for collection for subsequent characterization by other analytical techniques.

In view of the numerous requirements stated, quantitative measurement of proteins by GEMMA is restricted to samples with simple matrixes. To enable the determination of analyte proteins present at trace levels in highly complex samples such as urine, the integration of GEMMA with an MC-CE device designed for analyte protein isolation, sample cleanup, salt removal, and solvent switching is needed to expand its scope for real sample application.

16.5.3　Design and Fabrication of MC-CE Device

The design of an MC-CE device for online integration with CE/ultraviolet (UV) detection in the determination of urinary proteins was given in Chapter 7. However, some important protein biomarkers, such as hCG present in urine at trace levels, could not be detected due to insufficient detection sensitivity of the UV detector. Thus, the integration of an MC-CE device with highly sensitive GEMMA detection is investigated, with results presented here. The design of the MC-CE device is shown in Figure 16.7. An off-line integration is adopted to enable solvent switching, to assist in quantitative measurement using a fixed amount of GEMMA solvent for sample injection, and to avoid the various technical problems associated with making liquid and electrical contact to an enclosed GEMMA sample chamber under elevated pressure.

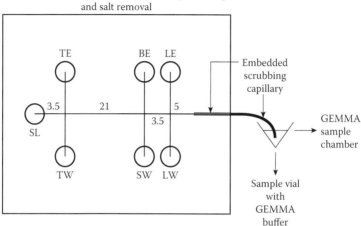

FIGURE 16.7

MC-CE device for off-line integration with GEMMA. The length of channel and capillary is shown in millimeters. The channels are ~120 μm in width and 110 μm in depth. SL, SW, SC, LE, LW, TE, TW, and BE represent the vials for sample loading, sample waste, sample collecting, leading electrolyte, leading electrolyte waste, terminating electrolyte, terminating electrolyte waste, and backing electrolyte, respectively. The pretreated urinary sample collected by SC is then transferred to the sample vial of electrospray aerosol generator of GEMMA for further analysis.

For the fabrication of the MC-CE device, the channels were created on a 1.5-mm-thick poly(methyl methacrylate) (PMMA) plate made by laser ablation from the CO_2 laser (V-series, Pinnacle V-series, Great Computer Corporation, Taipei) as described in the previous work [34]. The holes used for sample and buffer vials were also drilled by the CO_2 laser on another 1.5-mm-thick PMMA plate. The two PMMA plates and capillary were bonded under the pressure of 0.6 MPa at 95°C for 15 min. An eight-channel high-voltage system (0–4500 V, EMCO High Voltage Corporation, Sutter Creek, CA) was used to apply and switch high voltages under electronic control at designated vials in the MC-CE device.

16.5.4 Standards and Reagents

All chemicals used were analytical reagent grade. The leading electrolyte (LE) is composed of 25 mM Tris-HCl, 0.10 g/L Poly (ethylene oxide) (PEO), and 0.05% Hydroxyethyl Methyl Cellulose (HEMC), with pH adjusted to 3.6. The terminating electrolyte (TE) is composed of 25 mM acetic acid (HAc), 0.10 g/L PEO, and 0.05% HEMC at pH 3.0. Considering that the buffer used in GEMMA should be highly volatile, 20 mM NH_3Ac, with a conductivity of 0.2 S/m and pH adjusted to 7.5, was used as the background electrolyte as well as the buffer system for GEMMA to generate a stable "cone-jet" ES mode during ESI operation. All the solutions were prepared with ultrahigh-quality water and then filtered through 0.2-μm cellulose membrane filters (Pall Corp., East Hills, NY) and degassed in the ultrasonic bath prior to use.

Standard proteins included hCG, HSA, and β2-MG commercially available from Sigma Chemical (St. Louis, MO). All proteins were directly used, without further purification. The stock standard protein solutions of ~10 mg/mL were made up to given volumes by filtered DI water. Further dilution or mixture of the stock protein solutions with GEMMA buffer is necessary to prepare working solutions with different protein concentrations. All standard proteins were stored in refrigeration at −4°C.

16.5.5 Procedures and Operations

All channels of the MC-CE device were conditioned by loading 0.1 M NaOH, DI water, and LE from the open end of the capillary using a syringe pump. LE was then added to vial LE, urinary samples to vial SL, and TE to vial TE. In preparation of urinary samples, the stock solutions of standard proteins were mixed, diluted, and spiked into the collected normal human urine samples to prepare the working solution to the concentration range of 0.1–100 μg/ml for target urinary proteins.

Urine sample mixed with IEF buffer was introduced through SL; other vials were filled up with relevant electrolytes. High voltage of 600 V/cm was then applied from vial SL (HV) to SW (0 V) for 90 s to inject sample; high voltage of 600 V/cm was applied from vial BE (HV) to SW (0 V) for 30 s to load 20 mM NH_3Ac and high voltage of 600 V/cm was applied for 30 s across vial TE (HV) to TW (0 V) for loading of the TE [35,36]; all other vials were kept floating. HV of 1260 V/cm was then applied for 30 s from vial TE (HV) to LE (0 V) for salt removal, while other vials floated.

After salt removal, the NH_3Ac buffer solution from vial BE is pumped by electroosmotic flow to the scrubbing capillary together with the desalted sample by applying HV of 3000 V across BE and the sample vial with previously added GEMMA buffer at ~100 μL. To obtain enough concentration of urinary proteins, the same operation was repeated three to five times before transferring the sample vial to the GEMMA. The sample chamber was then used for sample injection via the ES aerosol generator of the GEMMA system.

16.5.6 Preparation of Urine Sample for Protein Determination

Three urinary proteins, hCG, HSA, β2-MG, are selected for assessing the MC-CE device with GEMMA integration. hCG is a small protein present at trace levels in urine of pregnant women. HSA and β2-MG were selected because they could provide important indicators for early diagnosis of diabetes mellitus and diabetic nephropathy and other urinary proteins in urine provide useful parameters as guidance for therapy treatment [23,29–32]. HSA is the most abundant protein in human blood plasma; it has a molecular mass of 67 kDa. The excretion of albumin in urine is related to the development of diabetes mellitus [37,38]. β2-MG is present on all nucleated cells and its molecular mass is 12 kDa. β2-MG is regarded as one of the most important markers for specific and accurate clinical analysis for diabetic nephropathy, and it has also been reported as an indicator of tubular and globular injuries in diabetic patients [23,30,31,39]. It is strongly proposed that the combination test of other marker proteins, such as β2-MG with HSA, can provide more accurate results than the albumin test alone [30,39,40]; thus, these two proteins are selected to spike into a normal urine sample to simulate the urine excreted from patients with diabetes mellitus or diabetic nephropathy.

To assess the performance of the MC-CE device integrated with GEMMA, two types of urine samples are prepared. The first type is spiked with the standard hCG protein into a normal urine sample from a human female to simulate pregnancy. The second type is spiked with the standard proteins HSA and β2-MG into normal urine to simulate pathologic urinary proteins. Typically, spiked urine samples are first centrifuged at 13.2 krpm for 10 min to remove suspending matter, such as cellular debris, before introducing to the MC-CE device for desalting and cleanup. Particles in the treated sample solutions were then removed by means of filtration through a 0.2-μm membrane. The pH and conductivity values of the filtered samples can be adjusted to desired values by the addition of HNO_3 or aqueous ammonia according to the required working conditions for the GEMMA procedure.

16.5.7 Quantitative Measurement of Urinary Proteins by Particle Counting and EMD Identification

The quantitative online measurement of the GEMMA system is the aerosol particle count by the UCPC for the singly charged monodisperse particles exiting from the DMA. To achieve the optimized conditions for delivering single molecule in each aerosol droplet generated and the isolation of required singly charged monodisperse protein from the exit of GEMMA for measurement by other integrated methods, all the relative optimized working conditions for the ES, DMA, and UCPC are summarized in Table 16.5.

Two types of information can be retrieved from the GEMMA system: the number of particles counted in a specified period and the gas-phase EMD calculated from the controlled parameters of DMA based on the gas flowrate and potential imposed on the central collection rod. Typical EMD spectra are shown in Figure 16.8 obtained from GEMMA measurement of two urine samples spiked with HSA standard proteins to concentrations of 10 μg/mL (high peak) and 5 μg/mL (low peak). Correspondingly, the aerosol concentrations obtained from these different spiked concentration urinary samples are plotted as the number of particles per cubic centimeter versus the EMD.

As shown in Figure 16.8, both HSA concentrations produce good profiles for the EMD spectra. The height/area of the peak from the 10 μg/mL HSA sample shows almost twice as much as that of the peak from the 5 μg/mL sample. From the spectra, the primary peaks are found at 6.85 nm EMD for the HSA protein, corresponding to the value of 69 kDa based on the linear

TABLE 16.5

Optimized GEMMA Operation Parameters for Urinary Proteins

Parameter	Optimized Condition
ES	
Buffer solution pH	7.5
Buffer solution conductivity (S/m)	0.2
Chamber pressure (psi)	3.7
Capillary length (cm)	25
Capillary inner diameter (μm)	25
Voltage range (kV)	2.0–2.5
Current range (nA)	200–300
Air flowrate (L·min^{-1})	1.0
CO_2 flowrate (L·min^{-1})	0.1
DMA	
Nano-DMA sheath flowrate (L·min^{-1})	15–20
Nano-DMA aerosol flowrate (L·min^{-1})	1–1.5
Nano-DMA scan voltage range (kV)	–10 to 0
UCPC	
Aerosol inlet flowrate (L·min^{-1})	1.5–2.0
Particle scan size range (nm)	2–150
Scan time (s)	130

Note: ES, electrospray; DMA, differential mobility analyzer; UCPC, ultrafine condensation particle counter.

relationship between EMD and the molecular size of proteins [4]. This value is consistent with experimentally measured results using other techniques. From the spectra, secondary peaks are shown at 8.51 nm EMD that correspond to the 138 kDa, the size of the dimer for the HSA protein [4]. The result is also consistent with the experimental values measured by other techniques.

16.5.8 Effect of Sample Concentration on Quantitative Measurement

Different concentration levels of β2-MG standard protein were spiked into a normal human urine sample to simulate urine suggesting diabetes. Figure 16.9 shows the scans obtained with urinary protein β2-MG at different concentrations: 2, 10, 25, 50, and 100 μg/mL. The five spectra obtained from different concentrations of β2-MG have the same x-axis to illustrate the same range of the size distribution of the β2-MG EMD spectra, with only the change in the concentration levels. With the same scan rate, compared with the EMD spectra obtained from the lower concentrations, 100 μg/mL β2-MG as given in Figure 16.9e produces a broad symmetrical peak, which is not resolved, indicating 100 μg/mL is saturated for the measurement. However, when the slower scan rate was used, the spectrum for 100 μg/mL β2-MG could be resolved. For the other four concentrations of β2-MG (Figure 16.9a–d),

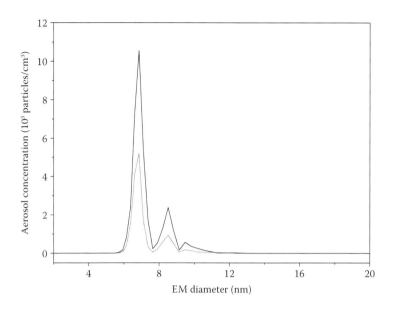

FIGURE 16.8

Electrophoretic mobility (EM) diameter spectrum of two different concentrations of HSA standard protein: 10 µg/mL (high peak) and 5 µg/mL (low peak), spiked into the normal human urinary samples.

the EMD spectra present three peaks with good resolution. The positions of the three peaks of different EMD spectra are in good agreement, and the three peaks correspond to 3, 4, and 4.96 nm, respectively.

The spectra indicate that the structure of the β2-MG is complicated. One possibility is that not only the dimer but also other multimers of β2-MG are coexisting with the monomer of this protein. With a decrease in concentration, the secondary peak decreases very fast, whereas the first peak is obviously the major peak, attributed to the monomer peak with a mass of 12 kDa, a value in consistent with the experimental value, obtained from the linear relationship between the EMD measurement to the molecular size/mass ratio of given protein molecule [4], which is consistent with the experimental values measured from other techniques. The secondary peak is attributed to formation of a dimer, with a mass of 24 kDa. The ratio of the dimer of β2-MG increases when the concentration of β2-MG is large enough as given in the 50 and 25 µg/mL diagram in Figure 16.9. The shoulder of the spectra yielding the third peak occurs at concentrations from 50 µg/mL down to 2 µg/mL, which could be formed by other multimers of β2-MG proteins.

16.5.9 Simultaneous Determination of Proteins as Internal Standard for Quantitative Measurement

To illustrate the capability for simultaneous determination of two proteins and to show that one can be used as an externally spiked protein acting as

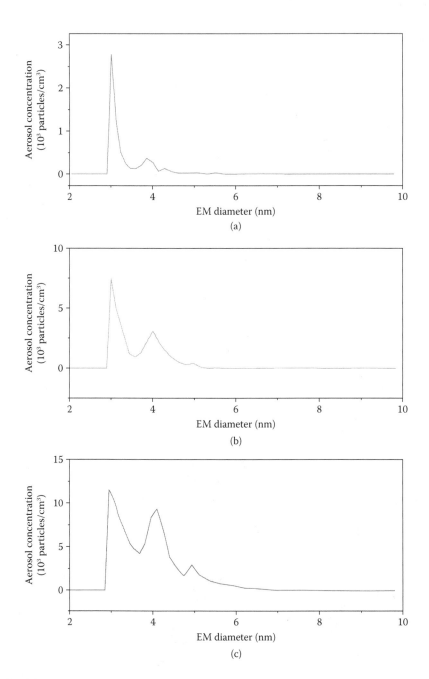

FIGURE 16.9
Electrophoretic mobility (EM) diameter spectrum of five concentrations (a–e) of standard β2-MG protein: from 2 to 100 μg/mL, spiked into the normal human urinary sample. [β2-MG]: A = 2, B = 10, C = 25, D = 50, and E = 100 μg/mL. *(Continued)*

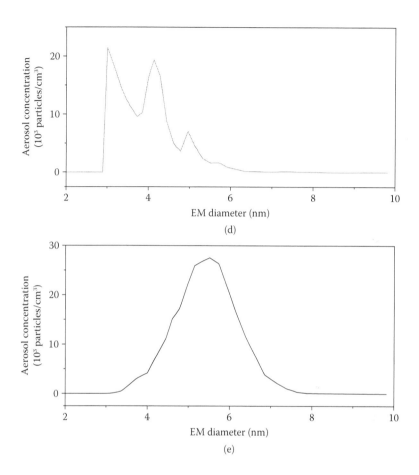

FIGURE 16.9 (Continued)
Electrophoretic mobility (EM) diameter spectrum of five concentrations (a–e) of standard β2-MG protein: from 2 to 100 μg/mL, spiked into the normal human urinary sample. [β2-MG]: A = 2, B = 10, C = 25, D = 50, and E = 100 μg/mL.

internal standard for quantitative measurement of the analyte protein, a normal urine sample was spiked with both HSA standard protein and β2-MG standard protein in a suitable concentration ratio. Figure 16.10 shows the scans obtained for a mixed standard solution containing 10 μg/mL HSA standard protein and 2 μg/mL β2-MG standard protein. According to the spectra, both HSA and β2-MG protein show good spectral profiles, and the dimers of both proteins can be observed. The EMD and mass of both HSA and β2-MG proteins obtained from these combination spectra are consistent with previous EMD spectra obtained by single HSA protein and single β2-MG protein. The HSA protein shows a higher peak than the β2-MG protein, which corresponds to the higher concentration HSA than that of the β2-MG protein.

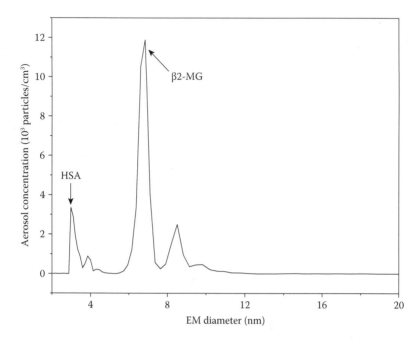

FIGURE 16.10
Electrophoretic mobility (EM) diameter spectrum obtained by spiking 10 μg/mL HSA standard protein and 2 μg/mL β2-MG standard protein into a normal urine sample.

The results indicate that no protein–protein interaction was observed that could change the EMD of a mixed standards protein spiked to a urine sample. Thus, the GEMMA system can be used for simultaneous determination of several nonoverlapping proteins in urine. In addition, another externally spiked protein can be used as an external standard to solve the quantitation problem due to the use of multiple instruments in the integration of MC-CE device with GEMMA for urinary protein determination. Details on quantitative measurement of the hCG protein, which is present in urine at trace levels, are given in Section 16.5.10.

16.5.10 Quantitative Measurement of Urinary Protein at Trace Levels

hCG is a very important glycoprotein hormone produced during pregnancy, and the detection and measurement of hCG levels from a urine sample are the bases of a pregnancy test, indicating the presence or absence of an implanted embryo. Testing for hCG may also be necessary when diagnosing or monitoring germ cell tumors and gestational trophoblastic disease, as high levels of hCG may be produced under these conditions, despite of the absence of pregnancy [41–43]. Furthermore, hCG levels are part of the triple test, a screening test for certain fetal chromosomal abnormalities.

hCG has dimensions of 7.5 nm × 3.5 nm × 2 nm, with a molecular mass of 36.7 kDa [44]. It consists of two noncovalently bonded heterodimeric subunits, α-hCG and β-hCG. The α-subunit is identical to all the four human glycoprotein hormones, and the β-subunit is unique to hCG.

Under the optimal experimental conditions, the DMA voltage was scanned for 135 s (120 s for increasing voltage and 15 s for the return) to cover the sample diameter range for each run. In the first series of experiments, a range of concentrations from 1 to 100 µg/mL of the standard proteins of hCG was spiked into a normal urine sample from a human female to simulate a pregnant woman's urine. After desalting and pretreatment by the MC-CE device, the pretreated hCG proteins were diluted in NH_3Ac buffer solution in the sample vial and transferred to the GEMMA system for further characterization.

Figure 16.11a shows the recorded GEMMA spectra. Several different concentrations of hCG (100, 75, 50, 25, 10, 5, and 1 µg/mL) are spiked into individual urine samples, with the same volume of 5 mL coming from the same person. According to the EMD spectra, $\sim 10^3$–10^4 of aerosol particles per cubic centimeter are coming from 1 mL of sample solution containing 1 µg of hCG per sample solution.

As shown in Figure 16.11b, the spectral profile shows a linear relationship between the concentrations of hCG protein in the sample solution and the number of particles counted by a UCPC at a given EMD under the normal experimental conditions. From the spectra, all the peaks are centered at 3.0 nm for the hCG protein, indicating that the EMD of hCG is ~3.0 nm, corresponding to 37 kDa, obtained from the linear relationship between EMD and the hCG protein molecular size [4], which is consistent with the experimental values measured from other techniques (36.7 kDa) [44].

According to Figure 16.11b, good linearity between the particle numbers counted (peak area/peak height) and the concentration level of the hCG proteins was observed in the range from 1 to 100 µg (hCG)/mL, which corresponds to hundreds of 10^3 particles per cubic centimeter, with a correlation coefficient of $R = 0.99724/0.99241$. This indicates that the working range of hCG protein using the GEMMA system varies from 1 to 100 µg/mL, a range that is sufficiently wide for the monitoring of hCG levels for pregnant women [42]. The linearity of the peak height to concentration level of hCG proteins is not as good as that of the peak area measurement. However, if the datum for the concentration of 100 µg/mL is excluded, the correlation coefficient is improved to 0.99251. Thus, the nonlinearity for peak height measurement may due to a large deviated datum.

To illustrate the effect after using an MC-CE device prior to GEMMA measurement, two test samples are prepared. The first test sample is prepared by spiking sufficient hCG standard proteins solution to a human male urine sample to reach a concentration of 8 µg/mL and treating with the MC-CE device for desalting and particle removal by filtering prior to GEMMA measurement. The second test sample is prepared by spiking the same concentration

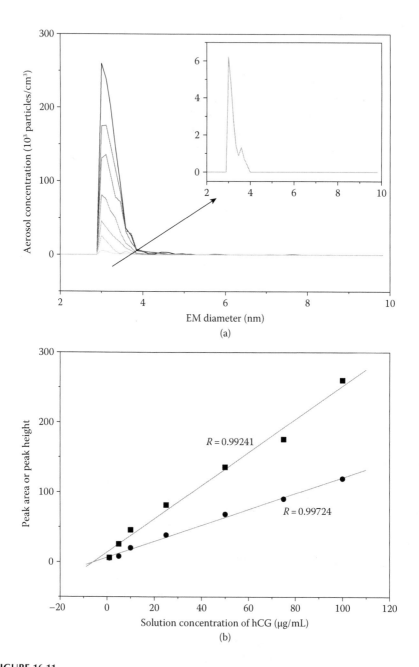

FIGURE 16.11
(a) Electrophoretic mobility (EM) diameter spectrum of seven concentrations of hCG: from 100 to 1 μg/mL, spiked into human female normal urine sample. (b) Linear relationship of peak area (square dots) and peak height (circle dots) with the concentrations of hCG proteins.

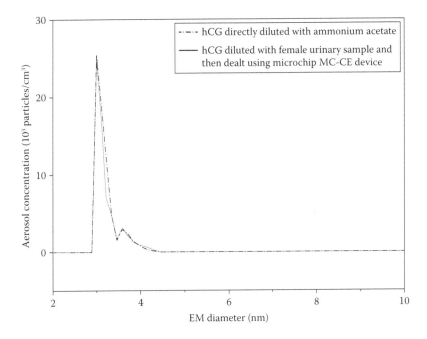

FIGURE 16.12
Comparison between two electrophoretic mobility (EM) diameter spectra of the same concentration levels of 8 μg/mL hCG standard proteins spiked into (1) human female normal urine sample that was then pretreated by MC-CE device and (2) GEMMA buffer solution of NH₃Ac that was then directly transferred to GEMMA.

of hCG standard protein to the GEMMA buffer solution consisting of NH₄Ac and directly using it for GEMMA measurement. The two GEMMA spectra obtained are shown in Figure 16.12, excellent matching in the peak profile and indicating a complete removal of sample matrix interference from the urine sample after sample cleanup and salt removal by the MC-CE device.

16.6 Summary

The use of GEMMA for protein characterization and quantitation has been illustrated in this chapter for two cases of application. The first case is characterization of proteins from fresh and imitation milk using GEMMA integrated with UCPC, NAS, and ICP-AES techniques coupled with impacted or acid scrubber method. The acid scrubber method offers a higher collection efficiency compared to the impacted plate method used in NAS for catching particles exiting from the DMA. The size information from GEMMA/UCPC shows that the number of fine particles is much higher than that of large particles for both samples. The results indicate that the addition of Ca to imitation milk led to larger particle formation for both fine and large particles. However, the ratio of casein micelles (large particles) compared to submicelles (fine particles) is kept constant. The model of a nonrigid structure of casein micelles with a core casein–micelle structure surrounded by dissociable submicelles agrees with the experimental results, as the large particle represents the casein–micelles and the fine particles the dissociated submicelles.

In the second case, the integration of the MC-CE device with GEMMA for the quantitation of urinary protein is explored. Three urinary proteins, HSA, β2-MG, and hCG, have been investigated for their quantitative measurement in urine. From the HSA study, dimers and impurities are found to be present in the standard proteins and β2-MG is found particularly prone for protein–protein interaction with the formation of dimers and aggregates at different protein concentrations, probably due to its structure. Thus, sample dilution until constant profile is necessary for protein measurement by GEMMA. Simultaneous determination of proteins using internal standards has been investigated using HSA and β2-MG. The results indicate no protein–protein interaction occurred between HSA and β2-MG and the GEMMA system can be used for simultaneous determination of several nonoverlapping proteins in urine and a second protein can be added to urine as an external standard to solve the quantitation problem due to the use of multiple instruments in the integration of MC-CE device with GEMMA for urinary protein determination. From the study on the determination of trace levels of hCG in urine, a linear range from 1 to 100 μg/mL was established between the concentration of hCG in urine and the number of particle counts by UCPC. For salt removal and sample cleanup, the MC-CE device was found to be capable of removing interference from the urine sample matrix prior to hCG determination by GEMMA.

References

1. Chen, D.R., Pui, D.Y.H. and Kaufman, S.L. Electrospraying of conducting liquids for monodisperse aerosol generation in the 4 nm to 1.8 pm diameter range. *J Aerosol Sci* 1995, 26, 963–977.
2. Kaufman, S.L., Skogen, J.W., Dorman, F.D. and Zarrin, F. Macromolecule analysis based on electrophoretic mobility in air: Globular proteins. *Anal Chem* 1996, 68, 1895–1904.
3. Mouradian, S., Skogen, J.W., Dorman, F.D., Zarrin, F., Kaufman, S.L. and Smith, L.M. DNA analysis using an electrospray scanning mobility particle sizer. *Anal Chem* 1997, 69, 919–925.
4. Kaufman, S.L., Kuchumov, A.R., Kazakevich, M. and Vinogradov, S.N. Analysis of a 3.6-MDa hexagonal bilayer hemoglobin from lumbricus terrestris using a gas-phase electrophoretic mobility molecular analyzer. *Anal Biochem* 1998, 259, 195–202.

5. Kaufman, S.L. Analysis of biomolecules using electrospray and nanoparticle methods: The gas-phase electrophoretic mobility molecular analyzer (GEMMA). *J Aerosol Sci* 1998, 29, 537–552.

6. Bacher, G., Szymanski, W.W., Kaufman, S.L., Zöllner, P., Blaas, D. and Allmaier, G. Charged-reduced nano electrospray ionization combined with differential mobility analysis of peptides, proteins, glycoproteins, noncovalent protein complexes and viruses. *J Mass Spectrom* 2001, 36, 1038–1052.

7. Loo, J.A., Berhane, B., Kaddis, C.S., Wooding, K.M., Xie, Y., Kaufman, S.L. and Chemnushevich, I.V. Electrospray ionization mass spectrometry and ion mobility analysis of the 20s proteasome complex. *J Am Soc Mass Spectrom* 2005, 16, 998–1008.

8. Kaddis, C.S., Lomeli, S.H., Yin, S., Berhane, B., Apostol, M.I., Kickhoefer, V.A., Rome, L.H. and Loo, J.A. Sizing large proteins and protein complexes by electrospray ionization mass spectrometry and ion mobility. *J Am Soc Mass Spectrom* 2007, 18, 1206–1216.

9. Allmaier, G., Laschober, C. and Szymanski, W.W. Nano ES GEMMA and PDMA, new tools for the analysis of nanobioparticles-protein complexes, lipoparticles, and viruses. *J Am Soc Mass Spectrom* 2008, 19, 1062–1068.

10. Laschober, C., Wruss, J., Blaas, D., Szymanski, W.W. and Allmaier, G. Gas-phase electrophoretic molecular mobility analysis of size and stoichiometry of complexes of a common cold virus with antibody and soluble receptor molecules. *Anal Chem* 2008, 80, 2261–2264.

11. Hogan, C.J., Kettleson, E.M., Ramaswami, B., Chen, D.R. and Biswas, P. Charge reduced electrospray size spectrometry of mega and gigadalton complexes: Whole viruses and virus fragments. *Anal Chem* 2006, 78, 844–852.

12. Wick, C.H. and McCubbin, P.E. Characterization of purified MS2 bacteriophage by the physical counting methodology used in the integrated virus detection system (IVDS). *Toxicol Methods* 1999, 9, 245–252.

13. Bernstein, S.L., Wyttenbach, T., Baumketner, A., Shea, J.E., Bitan, G., Teplow, D.B. and Bowers, M.T. Amyloid beta-protein: Monomer structure and early aggregation states of A beta 42 and its Pro(19) alloform. *J Am Chem Soc* 2005, 127, 2075–2084.

14. Koeniger, S.L. and Clemmer, D.E. Resolution and structural transitions of elongated states of ubiquitin. *J Am Soc Mass Spectrom* 2007, 18, 322–331.

15. Mora, J.F., Ude, S. and Thomson, B.A. The potential of differential mobility analysis coupled to MS for the study of very large single and multiply charged proteins and protein complexes in the gas phase. *Biotechnol J* 2006, 1, 988–997.

16. TSI Inc. 2000. Model *3480 Electrospray Aerosol Generator Instruction Manual*. Shoreview, MN: TSI Inc.

17. TSI Inc. 2002. Model *3080 Electrostatic Classifier Instruction Manual*. Shoreview, MN: TSI Inc.

18. TSI Inc. 2000. Model *3025A Ultrafine Condensation Particle Counter Instruction Manual*. Shoreview, MN: TSI Inc.

19. TSI Inc. 2001. Model *3089 Nanometer Aersol Sampler Instruction Manual*. Shoreview, MN: TSI Inc.

20. Kaufman, S.L. Electrospray diagnostics performed by using sucrose and proteins in the gas-phase electrophoretic mobility molecular analyzer (GEMMA). *Anal Chim Acta* 2000, 406, 3–10.

21. Reischl, G.P., Makela, J.M. and Necid, J. Performance of Vienna type differential mobility analyzer at 1.2-20 nanometer. *Aerosol Sci Technol* 1997, 27, 651–672.
22. Jabeen, R., Payne, D., Wiktorowicz, J., Mohammad, A. and Petersen, J. Capillary electrophoresis and the clinical laboratory. *Electrophoresis* 2006, 27, 2413–2438.
23. Hiratsuka, N., Shiba, K., Nishida, K., Iizima, S., Kimura, M. and Kobayashi, S. Analysis of urinary albumin, transferrin, N-acetyl-beta-D-glucosaminidase and beta(2)-microglobulin in patients with impaired glucose tolerance. *J Clin Lab Anal* 1998, 12, 351–355.
24. Wiedensohler, A. An approximation of the bipolar charge-distribution for particles in the submicron size range. *J Aerosol Sci* 1988, 19, 387–389.
25. Pelegrine, D.H.G. and Gasparetto, C.A.Whey proteins solubility as function of temperature and pH. *LWT Food Sci Technol* 2005, 38, 77–80.
26. Kuhn, R. and Hoffstetter-Kuhn, S. *Capillary Electrophoresis: Principles and Practice.* Springer-Verlag, Berlin, 1993.
27. Phadungath, C. Casein micelle structure: a concise review. *Songklanakarin J Sci Technol* 2005, 27, 201–212.
28. Montaser, A. and Golightly, D.W. *ICP in Analytical Atomic Spectrometry.* 2nd ed. VCH Publisher, Toledo, OH, 1992.
29. Kanauchi, M., Akai, Y. and Hashimoto, T. Validation of simple indices to assess insulin sensitivity and pancreatic beta-cell function in patients with renal dysfunction, *Nephron* 2002, 13, 190–193.
30. Hong, C.Y. and Chia, K.S. Markers of diabetic nephropathy. *J Diabet Compl* 1998, 12, 43–60.
31. Branten, A.J.W., Born, J.V.D., Jansen, J.L.J., Assmann, K.J.M., Wetzels, J.F.M. and Dijkman, H.B.P.M. Familial nephropathy differing from minimal change nephropathy and focal glomerulosclerosis. *Kidney Int* 2001, 59, 693–701.
32. Carroll, M.F. and Temte, J.L. Proteinuria in adults: A diagnostic approach. *Am Fam Physician* 2000, 62, 1333–1340.
33. Kolios, G., Bairaktari, E., Tsolas, O. and Seferiadis, K. Routine differential diagnosis of proteinurias by capillary electrophoresis. *Clin Chem Lab Med* 2001, 39, 784–788.
34. Nie, Z. and Fung, Y.S. Microchip capillary electrophoresis for frontal analysis of free bilirubin and study of its interaction with human serum albumin. *Electrophoresis* 2008, 29, 1924–1931.
35. Wainright, A., Nguyen, U.T., Bjornson, T. and Boone, T.D. Preconcentration and separation of double-stranded DNA fragments by electrophoresis in plastic microfluidic devices. *Electrophoresis* 2003, 24, 3748–3792.
36. Huang, H., Xu, F., Dai, Z. and Lin, B. On-line isotachophoretic preconcentration and gel electrophoretic separation of sodium dodecyl sulfate-proteins on a microchip. *Electrophoresis* 2005, 26, 2254–2260.
37. Pagana, K.D. and Pagana, T.J. *Mosby's Manual of Diagnostic and Laboratory Test-Urine Studies.* Mosby Elsevier, St. Louis, MO, 2010, pp. 946–1026.
38. Bessonova, E.A., Kartsova, L.A. and Shmukov, A.U. Electrophoretic determination of albumin in urine using on-line concentration techniques. *J Chromatogr A* 2007, 1150, 332–338.
39. Kazumi, T., Hozumi, T., Ishida, Y., Ikeda, Y., Kishi, K., Hayakawa, M. and Yoshino, G. Increased urinary transferrin excretion predicts microalbuminuria in patients with type 2 diabetes. *Diabetes Care* 1999, 22, 1176–1180.
40. Niwa, T. Biomarker discovery for kidney diseases by mass spectrometry. *J Chromatogr B* 2008, 870, 148–153.

41. Cole, L.A., Kardana, A., Park, S.Y. and Braunstein, G.D. The deactivation of HCG by nicking and dissociation. *J Clin Endoclin Metab* 1993, 73, 704–710.

42. O'Connor, J.F., Birken, S., Lustbader, J.W., Krichevsky, A., Chen, Y. and Canfield, R.E. Recent advances in the chemistry and immunochemistry of human chorionic—Gonadotropin—Impact on clinical measurements. *Endocrinol Rev* 1994, 15, 650–683.

43. Cole, L.A. New discoveries on the biology and detection of human chorionic gonadotropin. *Reprod Biol Endocrinol* 2009, 7, 1–37.

44. Goverman, J.M., Parsons, T.F. and Pierce, J.G. Enzymatic deglycosylation of the subunits of chrionic-gonadotropin-effects on formation of tertiary structure and biological-activity. *J Biol Chem* 1982, 257, 15059–15064.

Section V

Summary and Outlook

17

Current Achievements and
Future Perspectives

Ying Sing Fung

The University of Hong Kong
Hong Kong SAR, People's Republic of China

CONTENTS

17.1 Current Development and Achievements
of MC-CE Devices

There are two major areas for current development of integrated microfluidic chip-capillary electrophoresis (MC-CE) devices for intended applications. The first area is integrating CE with various on-chip sample and analyte pretreatment procedures, such as on-chip sample cleanup and concentration, manipulation of nanofluids at microchannels for analyte isolation and enrichment, and execution of desired procedures on-chip for mixing and other operations. The second area is integrating novel detection modes with

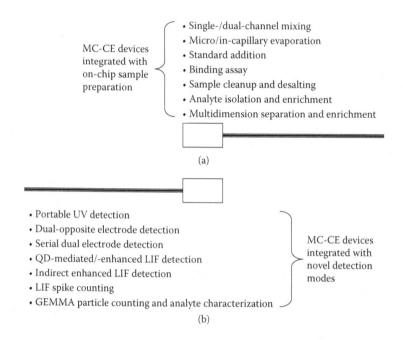

FIGURE 17.1
Current development of MC-CE devices integrated with (a) precapillary sample mixing and concentration, on-chip sample cleanup and desalting, on-channel analyte isolation and enrichment, and (b) postcapillary optical, electrochemical, and other novel detection modes.

the capillary column to enable sensitive and selective detection of analytes separated and eluted out from the separation capillary by various CE separation modes. A summary of these two areas is shown in Figure 17.1.

17.1.1 Areas for Current Development

For the first major area, the MC is normally placed before the CE separation capillary for carrying out the various sample preparation procedures on-chip prior to conducting a CE run for analyte separation and quantitation in capillary, mostly with an online and continuous integration of MC and CE operations. The on-chip operations include single- and dual-channel mixing for on-channel addition of standards to samples [1–4], micro- and in-capillary evaporation for in-capillary sample concentration (see Chapter 6), manipulation of nanofluids on-channel for sample cleanup and desalting [5–12], and analytical procedures on-chip for analyte isolation and enrichment as well as multidimensional separation and enrichment of analytes prior to CE quantitation [11–13].

In the second area, for integration with sensitive detection modes, MC is normally placed after the separation capillary to enable the use of a novel detection mode with high sensitivity for analyte detection. Various novel detection modes can be integrated with the MC-CE device to enable the incorporation of sensitive and selective detection of analytes that otherwise cannot be detected for the intended application. In general, postcapillary detection modes are avoided for CE detection due to the requirement for a postcapillary detector with an extremely small dead volume in the nanoliter region. Thus, most of the current CE detectors are operated for in-capillary detection with analytes detected within the separation capillary.

As the volume of the on-chip fabricated microchannel is about half that of the capillary for an equal length, it is possible to use a carefully designed microchannel segment pattern with acceptable dead volume for postcapillary reaction, such as the addition of reagent to react and convert nondetectable analytes to products that can be detected by novel, sensitive, optical or electrochemical detection modes [14–27]. This includes postcapillary analyte derivatization such as the addition of bromine for indirect determination of electroinactive analytes and the use of serial or opposite dual electrodes to enhance detection selectivity and to allow detection of nonelectroactive analysts [19,22,24,26].

The fabrication of MCs with carefully designed microchannel patterns enables easy alignment of microelectrodes for replacing the plugged capillary or fouled microelectrodes. In addition, it allows precise and repeatable positioning of microelectrodes very close to the exit of the separation capillary to minimize detector dead volume and to align dual microelectrodes in the desired arrangement. Detection mode with dual opposite microelectrode arrangement by holding the two opposite microelectrodes under suitable differential potentials enhances the detection selectivity by measuring the differential potential current profiles of the analyte and the comigrating impurity and by picking up comigrating impurities, as was demonstrated in the determination of polyphenols in wine (see Chapter 8). The capability of the indirect detection mode to detect nonelectroactive analytes was demonstrated using the serial dual-microelectrode arrangement with the upstream electrode to generate redox mediators that can react with the analyte, leaving behind residual mediators detected by the downstream electrode for nonelectroactive analyte quantitation (see Chapter 9).

For optical detector development, the integration of an MC-CE device with a portable ultraviolet (UV) detector was demonstrated in a field operation for assessing the quality of herbal medicine (see Chapter 12). Indirect analyte detection using the highly sensitive laser-induced fluorescence (LIF) detection mediated by quantum dots (QDs) and fluorescent dyes was shown for detecting analytes with no optical and electroactive groups, such as in the determination of pesticides in vegetables [28–38]. The detection mode is based on a suitable interaction of analytes with QDs as fluorescent dyes. The high thermal and irradiation stability of QDs and high detection sensitivity of LIF make the integration of QDs/LIF detection a promising detection mode for the determination of trace analytes in complex samples (see Chapters 10 and 11).

To achieve extreme high-sensitive detection for analytes with the absence of strong chromophores or electroactive groups, the integration of MC-CE device with particle counting detection mode was explored. Two detection modes based on LIF spike counting and gas-phase electrophoretic mobility molecular analyzer (GEMMA) nanoparticle counting were investigated (see Chapters 11, 15, and 16). LIF spike counting is less dependent on the sample matrix compared to GEMMA nanoparticle counting, but its scope of application is limited to analytes with selective and strong binding interaction with LIF dyes, whereas GEMMA is applicable to a wide range of analytes. Preliminary work on the integration of MC-CE device with GEMMA detection has shown promising results for the determination of proteins at trace levels in complex biofluids [39–42].

17.1.2 Achievements for Intended Applications

Two approaches have been adopted by MC-CE devices to meet the requirements for intended applications. The first approach is incorporation of tedious and time-consuming operations on-chip prior to online integration with CE separation to perform the desired analytical procedure to meet the requirements for the intended application. The second approach involves integration of a novel detection mode to enable quantitation of analytes without detectable functional groups. A summary of these two approaches is shown in Figure 17.2.

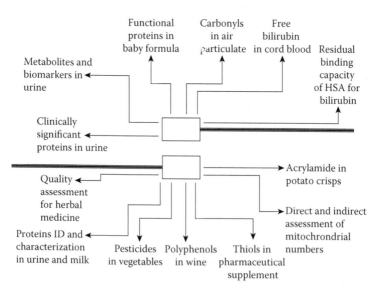

FIGURE 17.2
Schematic diagram showing the two approaches adopted by MC-CE devices based on integrating sample pretreatment procedures and novel detection modes on-chip with CE separation to meet the requirements for intended applications.

Success has been achieved by integrating sample pretreatment procedures on-chip in MC-CE devices for the determination of metabolites and bio-markers in urine [3,4,43–46], the assay of functional proteins in baby formula (see Chapter 3), the monitoring of carbonyl compounds in air particulates (see Chapter 6), and the determination of free bilirubin in cord blood, and assessment of the residual binding capacity of human serum albumin (HSA) for bilirubin [15,19,47–50]. Online integration between sample pretreatment procedures and CE separation have been shown to provide the most successful mode for integration in the above-mentioned applications, as various laboratory procedures can be performed on-chip and operated infield to widen the scope of application for the fabricated MC-CE device.

The integration of MC-CE devices with novel detection modes has been demonstrated in various application areas, such as quality assessment in herbal medicine (see Chapter 12); identification and characterization of proteins in urine and dairy products [5–10,12,13,51]; direct and indirect assessment of mitochondrial numbers [52–58]; and determination of pesticides in vegetables (see Chapter 11), polyphenols in wine (see Chapter 8), thiols in pharmaceutical supplement (see Chapter 9), and acrylamide in potato chips (see Chapter 10). Analyte detection modes can be divided into two major types: direct and indirect. For direct detection, various means have been adopted to enhance analyte detection sensitivity, including enrichment and isolation of analyte proteins by on-chip isoelectric focusing and integration with highly sensitive particle counting modes such as LIF-induced fluorescent spike counting (see Chapter 14) and GEMMA nanoparticle counting (see Chapter 16).

Indirect detection aims to detect analytes that cannot be detected by conventional detection mode due to the absence of detectable functional groups in the analyte. Electrochemical detection modes have been successfully applied using dual serial electrodes for detection of thiols via bromine as a mediator and dual opposite electrodes to differentiate polyphenols among comigration impurities in complex wine samples (see Chapters 8 and 9). Optical modes based on LIF indirect detection have been applied to acrylamide detection via interaction with QDs and indirect assessment of mitochondrial number using cardiolipin as a biomarker by indirect LIF-enhanced fluorescence detection mode. The indirect detection mode is made possible by a careful design of the operations for manipulating nanofluids in microchannels in MC-CE devices. It provides a general approach that can be adopted for other applications.

17.2 Problems and Obstacles

Obstacles to be addressed for further development of MC-CE devices fall into two major areas. The first area is the need for technological advancement of instrumentation and facilities to enable fast and easy fabrication and to assist in the control of MC-CE operation. The second area involves developing measures to deal with the application of MC-CE devices for routine analysis compared to currently used methods. These two areas are discussed in detail in the following sections.

In Sections III and IV of this book, success has been demonstrated with the application of MC-CE devices to different areas using a continuous microfluidic flow with online arrangement between MC and CE. However, for analytical procedures with a mismatch in flowrate between MC and CE, offline or flow-adjusting operations are needed to enable mixing and the use of slow chemical interaction in the detection process. Two- or three-way microvalves are required for such operations. Alternative methods for fabrication of valves with complex geometry and the need for instrumentation with flexible high-voltage (HV) control are discussed in the following sections.

17.2.1 Fabrication of MC-CE Devices

The fabrication of a desired microchannel pattern on an MC using computer-aided CO_2 laser ablation has been shown to enable flexible design by the users to meet the challenges of a given task. It is easy to engrave a desirable channel pattern onto an MC using CO_2 laser ablation under computer-aided manufacturing (CAD) simply by drawing the configuration required using user-friendly software (CorelDRAW 10, Corel Corporation, Ottawa, Ontario) prior to immediate implementation by CAD. The laser fabrication process is essentially a subtractive process by removing unwanted material from the MC to create the desirable channel pattern. However, subjects with complex shape and geometry, such as microvalve bodies, are difficult to fabricate. Although on-channel nanomagentic fluid activated by an external magnetic field had been used as a microvalve for MC-CE devices (see Chapters 3 and 13), the operations are slow, difficult to automate, and too fragile for field application.

As the manipulation of nanofluids on-chip is anticipated to be more complex to handle difficult tasks and integrate mismatching flow for MC and CE processes, the additive fabrication process may offer an alternative means for fabrication of microvalves for control of nanofluid in MC. The recent advances in three-dimensional (3D) printing for layer-to-layer buildup of a micro-object offer a promising means for fabrication of 3D microvalves. With the recent lowering in instrument prices to affordable levels and the availability of micronozzles with 0.3-mm diameter to print micro-objects with precision up to 20 μm [59], 3D printing is starting to show potential for fabrication of microvalves for MC-CE devices. However, 3D printing manufacturing requires considerable computing effort from well-trained programers familiar with the specialized CAD software to compile suitable software to print a given design by layer-to-layer deposition of nanopowders made up of metal or plastic materials. Hence, it is currently out of the scope that most analysts to handle. Given the rapid pace of development of 3D printing technology, user-friendly software will likely soon be developed and available for general use in the fabrication of complex objects required for operation of MC-CE devices.

17.2.2 Control and Operation of MC-CE Devices

For control and operation of MC-CE devices, hardware components for HV supply and HV switches are commercially available, with suppliers shown in Appendix II. In general, they offer sufficient capability for control and operation of MC-CE devices (see Chapter 3). However, at the system level for control and automation of HV, integration of multiple HV sources for MC-CE devices in designed sequence and specific duration is needed to program the required voltage sequences and time durations for an intended application. The commercially available HV control systems are limited in their capability (as discussed in Chapter 3), thus requiring the construction of the self-constructed HV switching circuit to enable ground voltage switching to direct the flow of nanofluids. The currently available two-pole switching system offered from Major Science Co. Ltd. (see Appendix II) presents HV control for simple systems, and the HV supply with eight programmable HV sources from EMCO High Voltage Corporation (Sutter Creek, CA) cannot switch ground voltage to direct nanofluid flow. The integration of an HV switching circuit with MC-CE devices, as shown in Chapter 3, offers the most flexible arrangement for HV control during operation.

The availability of a more user-friendly HV control system is preferable for new users and researchers who want to develop and operate their own MC-CE devices. The HV manufacturer [Spellman High Voltage Electronics (SIP) Co., Ltd., Suzhou, China] has developed multiple HV up to 6 kV in modular form to enhance application for microfluidic chip electrophoresis (MCE) to manipulate HVs at different electrodes. However, ground voltage cannot be switched, thereby limiting the scope of operations for HV control. In addition, voltages up to 6 kV are too low for CE run, limiting the separation for samples with less complicated sample matrixes. Thus, a more user-friendly system with integrated automation and control of HV is needed for the operation of MC-CE devices. Use of higher running voltages not only improve the analyte separation and but also allow for the implementation of micellar electrokinetic chromatography run that is needed for separation of neutral analytes.

17.3 Applicability Issues for MC-CE Devices

For application of MC-CE devices, two directions have been taken, with the first targeting the field or clinical application and the second focusing on the determination of analytes that otherwise cannot be detected. In this section, issues affecting the applicability of MC-CE devices for field or clinical analysis are discussed, whereas new application areas for the determination of previously nondetectable analytes are explored in Section 17.5. Based on the integration of CE with on-chip sample preparation and sensitive detection modes, various MC-CE devices have been fabricated and demonstrated for the separation and determination of urine metabolites and proteins [3,4,6,8,10,43–45]. Determination of urine metabolites and proteins is used to illustrate the various issues for the applicability of MC-CE devices.

To enable field and clinical application, various issues and performance parameters need to be considered for monitoring of newly discovered urinary proteins and metabolites clinically important for patients under critical conditions. New analytical methodologies developed have to meet the following requirements: (1) affordable capital cost, (2) low running cost, (3) applicability for real sample analysis, (4) validation for quantitative measurement, (5) multianalyte capability per assay, (6) high sample throughput, (7) fast turnaround of results, (8) negligible false-positive, (9) false-negative results, and (10) lab or on-site operable by clinical staff. The various currently used separation methods for the determination of urinary proteins and metabolites are compared to MC-CE devices, with assessments listed in Table 17.1.

The conventional laboratory-based techniques such as gas chromatography (GC), liquid chromatography (LC), and CE [60–68] have advantages such as previous cases for application to real samples and known validation of results with

TABLE 17.1

Separation Methods for Urinary Proteins and Metabolites

Considerations for Clinical Assay	Separation Method				
	GC	LC	CE	MCE	MC-CE
1. Capital cost	Medium high	Medium high	Medium	High	Medium
2. Running cost	Medium	Medium high	Low	Low	Low
3. Test using real sample	VC	NC	NC	NC	NC
4. Calibration for quantitative measurement	Yes	Yes	Yes	Yes	Yes
5. Multianalyte capability	Yes	Yes	Yes	Yes	Yes
6. Sample throughput	+	+	++	+++	++
7. Turnaround of results	+	+	++	+++	++
8. False-positive results	+	+	+	++	+
9. False-negative results	+	+	+	++	+
10. On-site operable by clinical staff	No	No	Yes	No	Yes
References	[60–62]	[63–66]	[67]	[68]	[3,4,10,46]

Note: GC, gas chromatography; LC, liquid chromatography; CE, capillary electrophoresis; MCE, microfluidic chip electrophoresis; microfluidic chip-capillary electrophoresis (MC-CE); VC, volatile compounds; NC, nonvolatile compounds; Low (+), medium (++) and high (+++) in considerations for clinical assay to techniques assessed.

acceptable false-positive and false-negative errors established for quantitative measurement. CE has matured to a stage as a stand-alone analyzer for target analytes, such as the commercialization of an automatic DNA analyzer for thalassemia disorder diagnosis [69] in hospitals. For other chip-based methods, such as MC and MC-CE, the advantages for high sample throughput, fast turnaround of results, and field operation capability have been demonstrated using various prototype devices, offering lab-on-a-chip operation for clinical, bedside, and personal monitoring applications.

In summary, most currently used laboratory-based separation methods are limited by low sample throughput, high expense for routine monitoring, and requirement for laboratory environment for their operation. With high sample throughput, fast turnaround of results, and field operation capability, the MC-CE–based separation methods can deliver urgently needed results on-site and in-time to assist in making decisions that are important for public health protection, such as diagnosis of infective diseases and on-site determination of toxic pesticides for food safety assessment. Thus, new application areas are expected from MC-CE devices upon further development and integration of enabling technology.

17.4 New Development of MC-CE Devices

17.4.1 Paths for New Development

The obvious path to be taken for future development of MC-CE devices is a full integration of both on-chip sample pretreatment procedures and novel sensitive detection methods in one MC-CE device with two MCs between the separation capillary to expand its scope of application to the determination of analytes at trace levels in complex samples (Figure 17.3). Challenging analytical tasks include the determination of low-abundance proteins and metabolites, infective pathogens, toxic environmental pollutants, and food contaminants at trace levels in biomedical, food, and environmental samples.

To deliver fast turnaround results on-site for difficult samples normally requires the use of tedious procedures for sample cleanup and expensive detection instrumentation. With a full integration of online sample cleanup and sensitive detection, the analytical performance of the MC-CE device is expected to be enhanced to meet the rising demand for determination of analytes at trace levels in complex sample matrixes encountered in food safety assessment, determination of toxins in biomedical substances, and monitoring of trace environmental pollutants with significant health impact.

Two major paths are expected to be taken for future development of MC-CE devices. The first path is further the development of a field-operable

Samples with complex matrixes requiring
use of tedious pretreatment procedures:
• Biomedical samples
• Clinical samples
• Food samples
• Environmental samples

Sample cleanup and analyte
enrichment for intended application

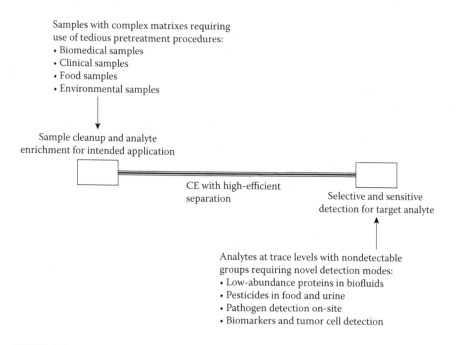

CE with high-efficient
separation

Selective and sensitive
detection for target analyte

Analytes at trace levels with nondetectable
groups requiring novel detection modes:
• Low-abundance proteins in biofluids
• Pesticides in food and urine
• Pathogen detection on-site
• Biomarkers and tumor cell detection

FIGURE 17.3

Integration of on-chip sample pretreatment and novel detection with high-efficiency CE separation for MC-CE devices to meet the demanding analytical tasks.

device for performing analytical tasks to obtain the required results for actions such as the determination of metabolites from infective agents in clinical and biomedical samples for their identification. One clear advantage of the MC-CE device is that the whole device can be completely sealed off from the external environment during its operation. Thus, the risk for exposure to infective agents during analysis is greatly reduced. The costs for a capillary and a plastic chip are low, and the whole setup can be disposed of after each analysis. Another example is the need for on-site monitoring of toxic organophosphorus pesticides in vegetables (see Chapter 11) using MC-CE devices as urgent results are needed before the contaminated vegetables enter the retail markets after their sampling at the border.

The second path is to develop on-chip quality assurance and quality control procedures for verification and confirmation of analytes to reduce false-positive or false-negative results. Various means for results verification have been described in this book such as on-chip standard addition to sample prior to CE run (see Chapter 5), use of dual opposite electrode detection to identify comigrating peaks from impurities present in complex wine samples (see Chapter 8), and the use of GEMMA for protein confirmation after sample cleanup by MC-CE devices (see Chapter 16). As the MC-CE device is further developed toward real-world applications, additional procedures are expected to be developed to validate results for MC-CE operations. These are important procedures required for acceptance of the MC-CE devices for intended applications.

17.4.2 New Application Areas

New application areas are being explored, together with technological advancement, which make the MC-CE device able to tackle difficult samples. A list of the expected technological advancements and new application areas is given in Table 17.2. The availability of a user-friendly automation system is essential for field application of the devices that are mostly going to be handled by nonlaboratory personnel during operation. Commercially available prototype MC-CE units can reduce the start-up time and cost for new users and hence speed up the application of MC-CE devices for use in new areas. The new application areas listed in Table 17.2 can be classified into the following three major groups: 1) monitoring of low-level analytes in samples collected in field, 2) determination of analytes at trace levels in complex samples, and 3) exploratory integration with selective cell capture for assay of circulating biomarkers and tumor cells.

For field application to deliver results in-time and on-demand, various MC-CE devices have been fabricated to demonstrate their applicability for various biomedical areas, such as assay of free bilirubin and binding capacity of HSA (see Chapter 13), determination of metabolites and proteins in urine (see Chapters 4, 5, and 7), monitoring of toxic pollutants in air (see Chapter 6), and assessment of harmful contaminants in food (see Chapters 10 and 11). Further development of the MC-CE device for field application requires automated control and sensitive and selective detection modes to meet the demanding requirements in new application areas.

TABLE 17.2

New Application Areas and Technological Advancement of MC-CE Devices

Technological Advancement
- Automation of MC-CE operation
- Fabrication of two- and three-way microvalves on-chip
- Integration with newly developed detection modes
- Prototyping unit for on-site assay
- Integration with selective cell capture modes

New Application Areas
- On-site monitoring of toxic and reactive environmental pollutants and contaminants
- Assay of regulated and harmful agents on-site for food safety and security monitoring
- Clinical and biomedical diagnosis of infective agents
- Protein assay in biomedical samples
- Assay of circulating biomarkers and tumor cells

For determination of analytes present at trace levels in complex samples, various techniques have been integrated with MC-CE devices to enhance the sensitivity and improve the selectivity for analyte detection. Typical CE techniques such as stacking and isotachophoresis (ITP) have been used for sample concentration [70,71] and analyte enrichment and sample cleanup [6,8,10]. In addition, sensitive detection modes such as dual serial microelectrode detection [17,19,21–26], QDs-assisted LIF detection mode [28–38], and particle counting modes [53–58] have been used with the MC-CE device for detection of analytes at trace levels in complex food and medical samples. The use of opposite dual opposite microelectrode detectors as described in Chapter 8 was shown to improve the detection selectivity to differentiate comigrating impurities from polyphenols in wine.

For new application areas requiring extremely high detection sensitivity, such as protein assay in biofluids, a particle counting detection method is needed, as a single protein particle could be counted if it appeared within the measuring period. However, thorough sample cleanup is required to avoid high background noise. The integration of GEMMA for protein particle counting with an MC-CE device for sample cleanup and desalting has demonstrated the applicability of the device for protein identification and its determination at trace levels in complex urine samples, with details provided in Chapter 16.

Recent interest has arisen for the isolation and capture of rare cells using an MC [72], as confirmation of a captured cell can be done by subsequent microscopic methods on unique cell morphology. In fact, the purpose for the incorporation of a capillary into an MC in the early work reported [73] is capturing of a cell through the capillary to an MC for microscopic investigation. The exploratory integration of MC-CE devices with selective cell capture has the following advantages. First, the previous experience accumulated for cell capture using an MC can be employed with similar procedures as the integration of sample preparation technique currently used in MC-CE devices. Second, dyes, drugs, or biomedically significant molecules can be added to the MC-CE device to investigate their effect on the captured cells. Third, the metabolites discharged from the captured cells in response to external stimulants can be separated and determined by MC-CE procedures integrated with a sensitive detection mode. The MC-CE device offers a promising tool for study and assay of circulating biomarkers and tumor cells.

17.5 Future Perspectives for Application

The physical form of MC-CE devices is visually simple: a capillary with poly(methyl methacrylate) chips bonded at either one or both of its open ends. It performs functions by integrating two or more analytical techniques in its operation to achieve the intended application, as described in Chapters 12–16. Compared to most analytical instruments developed for laboratory analysis of specific analytes in given samples, the scope of application using an MC-CE device is wide and flexible, with capability for on-site and laboratory analysis, based on the techniques and instruments to be integrated in its operation. Thus, it offers an open platform for integration of CE with techniques and instruments required to achieve an intended application.

To meet new and demanding requirements from recent analytical challenges, the task is often beyond the scope of existing methods; hence, the need to develop a new method(s) to tackle the required challenges is needed. New methods are often required for the determination of analytes at trace levels in complex samples containing high levels of interferents with similar chemical nature as the analyte. The MC-CE device offers a fast and low-cost solution to satisfy the required demand by integrating the existing detection systems with a suitable CE mode of separation. When the existing detection system does not have sufficient detection sensitivity to meet the analytical requirements, new detection modes or analyte enrichment schemes have to integrate with the MC-CE device to achieve the requirements for the intended application. There are currently two major directions for the development of MC-CE devices: (1) on-site assay of analytes with quick turnaround of results and (2) the determination of analytes at trace levels in complex samples, with details discussed in the following sections.

17.5.1 MC-CE Devices for On-Site Assay

For on-site assay of analytes with quick turnaround of results, MC-CE devices have demonstrated the following advantages, as described in this book. First, the integration of sample preparation procedures on-chip prior to online CE separation greatly reduces the sample preparation time due to efficient extraction and mixing in the microchannels. Second, sample cleanup is fast and efficient due to a large difference in the migration rates between analytes and impurities. Third, the risk for contamination during sample preparation is eliminated as the device is completely enclosed during operation. Fourth, the automated operations of the MC-CE device from sample injection, CE separation to analyte detection, and quantitation enable a fast sample throughput for a total analysis system operable in the field by non-laboratory personnel.

Successful cases for application of MC-CE devices for on-site assay have been demonstrated in clinical, environmental, and food safety applications, as given in Chapters 4–7 and 10–13. The use of CO_2 laser ablation provides the capability for flexibility in design of microchannel patterns on the MC to meet the requirements for integrating various sample preparation techniques with well-established CE modes to offer adequate separation and detection of analytes to meet the requirements for intended applications.

The noticeable success is to integrate multiple sample pretreatment procedures on-chip, such as isoelectric focusing and ITP, for online analyte enrichment and apply sample cleanup procedures prior to CE separation for the determination of urinary proteins, as described in Chapter 7, and the implementation of multidimensional analyte enrichment strategy to enable hourly determination of atmospheric carbonyl compounds, as described in Chapter 6.

To meet the requirements for on-site assay, fast, economic, sensitive, and portable detection modes are needed. Success with UV and electrochemical detection have been demonstrated in MC-CE devices for on-site assay. Apart from direct determination after CE separation, indirect methods have also been developed by integration of serial and opposite dual-microelectrode detection to enhance detection selectivity and to enable the detection of electroinactive analytes, such as for the determination of glutathione disulfide and glutathione in pharmaceutical supplement using serial dual microelectrodes (see Chapter 9). Dual opposite microelectrode detection has been shown capable to enhance detection selectivity by differentiating comigrating impurities from polyphenols in complex wine samples (see Chapter 8). With a rapid lowering in cost for the semiconductor laser, the new QDs/LIF detection mode as described in Chapters 10 and 11 provides a highly sensitive detection mode for quantitation of analytes with capability to interact with QDs.

17.5.2 MC-CE Devices for Quantitation of Analytes at Trace Levels in Complex Samples

Due to the recent concerns about public health protection, food safety, and toxic environmental pollutants, new methods are developed to meet the demand for the determination of analytes present at trace levels in complex sample matrixes. Expensive instrumentation is often employed to satisfy the demanding legal control requirements. However, the high capital and running costs have hindered the use of expensive equipment for routine and daily monitoring of controlled substances in clinical, food, and environmental samples.

With accumulated knowledge gained through the postgenome study of the interplay between DNA and proteins, new protein and metabolite biomarkers have been discovered for diseases and medical conditions that require new methods for their determination in various biofluids, such as urine and human serum. The determination of circulating biomarkers in human plasma, such as enzymes, DNA, proteins, and tumor cells, has recently become a hot topic for analytical method development to meet the challenge for detecting analytes at trace levels in very complex sample matrixes. New challenges have also arisen from the determination of infective pathogens in biofluids, newly discovered adulterants in food, and toxic pollutants in environmental samples. These are new areas of application to be explored using the MC-CE devices fabricated to deliver results on-demand for urgent analysis.

MC-CE devices offer a suitable option for the above-mentioned applications because of their capability for on-site determination of analytes at trace levels in complex samples. However, all technical features and full capability of the devices have to be utilized to satisfy the demanding requirements, such as the utilization of a large sample plug, application of an extensive sample cleanup and multistage analyte enrichment strategy, and the use of high CE separation efficiency and newly developed detection modes to achieve extremely high detection sensitivity at an affordable cost. To improve the separation for complex samples, relevant CE modes with higher running voltage and longer capillary columns are needed. For lowering the detection limits, multidimensional analyte enrichment should be used together with newly developed detection modes with extremely high detection sensitivity, such as QDs-LIF detection as described in Chapters 10 and 11, as well as the integration with the particle counting detection mode as described in Chapters 14 and 16.

In summary, success has been realized using MC-CE devices fabricated for the determination of metabolites and proteins in biofluids, pollutants, and harmful substances in environmental and food samples using portable detection modes such as UV-visible detector and dual serial and opposite microelectrode detection modes. For future application of MC-CE devices, the following new areas are expected to be explored: capture and identification of tumor cells; determination of circulating biomarkers; assay of proteins and metabolites in biofluids; and analysis for quantitation of trace analytes in complex samples using the full capability of the integrated MC-CE devices with multidimensional analyte enrichment, thorough sample cleanup and incorporation of newly developed detection modes to deliver on-site analysis for urgently required results.

References

1. Guo, W.P. and Fung, Y.S. Microfluidic chip-capillary electrophoresis for the determination of organic acids in biofluids. *Abstract, the 8th Chinese Symposium on Microscale Bioseparations & Methods for System Biology (CMSB 2008)*, Guilin, China, November 21–24, 2008, p. 197.
2. Guo, W.P., Lau, K.M. and Fung, Y.S. Microfluidic chip-capillary electrophoresis for emergent biomedical analysis in Hong Kong. *Abstract K27, the 9th Asia-Pacific International Symposium on Microscale Separation and Analysis (APCE 2009) and 1st Asian-Pacific International Symposium on Lab-on-Chip (APLOC 2009)*, Shanghai, China, October 28–31, 2009, p. 36.
3. Guo, W.P., Lau, K.M. and Fung, Y.S. Microfluidic chip-capillary electrophoresis for two orders extension of adjustable upper working range for profiling of inorganic and organic anions in urine. *Electrophoresis* 2010, 31, 3044–3052.

4. Guo, W.P. and Fung, Y.S. Microfluidic chip-capillary electrophoresis with dynamic multi-segment standards addition for rapidly identifying nephrolithiasis markers in urine. *Electrophoresis* 2011, 32, 3437–3445.

5. Wu, R.G., Fung, Y.S. and Yeung, W.S.B. Determination of lactoferrin and β-lactoglobulin in dairy products by microfluidic-chip capillary electrophoresis. *Abstract P4-10, 9th Asia-Pacific International Symposium on Microscale Separation and Analysis (APCE 2009) and 1st Asian-Pacific International Symposium on Lab-on-Chip (APLOC 2009)*, Shanghai, China, October 28–31, 2009, p. 238.

6. Wu, R.G., Fung, Y.S. and Yeung, W.S.B. Microfluidic-chip capillary electrophoresis for analysis of clinical urinary proteins. *Abstract P136, 25th International Symposium on Microscale Bioseparations (MSB 2010)*, Prague, Czech Republic, March 21–25, 2010, p. 129.

7. Wu, R.G., Fung, Y.S. and Yeung, W.S.B. Determination of lactoferrin and immunoglobulin G in baby formula by microfluidic-chip capillary electrophoresis. *Abstract PB8, the 10th Asian-Pacific International Symposium on Microscale Separations and Analysis (APCE 2010)*, Hong Kong, China, December 10–13, 2010, p. 99.

8. Wu, R.G., Fung, Y.S. and Yeung, W.S.B. Microfluidic-chip capillary electrophoresis for determination of clinical urinary proteins. *Abstract PB9, 10th Asian-Pacific International Symposium on Microscale Separations and Analysis (APCE 2010)*, Hong Kong, China, December 10–13, 2010, p. 100.

9. Wu, R.G., Fung, Y.S. and Yeung, W.S.B. Microfluidic chip-capillary electrophoresis for separation and determination of low abundance proteins. *Abstract, Microfluidic and Nanofluidic Devices for Chemical and Biochemical Experimentation Symposium, 2010 International Chemical Congress of Pacific Basin Societies (Pacifichem 2010)*, Honolulu, HI, December 15–20, 2010, TECH Paper 547, p. 1.

10. Wu, R.G., Yeung, W.S.B. and Fung, Y.S. 2D t-ITP/CZE determination of clinical urinary proteins using microfluidic-chip capillary electrophoresis device. *Electrophoresis* 2011, 32, 3406–3414.

11. Fung, Y.S. Enhancing analytical capability of capillary electrophoresis by integrated microfluidic chip-capillary devices. *Abstract, the 12th Asian-Pacific International Symposium on Microscale Separations and Analysis (APCE 2012)*, Singapore, December 16–19, 2012, SICC7-A-0238, p. 1.

12. Wu, R.G., Wang, Z.P., Zhao, W.F., Yeung, W.S.B. and Fung, Y.S. Multi-dimension microchip-capillary electrophoresis device for determination of functional proteins in infant milk formula. *J Chromatogr A* 2013, 1304, 220–226.

13. Wu, R.G., Wang, Z., Fung, Y.S., Seah, D.Y.P. and Yeung, W.S.B. Assessment of adulteration of soybean proteins in dairy products by 2D microchip-CE device. *Electrophoresis* 2014, 35, 1728–1734.

14. Nie, Z. and Fung, Y.S. Microchip capillary electrophoresis for frontal analysis of free bilirubin and study of its interaction with-human serum albumin. *Electrophoresis* 2008, 29, 1924–1931.

15. Du, F.Y. and Fung, Y.S. Determination of biologically important polyphenols by capillary electrophoresis with electrochemical detection. *Abstract P01-04, 8th Asia-Pacific International Symposium on Microscale Separation and Analysis (APCE 2008)*, Kaohsiung, Taiwan, November 2–5, 2008, p. 98.

16. Fung, Y.S. and Nie, Z. Microfluidic chip-capillary electrophoresis for biomedical analysis. *Sep Sci China* 2009, 1, 18–23.

17. Du, F.Y. and Fung, Y.S. Differential amperometric dual electrode detection for peak purity assessment of polyphenols in red wine after separation by microfluidic chip-capillary electrophoresis. *Abstract P3-01, 9th Asia-Pacific International Symposium on Microscale Separation and Analysis (APCE 2009) and 1st Asian-Pacific International Symposium on Lab-on-Chip (APLOC 2009)*, Shanghai, October 28–31, 2009, China, p. 212.

18. Fung, Y.S. and Nie, Z. Microfluidic chip-capillary electrophoresis for biomedical analysis. *Sep Sci* 2009, 1, 3–8.

19. Du, F.Y. and Fung, Y.S. Microfluidic-chip capillary electrophoresis device incorporating a new serial dual-microelectrode detector for the determination of amino acids and proteins in food. *Abstract PB12, the 10th Asian-Pacific International Symposium on Microscale Separations and Analysis (APCE 2010)*, Hong Kong, China, December 10–13, 2010, p. 103.

20. Du, F.Y. and Fung, Y.S. Determination of chlorophenols by capillary electrophoresis with dual electrode detection for peak purity assessment. *Abstract P056, 25th International Symposium on Microscale Bioseparations (MSB 2010)*, Prague, Czech Republic, March 21–25, 2010, p. 88.

21. Du, F.Y. and Fung, Y.S. Development of CE-dual opposite carbon-fibre micro-disk electrode detection for peak purity assessment of polyphenols in red wine. *Electrophoresis* 2010, 31, 2192–2199.

22. Fung, Y.S. and Du, F.Y. Microfluidic chip-capillary electrophoresis devices—Dual electrode detectors for direct and indirect determination of analytes in complicated sample matrixes. *Abstract IL2, the 11th Asian-Pacific International Symposium on Microscale Separations and Analysis (APCE 2011)*, Hobart, Australia, November 27–30, 2011, p. 21.

23. Du, F.Y. and Fung, Y.S. Dual opposite carbon-fiber micro-disk electrodes for detection and purity assessment of CE peaks from polyphenols in red wine. *Abstract, 11th Chinese National Symposium on Electroanalytical Chemistry (11NEAC)*, Liaocheng, Shandong, China, May 12–15, 2011, p. 1.

24. Du, F.Y., Chen, Q.D., Zhao, W.F. and Fung, Y.S. A new microfluidic-chip capillary electrophoresis device with a serial dual-electrode detector for biomedical analysis. *Abstract T21, the 4th International Symposium on Microchemistry and Microsystems (ISMM 2012)*, Zhubei City, Taiwan, June 10–13, 2012, pp. 326–327.

25. Du, F.Y. and Fung, Y.S. Capillary electrophoresis with dual-multiwalled carbon nanotube modified carbon fiber microelectrode detector for determination of methyl parathion metabolites in human urine. *Abstract, the 13th Asian-Pacific International Symposium on Microscale Separations and Analysis (APCE 2013)*, Jeju. Korea, November 3–6, 2013, Paper EA-TP61, p. 1.

26. Du, F.Y. and Fung, Y.S. A serial dual-electrode detector based on electrogenerated bromine for capillary electrophoresis. *Electrophoresis* 2014, 35, 3556–3563.

27. Nie, Z. and Fung, Y.S. Microfluidic chip-capillary electrophoresis for quality assessment of a complex herbal preparation. *Abstract, 7th Asia-Pacific International Symposium on Microscale Separation and Analysis (APCE 2007)*, Singapore, December 16–19, 2007, Paper 5:48, pp. 5:141.

28. Chen, Q.D. and Fung, Y.S. Quantum dots for detection of organophosphorus pesticides. *Abstract, 7th Asia-Pacific International Symposium on Microscale Separation and Analysis (APCE 2007)*, Singapore, December 16–19, 2007, Paper 5:60, pp. 5:145.

29. Chen, Q.D. and Fung, Y.S. Microfluidic-chip capillary electrophoresis with immobilized quantum dots detection for the separation and determination of organophosphorus pesticides in contaminated vegetables. *Abstract P01-07, 8th Asia-Pacific International Symposium on Microscale Separation and Analysis (APCE 2008)*, Kaohsiung, Taiwan, November 2–5, 2008, p. 101.

30. Chen, Q.D. and Fung, Y.S. Study of quantum dots immobilized at detection zone for capillary electrophoresis determination of organophosphorus pesticides. *Abstract ENV06, 34th International Symposium on High-Performance Liquid Phase Separations and Related Techniques (HPLC 2009)*, Dresden, Germany, June 28–July 2, 2009, p. 320.

31. Chen, Q.D. and Fung, Y.S. Determination of organophosphate pesticides in vegetables by integrating microextraction with microfluidic-chip capillary electrophoresis. *Abstract P4-01, 9th Asia-Pacific International Symposium on Microscale Separation and Analysis (APCE 2009) and 1st Asian-Pacific International Symposium on Lab-on-Chip (APLOC 2009)*, Shanghai, China, October 28–31, 2009, p. 226.

32. Chen, Q.D. and Fung, Y.S. Capillary electrophoresis with immobilized quantum dots fluorescence detection for rapid determination of organophosphorus pesticides in vegetables. *Electrophoresis* 2010, 31, 3107–3114.

33. Chen, Q.D., Mak, J.L.L. and Fung, Y.S. Coupling continuous micro-evaporator to capillary electrophoresis for determination of organophosphorus pesticides in vegetables. *Abstract P214, 25th International Symposium on Microscale Bioseparations (MSB 2010)*, Prague, Czech Republic, March 21–25, 2010, pp. 167–168.

34. Chen, Q.D. and Fung, Y.S. Determination of acrylamide in food by capillary electrophoresis with quantum dot-mediated LIF detection. *Abstract P-205, the 17th International Symposium on Capillary Electroseparation Techniques (ITP 2010)*, Baltimore, MD, August 29–September 1, 2010, p. 64.

35. Chen, Q.D., Mak, J.L.L. and Fung, Y.S. Continuous micro-evaporator method developed for pesticides analysis in vegetables. *Abstract PA18, the 10th Asian-Pacific International Symposium on Microscale Separations and Analysis (APCE 2010)*, Hong Kong, China, December 10–13, 2010, p. 90.

36. Chen, Q.D., Zhao, W.F. and Fung, Y.S. Determination of acrylamide in potato crisps by capillary electrophoresis with quantum dot-mediated LIF detection. *Electrophoresis* 2011, 32, 1252–1257.

37. Chen, Q.D., Zhao, W.F., Du, F.Y. and Fung, Y.S. Determination of pesticides by microchip-capillary electrophoresis device with magnetically immobilized QD for LIF detection. *Abstract T22, 4th International Symposium on Microchemistry and Microsystems (ISMM 2012)*, Zhubei City, Taiwan, June 10–13, 2012, pp. 328–329.

38. Chen, Q.D., Ma, T.M., Du, F.Y. and Fung, Y.S. Microchip-CE device with on-chip micro-evaporator for determination of organophosphorus pesticides in vegetable. *Abstract, the 12th Asian-Pacific International Symposium on Microscale Separations and Analysis (APCE 2012)*, Singapore, December 16–19, 2012, SICC7-A-0232, p. 1.

39. Ma, T.M., Wu, R.G. and Fung, Y.S. Direct determination of human urinary proteins by gas-phase electrophoretic mobility molecular analyzer. *Abstract, 11th Chinese National Symposium on Electroanalytical Chemistry (11NEAC)*, May 12–15, 2011, Liaocheng, Shandong, China, p. 1.

40. Ma, T.M., Wu, R.G. and Fung, Y.S. Hyphenation of microfluidic-chip capillary electrophoresis with gas-phase electrophoretic mobility molecular analyzer (GEMMA) for protein characterization. *Abstract PB10, the 10th Asian-Pacific International Symposium on Microscale Separations and Analysis (APCE 2010)*, Hong Kong, China, December 10–13, 2010, p. 101.

41. Ma, T.M., Zhao, W.F., Sze, K.L. and Fung, Y.S. Coupling gas-phase electrophoretic mobility molecular analyzer with atomic emission spectroscope for study of metal-protein binding in milk. *Abstract G53, the 14th Beijing Conference and Exhibition on Instrumental Analysis (BCEIA 2011)*, Beijing, China, October 13–16, 2011, pp. 922–923.

42. Ma, T.M. and Fung, Y.S. Hyphenating gas-phase electrophoretic mobility molecular analyzer with capillary electrophoresis for determination of proteins in human tear fluids. *Abstract 2773, 222nd Meeting of Electrochemical Society and Pacific Rim Meeting of Related Societies on Electrochemical and Solid-State Science (PRiME 2012)*, Honolulu, HI, October 7–12, 2012, 1 p.

43. Guo, W.P., Lau, K.M., Sun, H. and Fung, Y.S. Microfluidic chip-capillary electrophoresis for emergency onsite biomedical analysis. *Sep Sci* 2010, 2, 3–10.

44. Guo, W.P., Lau, K.M. and Fung, Y.S. Microfluidic chip-capillary electrophoresis for point-of-care analysis. *Proceedings, International Symposium on Micro-chemistry and Microsystems (ISMM 2010)*, Hong Kong, May 28–30, 2010, P31, pp. 144–145.

45. Guo, W.P. and Fung, Y.S. Microfluidic chip-capillary electrophoresis with adjustable on-chip sample dilution for profiling of urinary markers. *Proceedings, 14th International Conference on Miniaturized Systems for Chemistry and Life Sciences (μTAS 2010)*, Groningen, The Netherlands, October 3–7, 2010, Paper W11A, pp. 1481–1483.

46. Guo, W.P., Rong, Z.B., Li, Y.H., Fung, Y.S., Gao, G.Q. and Cai, Z.M. Microfluidic chip capillary electrophoresis coupled with electrochemiluminescence for enantioseparation of racemic drugs using central composite design optimization. *Electrophoresis* 2013, 34, 2962–2969.

47. Nie, Z. and Fung, Y.S. Multi-segment incremental sample injection for the determination of bilirubin binding capacity of albumin and study of drug interaction by microchip capillary electrophoresis. *Abstract, 22nd International Symposium on Microscale Bioseparations & Methods for System Biology (MSB 2008)*, Berlin, Germany, March 9–13, 2008, Paper P.284-Tu, p. 417.

48. Nie, Z. and Fung, Y.S. Determination of binding capacity of albumin for bilirubin by insitu titration at a circular ferrofluid driven micromixer in microfluidic chip-capillary electrophoresis. *Abstract P01-05, 8th Asia-Pacific International Symposium on Microscale Separation and Analysis (APCE 2008)*, November 2–5, 2008, Kaohsiung, Taiwan, p. 99.

49. Nie, Z., Sun, H. and Fung, Y.S. Determination of binding capacity of albumin for bilirubin by microfluidic chip-capillary electrophoresis based on insitu titration at a circular ferrofluid driven micromixer. *Abstract P2-02, 9th Asia-Pacific International Symposium on Microscale Separation and Analysis (APCE 2009) and 1st Asian-Pacific International Symposium on Lab-on-Chip (APLOC 2009)*, October 28–31, 2009, Shanghai, China, p. 155.

50. Sun, H., Nie, Z. and Fung, Y.S. Determination of free bilirubin and its binding capacity by albumin using a microfluidic chip-capillary electrophoresis device with a multi-segment circular-ferrofluid driven micromixing injection. *Electrophoresis* 2010, 31, 3061–3069.

51. Du, F.Y., Wu, R.G. and Fung, Y.S. Microfluidic chip-capillary electrophoresis devices for determination of proteins and amino acids in biofluids. *Abstract, the 13th Asian-Pacific International Symposium on Microscale Separations and Analysis (APCE 2013)*, Jeju, Korea, November 3–6, 2013, Paper LOC-WK02, p. 1.

52. Zhao, W.F., O, W.S. and Fung, Y.S. Quantitation of mitochondria separated by capillary electrophoresis. *Abstract, 7th Asia-Pacific International Symposium on Microscale Separation and Analysis (APCE 2007)*, Singapore, December 16–19, 2007, Paper 5:50, pp. 5:141.

53. Zhao, W.F., O, W.S. and Fung, Y.S. Microfluidic chip-capillary electrophoresis with laser induced fluorescence detection for assessing changes in mitochondria from HepG2 cells. *Abstract P01-06, 8th Asia-Pacific International Symposium on Microscale Separation and Analysis (APCE 2008)*, Kaohsiung, Taiwan, November 2–5, 2008, p. 100.

54. Zhao, W.F., Fung, Y.S. and O, W.S. Capillary electrophoresis with LIF detection for assessment of mitochondria based on the cardiolipin content. *Abstract P168, 25th International Symposium on Microscale Bioseparations (MSB 2010)*, Prague, Czech Republic, March 21–25, 2010, pp. 145–146.

55. Zhao, W.F., Fung, Y.S. and O, W.S. Assessment of the mitochondria number based on cardiolipin by capillary electrophoresis-LIF detection. *Abstract PA17, the 10th Asian-Pacific International Symposium on Microscale Separations and Analysis (APCE 2010)*, Hong Kong, China, December 10–13, 2010, p. 89.

56. Zhao, W.F., Fung, Y.S. and O, W.S. Evaluation of cardiolipin content changes in response to carbonyl cyanide p-(trifluoromethoxy) phenylhydrazone by capillary electrophoresis with laser-induced fluorescence detection. *Eur J Lipid Sci Technol* 2010, 112, 1058–1066.

57. Zhao, W.F., Fung, Y.S. and O, W.S. L-cysteine-capped CdTe quantum dots as a fluorescence probe for determination of cardiolipin. *Anal Sci* 2010, 26, 879–884.

58. Zhao, W.F., Chen, Q.D., Wu, R.G., Wu, H., Fung, Y.S. and O, W.S. Capillary electrophoresis with LIF detection for assessment of mitochondrial number based on the cardiolipin content. *Electrophoresis* 2011, 32, 3025–3033.

59. 3D printer from CEL Robox Co., Ltd. Available at http://www.cel-robox.com/technical/

60. Kvitvang, H.F.N., Andreassen, T., Adam, T., Villas-Boas, S.G. and Bruheim, P. Highly sensitive GC/MS/MS method for quantitation of amino and nonamino organic acids. *Anal Chem* 2011, 83, 2705–2711.

61. Kind, T., Tolstikov, V, Fiehn, O. and Weiss, R.H. A comprehensive urinary metabolomic approach for identifying kidney cancer. *Anal. Biochem.* 2007, 363, 185–195.

62. Kuhara, T. Diagnosis of inborn errors of metabolism using filter paper urine, urease treatment, isotope dilution and gas chromatography-mass spectrometry. *J Chromatogra B* 2001, 758, 3–25.

63. Roux, A., Xu, Y., Heilier, J.F., Olivier, M.F., Ezan, E., Tabet, J.C. and Junot, C. Annotation of the human adult urinary metabolome and metabolite identification using ultra high performance liquid chromatography coupled to a linear quadrupole ion trap-orbitrap mass spectrometer. *Anal Chem* 2012, 84, 6429–6437.

64. Lutz, U., Lutz, R.W. and Lutz, W.K. Metabolic profiling of glucuronides in human urine by LC-MS/MS and partial least-squares discriminant analysis for classification and prediction of gender. *Anal Chem* 2006, 78, 4564–4571.

65. Nielen, M.W.F., Bovee, T.F.H., van Engelen, M.C., Rutgers, P., Hamers, A.R.M., van Rhijn, J.H.A. and Hoogenboom, L.R.A.P. Urine testing for designer steroids by liquid chromatography with androgen bioassay detection and electrospray quadrupole time-of-flight mass spectrometry identification. *Anal Chem* 2006, 78, 424–431.

66. Kim, H., Ahn, E. and Moon, M.H. Profiling of human urinary phospholipids by nanoflow liquid chromatography/tandem mass spectrometry. *Analyst* 2008, 133, 1656–1663.

67. Ramautar, R., Somsen, G.W. and de Jong, G.J. CE-MS in metabolomics. *Electrophoresis* 2009, 30, 276–291.

68. Lin, C.C., Tseng, C.C., Chuang, T.K., Lee, D.S. and Lee, G.B. Urine analysis in microfluidic devices. *Analyst* 2011, 136, 2669–2688.

69. Chang, P.L., Kuo, I.T., Chiu, T.C. and Chang, H.T. Fast and sensitive diagnosis of thalassemia by capillary electrophoresis. *Anal Bioanal Chem* 2004, 379, 404–410.

70. Sun, H., Lai, J.P. and Fung, Y.S. Simultaneous determination of gaseous and particulate carbonyls in air by coupling micellar electrokinetic capillary chromatography with molecular imprinting solid-phase extraction. *J Chromatogr A* 2014, 1358, 303–308.

71. Sun, H. and Fung, Y.S. Integrating micellar electrokinetic capillary chromatography with molecular imprinting polymer- solid phase extraction for ambient carbonyl determination. *Abstract, the 12th Asian-Pacific International Symposium on Microscale Separations and Analysis (APCE 2012)*, Singapore, December 16–19, 2012, SICC7-A-0240, p. 1.

72. Smith, J.P., Barbati, A.C., Santana, S.M., Gleghorn, J.P. and Kirby, B.J. Microfluidic transport in microdevices for rare cell capture. *Electrophoresis* 2012, 33, 3133–3142.

73. Ocvirk, G., Tang, T. and Harrison, D.J. Optimization of confocal epifluorescence microscopy for microchip-based miniaturized total analysis systems. *Analyst* 1998, 123, 1429–1434.

Appendix I: Definition of Commonly Used Microfluidic Devices

Microfluidic Chip-Capillary Electrophoresis

The microfluidic chip-capillary electrophoresis (MC-CE) device integrates the capability of MC for on-chip sample pretreatment and enhanced analyte detection to capillary electrophoresis (CE) with high analyte separation efficiency and established separation modes to perform on-site lab-on-a-chip analysis of complex samples for multianalyte determination at trace levels.

Microfluidic Chip Electrophoresis or Microchip CE

Microfluidic chip electrophoresis (MCE) or microchip CE devices integrate all operations on-chip, including sample pretreatment, analyte separation by electrophoresis, and detection for analysis of multianalytes in a given sample.

Microfluidic Chip

Microfluidic chip (MC) offers the capability to move microfluids through microchannels to achieve a desired function, such as mixing, pumping, and allowing chemical reactions for direct detection of a given analyte in a sample.

Microarray Chip or Biochip

A microarray chip or biochip is a two-dimensional array fabricated on a solid substrate to detect analytes in a sample by direct chemical reaction between the samples delivered to each array containing specific reagents for a high throughput, multiplexed, and parallel detection screening method.

Appendix II: Suppliers for Multi-High-Voltage Power Systems, Portable CE Instrumentation, and Associated High-Voltage Accessories

1. Major Science Co. Ltd.

   Address: No. 7, Hwai-Der Street, Pan-Chao City, Taiwan, Republic of China

   Telephone: (886) 2-2256-6980

   Website: http://www.major-sci.com

   Information: Supplier for power supply with eight switchable high-voltage channels operated by one high-voltage source

2. CE Resources Pte Ltd. (CE)

   Address: Block 51, Ayer Rajah Crescent, No. 06-08/09, Ayer Rajah, Singapore 139948

   Telephone: (65) 6779-8208

   Fax: (65) 6778-2987

   E-mail: salessupport@ce-resources.com

   Information: Supplier for portable CE and associated equipment

3. EMCO High Voltage Corporation

   Address: 1 EMCO Court, Sutter Creek, CA 95685, USA

   Telephone: (209) 267-1630

   Website: http://www.emcohighvoltage.com/

   E-mail: sales@emcohv.com

   Information: Supplier for power supply with eight programable high-voltage channels independently operated by eight high-voltage sources

4. Spellman High Voltage Electronics (SIP) Co., Ltd.

   Address: Block D, No. 16 SuTong Road, Suzhou Industrial Park, China 215021

   Telephone: (86) 512-6763-0010

   Fax: (86) 512-6763-0030

   E-mail: sales@spellmanhv.cn

   Information: Supplier for high-voltage sources and accessories

5. Bertan High Voltage Corporation

   Address: 121 New South Road, Hicksville, NY 11801, USA

   Telephone: (1) 516-433-3110

   Fax: (1) 516-935-1766

   Information: Supplier for high-voltage sources and accessories

6. RS Components Ltd.

   Address: 13/F, Wyler Centre, 2-Unit 1, 200 Tai Lin Pai Road, Kwai Chung, New Territory, Hong Kong Special Administrative Region, China

   Telephone: (852) 2421-9898

   Fax: (852) 2421-0339

   Website: http://hken.rs-online.com/web/

   Information: Supplier for high-voltage cables and switches

7. LabSmith, Inc.

   Address: 6111 Southfront Road, Suite E, Livermore, CA 94551, USA

   Telephone: +1 925-292-5161

   Fax: +1 925-454-9487

   E-mail: info@labsmith.com

   Website: http://labsmith.com/

   Information: Supplier for high-voltage sources and accessories

8. GIGAVAC Advanced Switching Solutions

   Address: 6382 Rose Lane, Carpinteria, CA 93013, USA

   Telephone: +1 805-684-8401/805-766-2000

   Fax: +1 805-684-8402

   E-mail: info@gigavac.com

   Website: http://www.gigavac.com/

   Information: Supplier for high-voltage switches and relays

9. ELAND Cables

   Address: 53-79 Highgate Road, London NW5 1TL, UK

   Telephone: (UK) + 44 20-7241-8787

   (International) +44 20-7241-8740

   E-mail: sales@elandcables.com

   Website: http://www.elandcables.com/

   Information: Supplier for high-voltage cables

10. The Okonite Company
    Address: 102 Hilltop Road, Ramsey, NJ 07446, USA
    Telephone: +1 201-825-0300
    Fax: +1 201-825-3524
    E-mail: info@okonite.com
    Website: http://www.okonite.com/
    Information: Supplier for high-voltage cables

Index

A

AA, *see* acrylamide (AA)
Acetic acid, 329
Acetonitrile (ACN), 296–300
ACN, *see* Acetonitrile (ACN)
Active ingredience, 19
Active mixing, 57–59
Acrylaminde (AA), 17, 19, 172–174, 178–180, 184–185
AD, *see* Detection, amperometric
Adenosine triphosphate (ATP), 290, 293
Agglomeration, 44
AIBN, *see* 2,2′-Azobisisobutyronitrile (AIBN)
Albumin, 248–249
Albumin reservoir (AR), 252–253
Amino acids, 17
Ammonium Acetate, 312, 318, 320, 322, 329, 330, 337, 339
Amperometric detection (AD), *see* Detection, amperometric
Analyte
    enrichment, 6, 9, 15, 96, 99, 122–123, 194, 348
    identification, 17, 62
    isolation, 62, 347–348
    separation, 7, 62
AR, *see* Albumin reservoir (AR)
ATP, *see* Adenosine triphosphate (ATP)
Authenification, 211, 232
2,2′-Azobisisobutyronitrile (AIBN), 100–101

B

Baby formula, 15
Background electrolyte, *see* Electrolyte, background
BE, *see* Electrolyte, backing
1,3,5-benzenetricarboxylic acid (BTA), 76
BGE, *see* Electrolyte, background
Bilirubin, 15, 19, 251–262, 350–351, 357
bilirubinemia, 248

Binding, 15, 19, 198, 350–351
Biochip, *see* Microarray chip
Biomarkers, *see* Urine biomarkers
Booth-Overbeek theory, 281
BR, *see* Buffer reservoir (BR)
Bromine generation, 158–163
BTA, *see* 1,3,5-benzenetricarboxylic acid (BTA)
Buffer reservoir (BR), 73, 181, 197, 234, 245, 269
Buffer vial (BV), *see* Buffer reservoir (BR)
Buffer waste (BW), 234
BV, *see* Buffer vial (BV)
BW, *see* Buffer waste (BW)

C

CA, *see* Citrate (CA)
CAD, *see* Computer-aided Design (CAD)
Capillary electrochromatography, 98
Capillary electrophoresis (CE), 7–8, 14–16, 83, 98, 135–137, 174, 193, 266–271, 294–302, 308, 327, 328, 353
Capillary liquid chromatography, 98
Capillary zone electrophoresis (CZE), 8, 60, 98, 138, 173, 354–355
Carbonyls, 15, 95–100, 102–107, 109–113, 350–351
Carbonyl-DNPH derivatives, 96, 99, 100, 102–107, 109, 113
Cardiolipin (CL), 292
CCD, *see* Charge-coupled Device (CCD)
β-CD, *see* β-Cyclodextrin (β-CD)
CE, *see* Capillary electrophoresis (CE)
Cetyltrimethylammonium bromide (CTAB), 86–91
Charge-coupled Device (CCD), 269, 270
CIM, *see* Cimaterol
Cimaterol, 137
Citrate (CA), 83, 86–91
CL, *see* Cardiolipin (CL)
cLOD, *see* Concentration Detection Limit (cLOD)